9th Biennial Conference on Classical and Quantum Relativistic Dynamics of Particles and Fields

(IARD 2014)

AA009062

Journal of Physics: Conference Series Volume 615

Storrs, Connecticut, USA
9 – 13 June 2014

Editor:

L. P. Horowitz

ISBN: 978-1-5108-0582-8
ISSN: 1742-6588

Printed from e-media with permission by:

Curran Associates, Inc.
57 Morehouse Lane
Red Hook, NY 12571

Some format issues inherent in the e-media version may also appear in this print version.

Printed by Curran Associates, Inc. (2015)

For permission requests, please contact the Institute of Physics
at the address below.

Institute of Physics
Dirac House, Temple Back
Bristol BS1 6BE UK

Phone: 44 1 17 929 7481
Fax: 44 1 17 920 0979

techtracking@iop.org

Additional copies of this publication are available from:

Curran Associates, Inc.
57 Morehouse Lane
Red Hook, NY 12571 USA
Phone: 845-758-0400
Fax: 845-758-2634
Email: curran@proceedings.com
Web: www.proceedings.com

TABLE OF CONTENTS

Journal of Physics: Conference Series **615** (2015) 011001 doi:10.1088/1742-6596/615/1/011001

Preface

The International Association for Relativistic Dynamics was organized in February 1998 in Houston, Texas, with John R. Fanchi as president.

Although the subject of relativistic dynamics has been explored, from both classical and quantum mechanical points of view, since the work of Einstein and Dirac, its most striking development has been in the framework of quantum field theory. The very accurate calculations of spectral and scattering properties, for example, of the anamolous magnetic moment of the electron and the Lamb shift in quantum electrodynamics, and many qualitative features of the strong and electroweak interactions, demonstrate the very great power of description achieved in this framework. Yet, many fundamental questions remain to be clarified, such as the structure of classical relativistic dynamical theories on the level of Hamilton and Lagrange in Minkowski space as well as on the curved manifolds of general relativity. There, moreover remains the important question of the covariant classical description of systems at high energy for which particle production effects are not large, such as discussed in Synge's book, *The Relativistic Gas,* and in Balescu's book on relativistic statistical mechanics. In recent years, the study of high energy plasmas and heavy ion collisions has emphasized the importance of developing the techniques of relativistic mechanics. In recent years, the very high accuracy of telescopes and advanced facilities for computation have brought a high level of interest in cosmological problems such as the structure of galaxies (dark matter) and the apparently anomalous expansion of the universe (dark energy). Some of the papers reported here deal with these problems, as well as other fundamental related issues.

The results of Linder et al [Physical Review Letters **95**, 0040401 (2005)], the more recent proposed experiment of Palacios et al [Physical Review Letters **103**, 253001 (2009)], the calculation of A. Bennett of the anomalous moment of the electron in this framework without the use of quantum field theory [A. Bennett, Jour. Phys. A:Math. Theor. **45**, 285302 (2012)], as well the study of M. Davidson, reported here, showing that low energy nuclear reactions associated with He production (indicated by experiments in many laboratories), have shown that there must be a quantum theory with coherence in time. The results, and the consequent variablility of mass as required in Davidson's analysis, suggest the applicability of a theory, manifestly covariant under the transformations of special relativity with an invariant evolution parameter, such as that of Steuckelberg [Helv. Phys. Acta **14**, 322, 588 (1941); **15**, 23 (1942); see also R.P. Feynman, Physical Review **80**, 4401 and J.S. Schwinger, Phys. Rev. **82**, 664, (1951)] that could provide a suitable basis for the study of such questions, as well as many others for which the application of the standard methods of quantum field theory are difficult to manage, involving, in particular, local properties of space time structure.

It was for this purpose, to bring together researchers from a wide variety of fields, such as particle physics, astrophysics, cosmology, heavy ion collisions, plasma research, and mathematical physics, with a common interest in relativistic dynamics, to investigate fundamental questions of this type, that this Association was founded.

The second meeting took place, in 2000, at Bar Ilan University in Ramat Gan, Israel, the third, in 2002, at Howard University in Washington, D.C., and the fourth, on June 12-19, 2004, in Saas Fee, Switzerland. In 2006, the meeting took place at the University of Connecticut campus in Storrs, Connecticut, and the sixth meeting, some of which is recorded in the papers published here, in Thessaloniki, Greece, in 2008, with the significant guidance of Professor Ioannis Antoniou, the help of his institution, the Aristotle University of Thessaloniki, the Municipality of the Heroic City of Naousa and the Euxeinos Club of Naousa. The meeting took place at the Teleglion Foundation of Art at AUTH.

The most recent meeting took place at the University of Connecticut, Storrs, on June 9-13, 2014.

This meeting forms the basis for the Proceedings that are recorded in this issue of the Journal of Physics:Conference Series. Along with the work of some of the founding members of the Association, we were fortunate to have lecturers from application areas that provided strong challenges for further developments in quantum field theory, cosmological problems, and in the dynamics of systems subject to accelerations and the effects of general relativity.

Topics treated in this issue include studies of the dark matter problem, rotation curves, and, in particular, for the (relatively accessible) Milky Way galaxy, compact stellar objects, a composite particle model, and the properties of a conformally invariant theory with spontaneous symmetry breaking. The Stueckelberg theory is further investigated for its properties in producing bremsstrahlung and pair production and apparent superluminal effects, and, as mentioned above, the implications of low energy nuclear reactions for such off-shell theories. Other "proper time" theories are investigated as well, and a study of the clock synchronization problem is presented.

A mathematical study of to quantum groupo associated with the Toda lattice and its implications for quantum field theory, as well as a phenomenological discussion of supernova mechanics as well as a semiclassical discussion of electron spin and the question of the compatibility of special relativity and the quantum theory. A careful analysis of the covariant Aharonov-Bohm effect is given as well. The quantization of massless fields and the relation to the Maxwell theory is also discussed,

We wish to thank the participants who contributed very much through their lectures, personal discussions, and these papers, to the advancement of the subject and our understanding.

We thank the Scientific Advisory Committee for their invaluable guidance and advice:

Stephen Adler (Institute for Advanced Study)
Itzhak Bars (University of Southern California)
Gordon Baym (University of Illinois)
Jacob Bekenstein (Hebrew University)
Fred Cooper (Los Alamos National Laboratory)
Bei-Lok Hu (University of Maryland)
Werner Israel (University of Victoria)
E.V. Shuryak (Brookhaven National Laboratory)
L.S. Shulman (Clarkson University)
William Unruh (University of British Columbia)
Luca Lusanna (National Institute for Nuclear Physics, INFN)
Benoit Famaey (Observatoire Astronomique de Strasbourg, CNRS)

The organizers express their gratitude to the University of Connecticut at Storrs, the UConn Department of Physics, and Prof. Nora Berrah for their support and the use of its excellent facilities.

Finally, we thank the participants who contributed through their lectures, personal discussions, and these papers, to the advancement of the subject and our understanding.

For the Editors and Organizing Committee,

L.P. Horwitz (Tel-Aviv University, Bar Ilan University), Editor-in-Chief
Philip Mannheim, Chair of the Organizing Committee
Tepper Gill, IARD Treasurer
Martin Land, IARD President
James O'Brien, IARD Vice-President
Benoit Famaey, IARD SAC
Luca Lusanna, IARD SAC

Peer review statement

All papers published in this volume of *Journal of Physics: Conference Series* have been peer reviewed through processes administered by the proceedings Editors. Reviews were conducted by expert referees to the professional and scientific standards expected of a proceedings journal published by IOP Publishing.

Modeling wave dark matter in dwarf spheroidal galaxies

H L Bray[1] and A R Parry[2]

[1]Department of Mathematics and Department of Physics, Duke University, Box 90320, Durham, NC 27708, USA

[2]Department of Mathematics, University of Connecticut, 196 Auditorium Road, Unit 3009, Storrs, CT 06269, USA

E-mail: [1]`bray@math.duke.edu`, [2]`alan.parry@uconn.edu`

Abstract. This paper studies a model of dark matter called wave dark matter (also known as scalar field dark matter and boson stars). Wave dark matter describes dark matter as a scalar field which satisfies the Einstein-Klein-Gordon equations. These equations rely on a fundamental constant Υ (also known as the "mass term" of the Klein-Gordon equation), which can be interpreted physically as a characteristic frequency of the scalar field. In this work, we compare the wave dark matter model to observations to obtain an estimate of Υ. Specifically, we compare the mass profiles of spherically symmetric static states of wave dark matter to certain Burkert mass profiles recently shown to predict well the velocity dispersion profiles of the eight classical dwarf spheroidal galaxies. We outline a procedure for estimating Υ in these circumstances and show that under precise assumptions the value of Υ can be bounded above by 1000 yr^{-1}. We also show that a reasonable working value for this constant is $\Upsilon = 50\ \mathrm{yr}^{-1}$.

1. Introduction

Ever since the first postulation of dark matter in the 1930's by Oort [1] and then Zwicky [2], much evidence for the existence of dark matter has been accumulated including the unexpected behavior in the rotation curves of spiral galaxies [3, 4], the velocity dispersion profiles of dwarf spheroidal galaxies [5, 6, 7, 8], and gravitational lensing [9]. These and other observations support the idea that most of the matter in the universe is not baryonic, but is, in fact, some form of exotic dark matter and that almost all astronomical objects from the galactic scale up contain a significant amount of this dark matter. Describing dark matter, its nature, and effects is currently one of the biggest open problems in astrophysics [10, 11, 12, 13, 14, 15, 16].

Essentially all of the observations indicating the presence of dark matter are gravitational observations, that is, we have only observed the gravitational effects of dark matter. Since, from the point of view of general relativity, gravity is an effect of the curvature of spacetime, this indicates that, when it comes to dark matter, what is being observed is some kind of unexplained curvature of the spacetime. This suggests the possibility that the key to understanding the nature of dark matter already lies within the fraemwork of general relativity. Recently, Bray has geometrically motivated the study of a scalar field satisfying the Einstein-Klein-Gordon equations as a viable dark matter candidate via constructing axioms for general relativity [17, 18]. This geometric motivation shows that modeling dark matter with a scalar field is perhaps the most natural large scale point of view for the dark matter problem.

Journal of Physics: Conference Series **615** (2015) 012001 doi:10.1088/1742-6596/615/1/012001

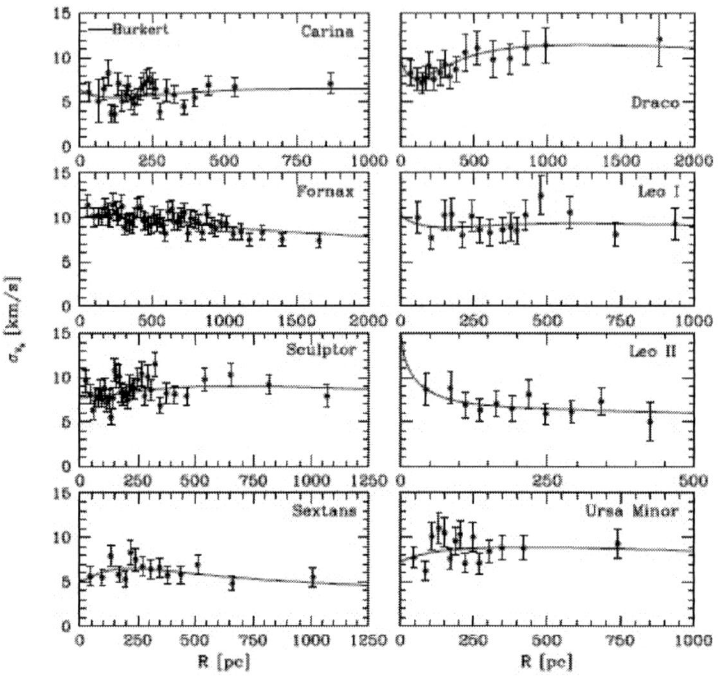

Figure 1. Observed velocity dispersion profiles of the eight classical dwarf spheroidal galaxies are denoted by the points on each plot with its associated error bars. The solid lines overlayed on these profiles are the best fit velocity dispersion profiles induced by the Burkert mass profile. This figure is directly reproduced from Figure 1 of the paper "Dwarf spheroidal galaxy kinematics and spiral galaxy scaling laws" by Salucci et al. appearing in the Monthly Notices of the Royal Astronomical Society volume 420 issue 3 [6]. The reader is referred to their paper for a complete description of how these models were computed.

The idea of using a scalar field to describe dark matter is not new. In fact, a scalar field model has been seriously considered as a candidate for dark matter for more than two decades and has been shown to be in agreement with many cosmological observations [17, 19, 20, 21, 22, 23, 24, 25, 26, 27, 28, 29, 30, 31]. In most of these settings, these scalar fields are considered from a quantum mechanical motivation for the same Einstein-Klein-Gordon equations and go by the name scalar field dark matter or boson stars. However, due to the fact that the Klein-Gordon equation is a wave-type partial differential equation, we prefer the name *wave dark matter* and will refer to the scalar field model as such in the remainder of this paper.

Specifically, wave dark matter is characterized by a single scalar field permeating all of a spacetime (N, g) whose metric g has signature $(- + ++)$. The scalar field, which is either a real or complex-valued function, and the metric together form a solution to the Einstein-Klein-Gordon equations. In this paper, we consider the point of view of a complex-valued scalar

Journal of Physics: Conference Series **615** (2015) 012001 doi:10.1088/1742-6596/615/1/012001

field$f : N \to \mathbb{C}$. In this case, the Einstein-Klein-Gordon equations are given by

$$G = 8\pi\mu_0 \left(\frac{df \otimes d\bar{f} + d\bar{f} \otimes df}{\Upsilon^2} - \left(\frac{|df|^2}{\Upsilon^2} + |f|^2 \right) g \right) \tag{1a}$$

$$\Box_g f = \Upsilon^2 f \tag{1b}$$

where $\Box_g$ is the Laplacian with respect to the metric g, and $\mu_0 \geq 0$ and $\Upsilon > 0$ are constants. The constant μ_0 is for convenience and is related to the value of the energy density of the scalar field at the origin. If desired, μ_0 can be completely absorbed into f and hence its value does not affect the solutions qualitatively. However, Υ is a fundamental constant of this system and its value affects the qualitative behavior of the solutions. Physically, Υ can be interpreted as a characteristic frequency of the scalar field. From a particle physics point of view, the constant Υ is related to the mass m^* of the dark matter particle. Specifically,

$$m^* = \frac{\hbar\Upsilon}{c} = 2.09 \times 10^{-23} \text{ eV} \left(\frac{\Upsilon}{1 \text{ yr}^{-1}} \right) \tag{2}$$

In the wave dark matter model, the value of Υ is the same throughout all of spacetime since the scalar field permeates all of spacetime, making Υ a fundamental constant of the universe. Thus an important check for the validity of this model is to see if a single constant value of Υ can successfully describe all of the observed gravitational effects of dark matter in all contexts. Moreover, we can use observations of dark matter in any context to bound this parameter since its value is the same everywhere. This is positive since the more estimates of Υ that we can obtain, the more clearly we will be able to determine if this model is valid.

One of the simplest models defined by wave dark matter are the spherically symmetric static states. These models are well known and are sometimes referred to as standing waves of the Einstein-Klein-Gordon equations. These models are shown by Parry in [32] to have certain useful properties. These properties, in fact, present an opportunity to estimate Υ. The purpose of this paper is to use these properties of spherically symmetric static states and recent observations of dwarf spheroidal galaxies to provide an estimate of and bounds on Υ. Additionally, this paper gives support to the notion that a single value of Υ is sufficient to describe dark matter at all levels by showing that a single value is at least sufficient to describe the dark matter content in the dwarf spheroidal galaxies considered here and that, in fact, many values work in this case.

In particular, Salucci et al. recently used the cored Burkert profile [33] to model the dark matter energy density profiles of the eight classical dwarf spheroidal galaxies orbiting the Milky Way. They found excellent agreement between the observed velocity dispersion profiles of these galaxies and those velocity dispersion profiles induced by the Burkert profile [6]. This can be seen in Figure 1, which we have reproduced exactly as it appears in the paper by Salucci et al. In what follows, we will use the properties of spherically symmetric static states of wave dark matter to show that under precise assumptions, comparisons to these Burkert profiles can be used to bound the value of Υ above by 1000 yr^{-1}. We will also show that a value of $\Upsilon = 50 \text{ yr}^{-1}$ produces wave dark matter mass models that are qualitatively similar to the Burkert mass models found by Salucci et al. as an example that a single value of Υ can be used to describe the dark matter mass content of all of the dwarf spheroidal galaxies considered here.

2. Burkert Mass Profiles

The Burkert energy density profile models the energy density of a spherically symmetric dark matter halo using the function

$$\mu_B(r) = \frac{\rho_0 r_c^3}{(r + r_c)(r^2 + r_c^2)} \tag{3}$$

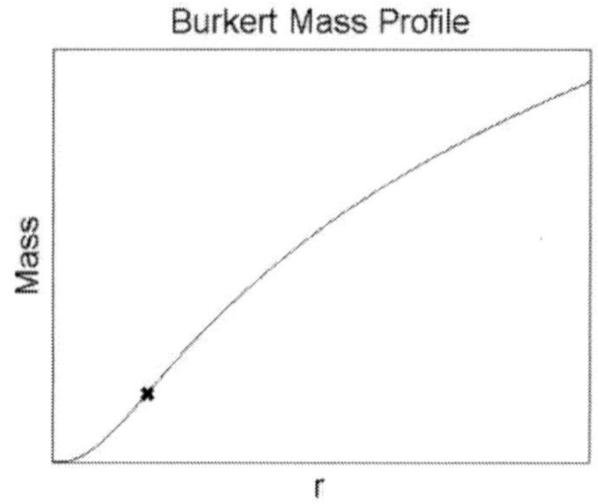

Figure 2. Plot of a Burkert mass profile. The inflection point is marked with an $\times$.

where ρ_0 is the central density and r_c is the core radius. Integrating this function over the ball of radius r centered at the origin, $B_r(0)$, with respect to the standard spherical volume form yields the Burkert mass profile as follows.

$$\begin{aligned} M_B(r) &= \int_{B_r(0)} \mu_B(s)\, dV_{\mathbb{R}^3} \\ &= 4\pi \int_0^r s^2 \mu_B(s)\, ds \\ &= 4\pi \int_0^r \frac{s^2 \rho_0 r_c^3}{(s+r_c)(s^2+r_c^2)}\, ds \\ M_B(r) &= 2\pi\rho_0 r_c^3 \Bigg(\ln\left(\frac{r+r_c}{r_c}\right) \\ &\qquad + \frac{1}{2}\ln\left(\frac{r^2+r_c^2}{r_c^2}\right) - \arctan\left(\frac{r}{r_c}\right)\Bigg) \end{aligned} \tag{4}$$

A generic plot of a Burkert mass profile, $M_B(r)$, defined to be the dark matter mass in the ball of radius r, is shown in Figure 2. We make a few remarks about the behavior of this mass function.

Note that the behavior of the graph changes concavity at the inflection point $r = r_{ip}$, which we have marked on the plot in Figure 2 with an $\times$. Recalling from equation (4) the fact that $M_B(r)$ is the integral over the interval $[0, r]$ of the function $4\pi r^2 \mu_B(r)$, we can compute this inflection point as follows.

$$M_B'(r) = 4\pi r^2 \mu_B(r) = \frac{4\pi\rho_0 r_c^3 r^2}{(r+r_c)(r^2+r_c^2)} \tag{5}$$

Differentiating again yields

$$M_B''(r) = \frac{-4\pi\rho_0 r_c^3 (r^4 - r^2 r_c^2 - 2r r_c^3)}{(r+r_c)^2(r^2+r_c^2)^2}, \tag{6}$$

which has two complex zeros and two real zeros. The two real zeros are $r = 0$ and

$$r_{ip} = \left(\frac{3 + (27+3\sqrt{78})^{2/3}}{3(27+3\sqrt{78})^{1/3}}\right) r_c \approx 1.52 r_c, \tag{7}$$

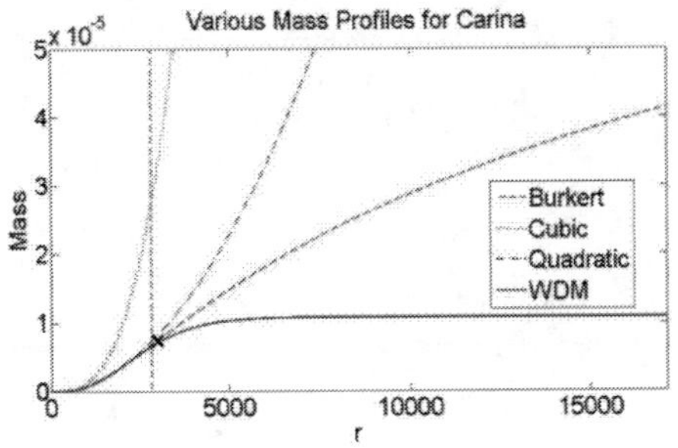

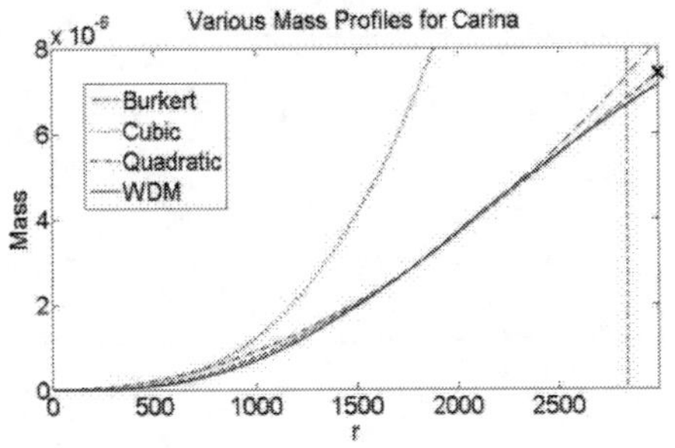

Figure 3. Left: Plot of the Burkert mass profile for the Carina galaxy found by Salucci et al. [6] along with a mass plot of a wave dark matter static ground state, the cubic function which is the leading term of the Taylor expansion of the Burkert mass profile, and the quadratic power function $\frac{M_B(r_c)}{r_c^2}r^2$ where r_c is the core radius of the Carina galaxy. The $\times$ marks the location of the inflection point of the Burkert mass profile, while the vertical line denotes the location of the outermost data point for the Carina galaxy and is presented for reference purposes only. Right: Closeup of the plot on the left over the r interval $[0, r_{ip}]$.

the latter being the inflection point of the mass model.

For $r \gg r_{ip}$, the plot grows logarithmically due to the fact that the arctan term in equation (4) approaches a constant value as $r \to \infty$. To describe the behavior when $r \ll r_{ip}$, we note that the Taylor expansion of $M_B(r)$ centered at $r = 0$ is as follows,

$$M_B(r) = \frac{4}{3}\pi\rho_0 r^3 + O(r^4). \tag{8}$$

Thus for $r \ll r_{ip}$, $M_B(r)$ is dominated by an r^3 term making the initial behavior cubic.

In fact, several other models for dark matter mass profiles have similar initial behavior to the Burkert profile including a quadratic mass profile (which is not physical and is only included for the sake of comparison) and wave dark matter mass profiles. In Figure 3, we have collected several mass models that have similar behavior inside $r = r_{ip}$ to the Burkert mass profile computed by Salucci et al. for the Carina galaxy [6]. While these models have similar behavior inside $r = r_{ip}$, they are very different outside $r = r_{ip}$.

We have computed the inflection points of each of the Burkert mass profiles computed by Salucci et al. for the eight classical dwarf spheroidal galaxies [6] and have marked these points on a plot of each Burkert mass profile in Figure 4. We have constrained the viewing window of each plot to the range of data points collected. That is, we plot the Burkert mass profiles on the interval $[0, r_{last}]$, where r_{last} is the radius of the outermost data point given by Walker et al. [7, 8] for the observed velocity dispersion profiles. We have presented them in order from greatest to least according to the ratio of r_{last}/r_{ip}.

In Table 1, we have collected the defining parameters, ρ_0 and r_c, computed by Salucci et al. for the Burkert mass profiles which best predict the velocity dispersion profiles of each galaxy [6]. We have also collected the outermost data point, r_{last}, of these velocity dispersion profiles [7, 8], as well as our computations of the inflection point, r_{ip}, and the ratio r_{last}/r_{ip} for each of the classical dwarf spheroidal galaxies. All quantities have been converted to geometrized units (the universal gravitational constant and the speed of light set to one) of (light)years for mass, length, and time.

Journal of Physics: Conference Series **615** (2015) 012001 doi:10.1088/1742-6596/615/1/012001

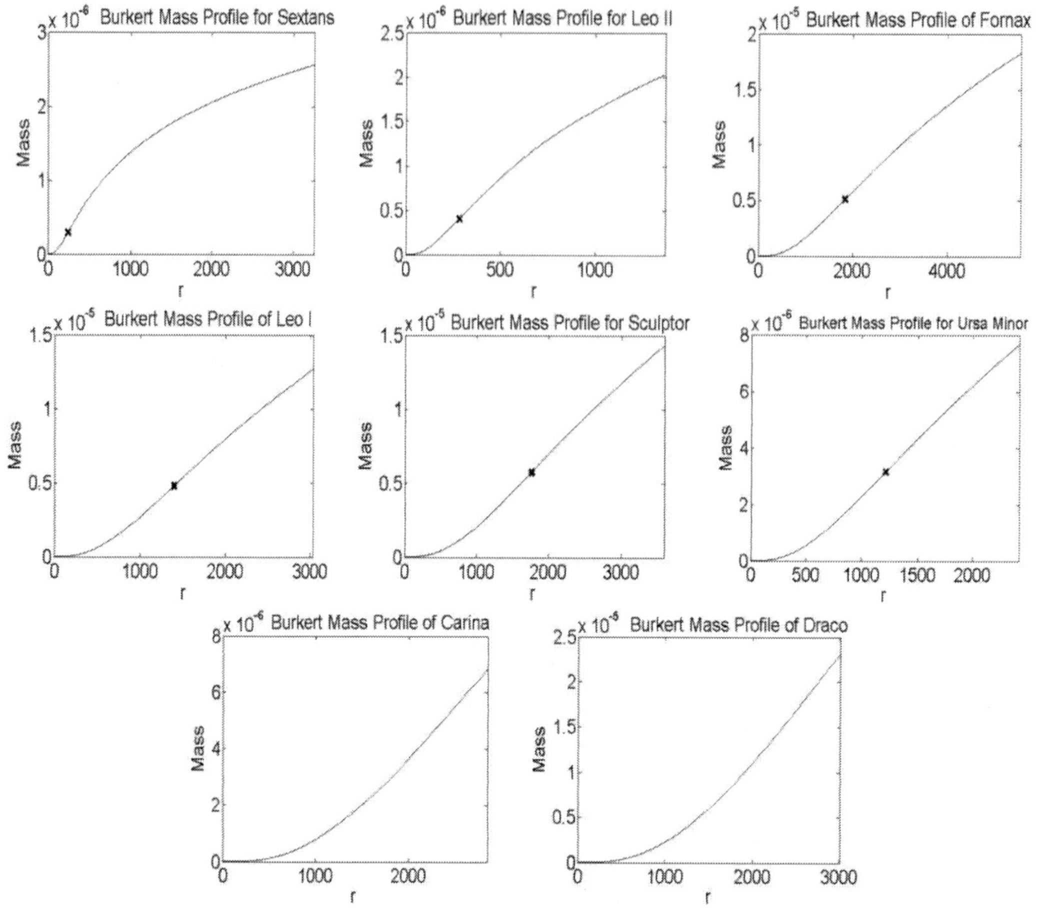

Figure 4. Plots of the Burkert mass profiles computed by Salucci et al. of the eight classical dwarf spheroidal galaxies within the range of observable data. The inflection point is marked on each plot by an $\times$. Carina and Draco have no inflection point marked because the inflection point for their Burkert mass profiles occurs outside the range of observable data.

3. Static States of Wave Dark Matter

Now that we have presented mass profiles which model actual data well, we need to describe the wave dark matter models we will use to make our comparison. In the following, we present only the basic background information required to understand the model we use and refer the reader to [17, 18] for more discussion on its motivation and successes thus far.

Since we wish to compare solutions of (1) to the spherically symmetric Burkert mass profiles computed by Salucci et al. [6], we choose to work in spherical symmetry. In [34], Parry surveyed the well-known form of the metric of a general spherically symmetric spacetime in polar-areal coordinates, namely,

$$g = -e^{2V(t,r)}\,dt^2 + \left(1 - \frac{2M(t,r)}{r}\right)^{-1} dr^2 + r^2\,d\sigma^2, \tag{9}$$

for real valued functions V and M and where $d\sigma^2 = d\theta^2 + \sin^2\theta\,d\varphi^2$ is the standard metric

Journal of Physics: Conference Series **615** (2015) 012001 doi:10.1088/1742-6596/615/1/012001

Galaxy Name	ρ_0 (yr^{-2})	r_c (yr)	r_{last} (yr)	r_{ip} (yr)	r_{last}/r_{ip}
Sextans	2.47×10^{-14}	1.53×10^{2}	3.26×10^{3}	2.32×10^{2}	14.05
Leo II	1.83×10^{-14}	1.88×10^{2}	1.37×10^{3}	2.86×10^{2}	4.80
Fornax	8.57×10^{-16}	1.21×10^{3}	5.54×10^{3}	1.84×10^{3}	3.01
Leo I	1.83×10^{-15}	9.19×10^{2}	3.03×10^{3}	1.40×10^{3}	2.17
Sculptor	1.10×10^{-15}	1.16×10^{3}	3.59×10^{3}	1.76×10^{3}	2.04
Ursa Minor	1.83×10^{-15}	8.01×10^{2}	2.41×10^{3}	1.22×10^{3}	1.98
Carina	2.90×10^{-16}	1.97×10^{3}	2.84×10^{3}	2.99×10^{3}	0.95
Draco	8.19×10^{-16}	2.11×10^{3}	3.00×10^{3}	3.20×10^{3}	0.94

Table 1. Burkert mass profile data for the eight classical dwarf spheroidal galaxies converted to units of years for mass, length, and time. The parameters ρ_0 and r_c are those found by Salucci et al. for the best fit Burkert profiles [6], and r_{last} is the radius of the outermost data point given by Walker et al. [7, 8]. Also included is the value of the inflection point, r_{ip}, of the Burkert mass profile for each galaxy and the ratio of r_{last} to r_{ip}.

on the unit sphere. This metric has the following useful properties. The function $M(t,r)$ is the Hawking mass of the metric sphere of radius r and time t. Under the Einstein equation, $G = 8\pi T$, M is also the flat volume integral of the energy density term in the stress energy tensor. This motivates interpreting the function $M(t,r)$ as the mass inside the metric sphere of radius r at time t. Finally, given the Einstein equation, in the low field limit, V is approximately the gravitational potential of the system. We refer the reader to [34] for detailed proofs of these facts.

It is also shown in [34] that in spherical symmetry and using the metric (9), solving the Einstein-Klein-Gordon system (1) reduces to solving the system

$$M_r = 4\pi r^2 \mu_0 \left(|f|^2 + \left(1 - \frac{2M}{r}\right) \frac{|f_r|^2 + |p|^2}{\Upsilon^2} \right) \tag{10a}$$

$$V_r = \left(1 - \frac{2M}{r}\right)^{-1} \left(\frac{M}{r^2} - 4\pi r \mu_0 \left(|f|^2 - \left(1 - \frac{2M}{r}\right) \frac{|f_r|^2 + |p|^2}{\Upsilon^2} \right) \right) \tag{10b}$$

$$f_t = p e^V \sqrt{1 - \frac{2M}{r}} \tag{10c}$$

$$p_t = e^V \left(-\Upsilon^2 f \left(1 - \frac{2M}{r}\right)^{-1/2} + \frac{2 f_r}{r} \sqrt{1 - \frac{2M}{r}} \right) + \partial_r \left(e^V f_r \sqrt{1 - \frac{2M}{r}} \right). \tag{10d}$$

To solve this system, we need boundary conditions. At the central value, we require all of the functions to be smooth. Since all of the functions are spherically symmetric, this implies that M_r, V_r, f_r, and p_r all vanish at $r = 0$ for all t. We will also require that the spacetime be asymptotically Schwarzschild, that is, it approaches a Schwarzschild metric as $r \to \infty$.

Journal of Physics: Conference Series **615** (2015) 012001 doi:10.1088/1742-6596/615/1/012001

Specifically, this implies that

$$e^{2V} \to \kappa^2 \left(1 - \frac{2M}{r}\right) \quad \text{as } r \to \infty \tag{11}$$

$$\Box_{g_S} f \to \Upsilon^2 f \quad \text{and} \quad f \to 0 \quad \text{as } r \to \infty \tag{12}$$

where $\kappa > 0$ and g_S is the appropriate Schwarzschild metric. Since $f \to 0$ as $r \to \infty$, M approaches a constant value m, which is the total mass of the system.

Note that these boundary conditions ensure that as $r \to \infty$, the metric g in equation (9) becomes the Schwarzschild metric

$$g_S = -\kappa^2 \left(1 - \frac{2m}{r}\right) dt^2 + \left(1 - \frac{2m}{r}\right)^{-1} dr^2 + r^2\, d\sigma^2. \tag{13}$$

Thus κ represents a scaling of the t coordinate in the standard Schwarzschild metric. The effect of this on our discussion is that $V \to \ln \kappa$ as $r \to \infty$.

We have to numerically solve these equations and so in practice, we will impose these boundary conditions at an artificial right hand boundary point r_{max} and solve the system on the r-interval $[0, r_{max}]$.

One of the simplest solutions to this system are those where the scalar field is of the form

$$f(t, r) = e^{i\omega t} F(r) \tag{14}$$

where $\omega \in \mathbb{R}$ is a constant and F is real valued.

Note that for f of this form, solving equation (12) for large r and requiring the solution to decay to 0 yields that, for large r and small total mass, F must satisfy

$$F' + \left(\sqrt{\Upsilon^2 - \frac{\omega^2}{\kappa^2}} + \frac{1}{r}\right) F \approx 0. \tag{15}$$

Requiring this condition on our system ensures that f appropriately decays to 0 as $r \to \infty$. See [35] for more details on this computation.

Solutions of the form in equation (14) produce static metrics and, once substituted into the system (10), yield the following set of ODEs [32],

$$M' = 4\pi r^2 \mu_0 \left[\left(1 + \frac{\omega^2}{\Upsilon^2} e^{-2V}\right) |F|^2 + \left(1 - \frac{2M}{r}\right) \frac{|H|^2}{\Upsilon^2}\right] \tag{16a}$$

$$V' = \left(1 - \frac{2M}{r}\right)^{-1} \left\{\frac{M}{r^2} - 4\pi r \mu_0 \left[\left(1 - \frac{\omega^2}{\Upsilon^2} e^{-2V}\right) |F|^2 - \left(1 - \frac{2M}{r}\right) \frac{|H|^2}{\Upsilon^2}\right]\right\} \tag{16b}$$

$$F' = H \tag{16c}$$

$$H' = \left(1 - \frac{2M}{r}\right)^{-1} \left[\left(\Upsilon^2 - \frac{\omega^2}{e^{2V}}\right) F + 2H \left(\frac{M}{r^2} + 4\pi r \mu_0 |F|^2 - \frac{1}{r}\right)\right] \tag{16d}$$

with boundary conditions

$$F(0) = 1, \qquad H(0) = 0, \qquad M(0) = 0, \qquad V(0) = V_0, \tag{17}$$

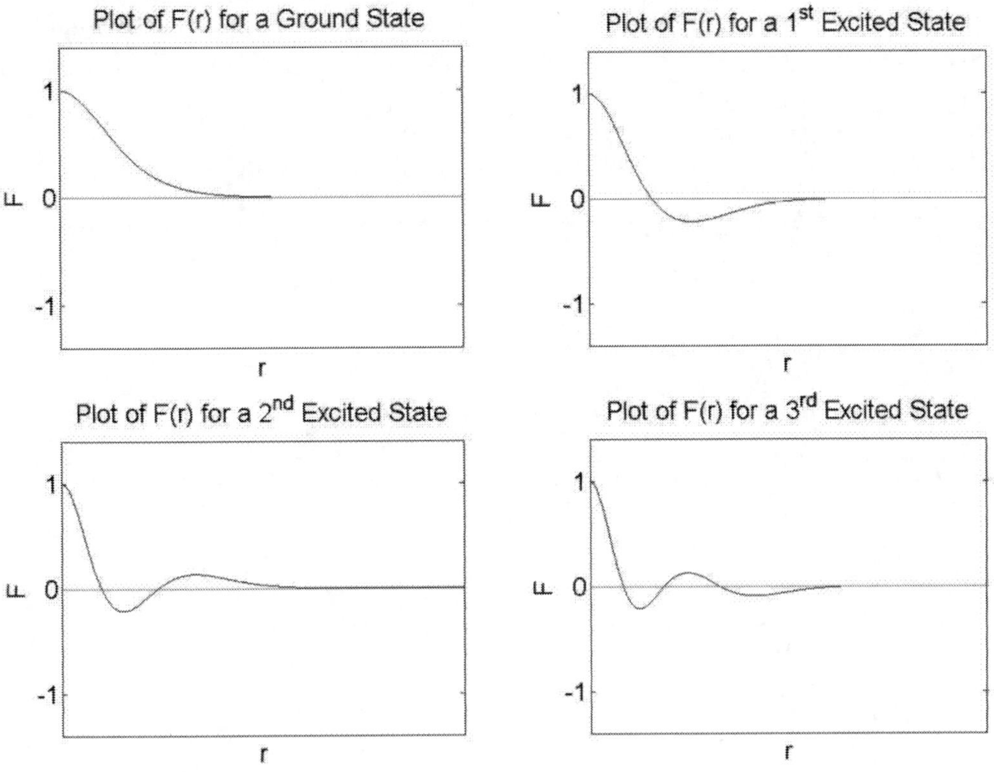

Figure 5. Plots of spherically symmetric static state scalar fields (specifically, the function $F(r)$ in equation (14)) in the ground state and first, second, and third excited states. Note the number of nodes (zeros) of each function.

$$F'(r_{max}) + \left(\sqrt{\Upsilon^2 - \frac{\omega^2}{\kappa^2}} + \frac{1}{r_{max}}\right) F(r_{max}) \approx 0, \tag{18}$$

$$V(r_{max}) - \frac{1}{2}\ln\left(1 - \frac{2M(r_{max})}{r_{max}}\right) - \ln\kappa \approx 0, \tag{19}$$

by equations (11) and (12). For simplicity, we set $\kappa = 1$, which corresponds to the assumption on our choice of t coordinate that V goes to zero at infinity and that the metric goes to the standard Schwarzschild metric as $r \to \infty$. A solution to these equations depends on the choice of the parameters Υ, μ_0, ω, and V_0. We solve a shooting problem for ω and V_0 to satisfy (18) and (19) leaving Υ and μ_0 freely selectable.

For each choice of Υ and μ_0, there are an infinite number of discrete finite mass solutions characterized by the number of zeros that F exhibits [32, 36]. These are called static states. A static state with no zeros is called a ground state. With n zeros for $n > 0$, it is called an nth excited state. In Figure 5, we have plotted examples of F for a ground through third excited state. In Figure 6, we have presented the plots of the mass, M, corresponding to the plots of F in Figure 5.

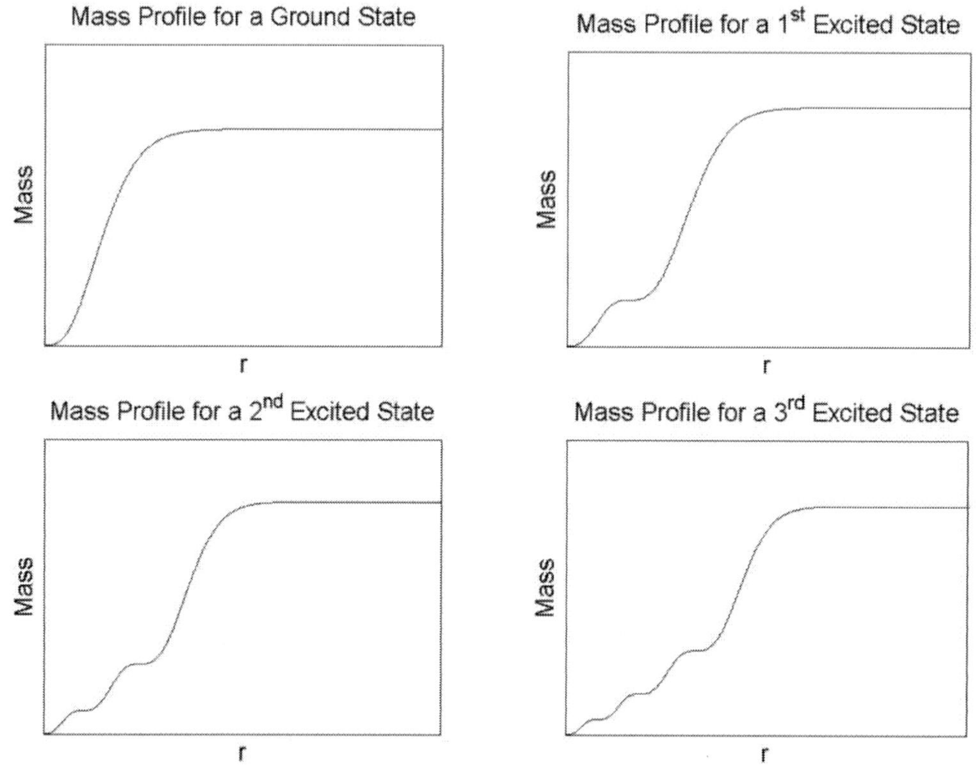

Figure 6. Mass profiles for a static ground state and first, second, and third excited states of wave dark matter.

3.1. Hyperbolas of Constant Υ

Parry showed in [32] how the parameters which define an nth excited state, Υ, μ_0, ω, and V_0, as well as the values of the total mass, m, and the radius, r_h, called the half mass radius, for which $M(r_h) = m/2$, are related to each other. The approximations which follow only apply to solutions in the long wavelength, low field limit, which is what is relevant for modeling galaxies. In particular, for constant Υ, there is a one parameter family of solutions for each nth excited state defined by the equations

$$\omega^n(\Upsilon, \mu_0) \approx \Upsilon \exp\left(C^n_{frequency} \frac{\sqrt{\mu_0}}{\Upsilon}\right) \tag{20}$$

$$V_0^n(\Upsilon, \mu_0) \approx C^n_{potential} \frac{\sqrt{\mu_0}}{\Upsilon} \tag{21}$$

$$m^n(\Upsilon, \mu_0) \approx C^n_{mass} \Upsilon^{-3/2} \mu_0^{1/4} \tag{22}$$

$$r_h^n(\Upsilon, \mu_0) \approx C^n_{radius} \Upsilon^{-1/2} \mu_0^{-1/4}. \tag{23}$$

where the constants C^n_* depend upon which state we wish to consider (i.e. they depend on n). Thus for constant Υ, the different possible excited states satisfying equations (17), (18), and (19) are defined entirely by the value of μ_0, which corresponds to the value of the energy density

n	C^n_{mass}	C^n_{radius}
0	4.567 ± 0.05	0.8462 ± 0.004
1	10.22 ± 0.10	2.2894 ± 0.009
2	15.81 ± 0.16	3.8253 ± 0.014
3	21.37 ± 0.22	5.3994 ± 0.018
4	26.91 ± 0.27	6.9860 ± 0.022
5	32.42 ± 0.33	8.5606 ± 0.026
10	60.32 ± 1.18	15.1357 ± 0.039
20	116.62 ± 2.57	29.6822 ± 0.107

Table 2. Values of the constants in equations (22) and (23) for the ground through fifth excited states as well as the tenth and twentieth excited states. We have given these values error ranges which encompass the interval we observed in our experiments, but it is possible that values outside our ranges here could be observed. However, we do not expect them to be outside by much if the discretization of r used in solving the ODEs is sufficiently fine. Note also that our values have less precision as we increase n. This is because as n increases, it becomes more difficult to compute the states with as much precision.

function, μ, as defined in [34] at the origin via the equation

$$\mu(0) = \mu_0 \left(1 + \frac{\omega^2}{\Upsilon^2} e^{-2V_0}\right). \tag{24}$$

The equations relevant to our discussion here are equations (22) and (23). We have collected the values of C^n_{mass} and C^n_{radius} from these two equations for the ground through fifth excited states as well as for the tenth and twentieth excited states in Table 2. A complete table of the values of all the constants C^n_* for these excited states can be found in [32].

As explained in [32], equations (22) and (23) imply that the product of m and r_h does not depend on the value of μ_0, but only on Υ. Specifically,

$$mr_h = \frac{C_{mass} C_{radius}}{\Upsilon^2}, \tag{25}$$

where we have suppressed the notation of n. If Υ is constant, then, because both C_{mass} and C_{radius} are positive, the right hand side of this equation is some positive constant, k, and we have

$$mr_h = k \tag{26}$$

which defines a hyperbola. Thus, for a given n^{th} excited state, all of the possible mass profiles for a constant value of Υ lie along a hyperbola. We illustrate this phenomenon in Figure 7.

3.2. Fitting Burkert Mass Profiles

With these properties of static state mass profiles in mind, we turn our attention to finding static state mass profiles that best fit the Burkert mass profiles computed by Salucci et al. [6]. Given a Burkert mass profile, M_B, a value for Υ, and a specific state (i.e. value for n), we define for our purposes the best fit wave dark matter static state mass profile, M_W, as the one which minimizes the L^2 norm of the difference between these profiles, E, given by

$$E = \|M_B - M_W\|^2_{L^2} = \int_0^{r_{last}} (M_B - M_W)^2 \, dr. \tag{27}$$

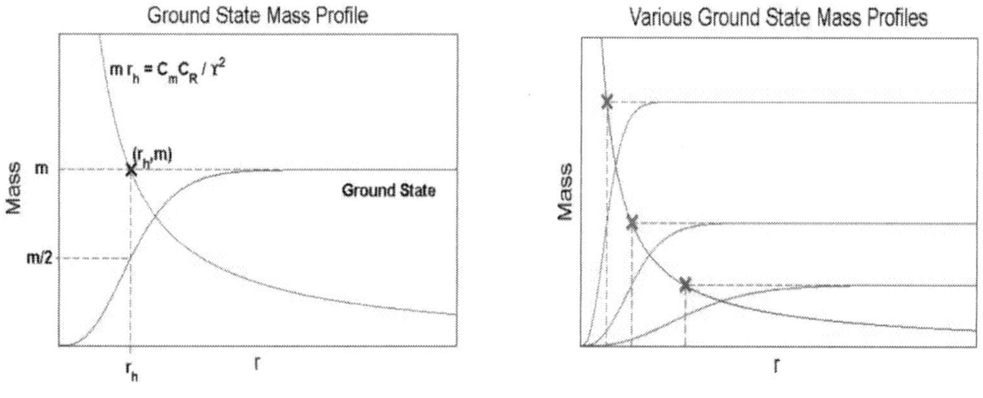

Figure 7. Left: Plot of the mass profile of a ground state with its corresponding hyperbola of constant Υ overlayed. Any ground state mass profile that keeps the presented relationship with this hyperbola corresponds to the same value of Υ. Right: Examples of different ground state mass profiles corresponding to the same value of Υ. The corresponding hyperbola of constant Υ is overlayed. Notice that all three mass profiles have the same relationship with the hyperbola.

Of course, since we compute the static states numerically, we have to approximate this norm by an appropriate Riemann sum defined on a discretization of the interval $[0, r_{last}]$.

To find this minimum, we first note that since Υ is fixed, we can write the total mass m and the value of μ_0 in terms of a choice of r_h via equations (25) and (23) respectively. Thus, we parameterize the different mass profiles of constant Υ, and hence E by r_h, that is, $E = E(r_h)$. Furthermore, since all of the static state mass profiles of constant Υ lie on a hyperbola, there will be a value of r_h that yields the minimum of $E(r_h)$.

To make computing the best fits more uniform from galaxy to galaxy, we make the choice $r_h = br_c$, where $b > 0$ and r_c is the core radius of the Burkert profile we wish to match, and vary the free parameter b. To compute which value of b produces a minimum value of $E(r_h)$, we create a grid of r_h values around an initial choice of b of the form $[(b - step)r_c, br_c, (b + step)r_c]$ for some $step > 0$. Next we compute $E(r_h)$ for each of the values of r_h and shift the grid, if necessary, so that it is centered on the r_h value which yielded the smallest value of $E(r_h)$. If the grid shifts, we recompute $E(r_h)$ on the new grid and continue to shift, if necessary. Once the minimum $E(r_h)$ value occurs at the center of the grid, we keep that point as the center, but cut the step size in half. We then run this shifting procedure again for this smaller grid until the minimum is at the center and then we shrink again. We continue to shrink the step size until we get to a predetermined terminal value. We generally would run the procedure until the step size was less than or equal to 2^{-10}.

This best fitting procedure provides a method of finding values of Υ, for Υ sufficiently large, which produce untenable matches to the Burkert profiles of Salucci et al. [6]. This is the topic of the next section.

3.3. Upper Bound for Υ

To find an upper bound for Υ, we first need to explain how the static states change as Υ gets large. Equation (25) implies that for a given n^{th} excited state, as Υ increases, the product mr_h decreases. The hyperbolas corresponding to smaller values of mr_h are those that lie closer to the mass and radius axes.

Now consider the nth excited state mass profile that is the best fit to a Burkert mass profile

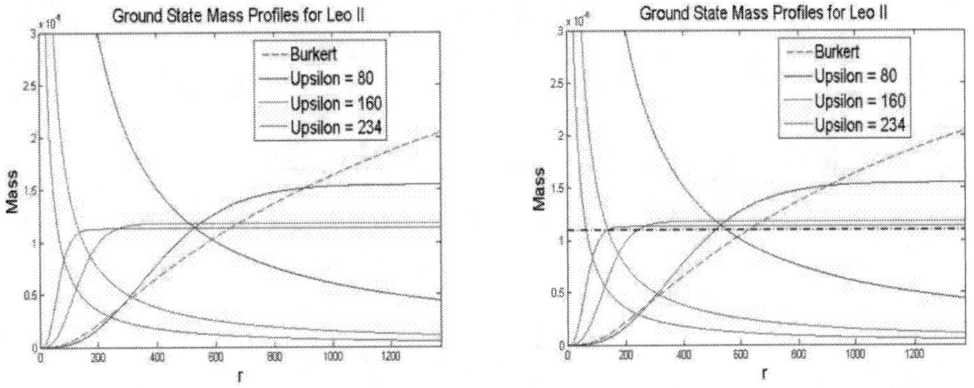

Figure 8. Left: Ground state mass profiles of various values of Υ that are best fits to the Burkert mass profile found by Salucci et al. [6] for the Leo II galaxy. The corresponding hyperbolas of constant Υ on which these profiles lie are also plotted. Ground states and their corresponding hyperbolas are drawn in the same color. Right: The same plots as in the left frame, but with the constant function which best fits the Burkert profile also plotted. Note that the best fit mass profiles approach this constant mass profile as Υ increases.

for a given Υ. As Υ increases, the hyperbola to which this static state mass profile corresponds will get closer to the mass and radius axes, but since the mass profile must also minimize $E(r_h)$, the value of its total mass will not tend to 0. Since mr_h tends to zero as $\Upsilon \to \infty$, it must be instead that $r_h \to 0$ as $\Upsilon \to \infty$. This implies that, as Υ increases, more of the constant portion of the best fitting n^{th} excited state mass profile will be compared to the Burkert profile. Thus, as $\Upsilon \to \infty$, the best fitting n^{th} excited state mass profile will limit to the constant function of r that best fits the Burkert profile under the same fitting criteria used for the static states. We illustrate this phenomenon in Figure 8.

As $\Upsilon \to \infty$, the initial increasing region (i.e. the region before the constant portion) of the nth excited state mass profile that best fits a Burkert mass profile becomes more compressed. This initial region is where all of the dark matter mass is located. Thus as Υ increases, the dark matter corresponding to the best fit n^{th} excited state extends out to smaller radii. However, observations suggests that dwarf spheroidal galaxies are dark matter dominated at all observable radii [37]. Thus the best fit n^{th} excited state mass profiles for large Υ do not represent observations well and can be rejected. The question then is exactly when should we reject them.

Since every static state has the initial increasing region just described, the best fit n^{th} excited state for any value of Υ will be a better fit than the best fit constant function. Moreover, since this initial region becomes more compressed as $\Upsilon \to \infty$, for large Υ, the value of $E(r_h)$ for the best fit n^{th} excited state mass profile increases monotonically as $\Upsilon \to \infty$ approaching the value of E for the best fitting constant function.

This suggests a criteria for when to reject values of Υ. We will reject a best fit n^{th} excited state mass profile, and hence its corresponding value of Υ, as an untenable model of the dark matter mass if its value of $E(r_h)$ is greater than some prescribed fraction of the value of E for the best fitting constant function. We choose to use 80%. Explicitly, we use the following rejection criteria.

Rejection Criteria 3.1 *Given Υ, n, and a Burkert mass profile M_B, let M_W be the spherically symmetric n^{th} excited state mass profile corresponding to Υ that best fits M_B, that is, that minimizes E from equation (27) along the hyperbola defined by the value of Υ and equation*

Journal of Physics: Conference Series **615** (2015) 012001 doi:10.1088/1742-6596/615/1/012001

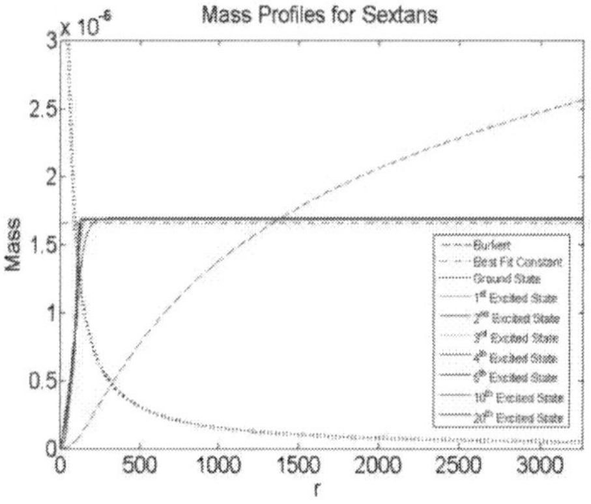

Figure 9. The Burkert mass profile found by Salucci et al. [6] for the Sextans galaxy. The best fit static state mass profiles for a ground through fifth excited state, tenth excited state, and twentieth excited state all lying on the same hyperbola are overlayed on the plot. The hyperbola here satisfies the rejection criteria for all of the different static states represented in the plot, thus all of these static states correspond to an upper bound on the value of Υ for Sextans for their respective value of n (i.e. the set of nth excited states). Note how close together all of the states are. This is due to the fact that the majority of their profiles which are being compared to the Burkert mass profile is the common and constant portion of the profiles.

(25). Let E_W be the value of E for this mass profile. Furthermore, let M_C be the constant function which best fits M_B, also by minimizing the corresponding function E, and let E_C be the value of E for the constant function. Reject the given value Υ as a tenable value for this fundamental constant if

$$E_W \geq .8E_C.$$

In other words, any fit that is less than 20% better than the best fitting constant function of r is rejected as a bad fit.

For each of the eight classical dwarf spheroidal galaxies and $n \in \{0, 1, 2, 3, 4, 5, 10, 20\}$, we computed values of Υ that yielded n^{th} excited state mass profiles that best fit that galaxy's Burkert profile which were rejected by the above criteria. All of the values of Υ above those computed are also rejected because they produce mass profiles even closer to the constant function. In Table 3, we have collected these upper bounds of Υ. In Figure 9, for the galaxy Sextans, we present best fit static state mass profiles for the ground through fifth, tenth, and twentieth excited states for which $E(r_h)$ is more than 80% of the value of E for the best fit constant function.

We observe from Table 3 that the upper bound values of Υ increase as we increase the state we consider. This is due to the following. The rejected values of Υ correspond to rejected hyperbolas of constant Υ, and hence constant mr_h. Furthermore, the only qualitative difference between any two n^{th} excited state mass profiles is the number of ripples in the initial increasing region of the profile. For large Υ, the majority of a best fit n^{th} excited state mass profile that is compared to the Burkert profile is the constant region which is shared by static state mass profiles for any n. Thus the hyperbola corresponding to a rejected best fit ground state is close to the hyperbola corresponding to a rejected best fit n^{th} excited state for any n. In particular, there is a hyperbola of constant mr_h, for which the corresponding best fit n^{th} excited state mass profiles for any n are rejected by the above criteria. Then, since C^n_{mass} and C^n_{radius} appear to monotonically increase as n increases (see Table 2), by equation (25), we would expect the same

Galaxy \ State	0	1	2	3
Sextans	$\Upsilon < 160$	$\Upsilon < 394$	$\Upsilon < 633$	$\Upsilon < 875$
Leo II	$\Upsilon < 234$	$\Upsilon < 576$	$\Upsilon < 926$	$\Upsilon < 1279$
Fornax	$\Upsilon < 35$	$\Upsilon < 87$	$\Upsilon < 139$	$\Upsilon < 192$
Leo I	$\Upsilon < 57$	$\Upsilon < 141$	$\Upsilon < 226$	$\Upsilon < 312$
Sculptor	$\Upsilon < 49$	$\Upsilon < 121$	$\Upsilon < 194$	$\Upsilon < 268$
Ursa Minor	$\Upsilon < 82$	$\Upsilon < 202$	$\Upsilon < 325$	$\Upsilon < 449$
Carina	$\Upsilon < 84$	$\Upsilon < 207$	$\Upsilon < 333$	$\Upsilon < 459$
Draco	$\Upsilon < 45$	$\Upsilon < 111$	$\Upsilon < 179$	$\Upsilon < 246$

Galaxy \ State	4	5	10	20
Sextans	$\Upsilon < 1116$	$\Upsilon < 1356$	$\Upsilon < 2460$	$\Upsilon < 4789$
Leo II	$\Upsilon < 1632$	$\Upsilon < 1983$	$\Upsilon < 3597$	$\Upsilon < 7003$
Fornax	$\Upsilon < 245$	$\Upsilon < 297$	$\Upsilon < 538$	$\Upsilon < 1048$
Leo I	$\Upsilon < 398$	$\Upsilon < 484$	$\Upsilon < 877$	$\Upsilon < 1706$
Sculptor	$\Upsilon < 342$	$\Upsilon < 416$	$\Upsilon < 754$	$\Upsilon < 1467$
Ursa Minor	$\Upsilon < 572$	$\Upsilon < 695$	$\Upsilon < 1261$	$\Upsilon < 2455$
Carina	$\Upsilon < 586$	$\Upsilon < 712$	$\Upsilon < 1292$	$\Upsilon < 2514$
Draco	$\Upsilon < 314$	$\Upsilon < 382$	$\Upsilon < 692$	$\Upsilon < 1347$

Table 3. Upper bound values for Υ corresponding to poor best fits of the Burkert mass profiles for each of the classic dwarf spheroidal galaxies. The values in each column for each galaxy should be interpreted as an upper bound on the value of Υ, under the approximations explained in the paper, if that galaxy is best modeled by an nth excited state. The units on Υ are yr^{-1}.

behavior for the value of Υ in order for mr_h to remain constant, which is what we observe in Table 3.

Thus if a dwarf spheroidal galaxy is correctly modeled by a twentieth excited state or less, then an overall upper bound on the value of Υ would be the upper bound corresponding to the twentieth excited state. The least upper bound corresponding to the twentieth excited state over all eight galaxies is that value for the Fornax galaxy, which yields approximately that

$$\Upsilon < 1000 \text{ yr}^{-1}. \tag{28}$$

3.4. Working Value of Υ

The last objective of this paper is to provide an example of a single value of Υ being qualitatively consistent with the dark matter mass profiles of each of the eight classical dwarf spheroidal galaxies. As shown in the last section, all of the values $\Upsilon < 1000$ are such examples to varying levels of tolerance. In this section, we present an example where the qualitative similarity is readily apparent.

For $\Upsilon = 50 \text{ yr}^{-1}$, there is at least one wave dark matter static state mass profile, which is a third excited state or less, for each of the Burkert mass profiles of the eight classical dwarf spheroidal galaxies that matches reasonably well. We have plotted such matches in Figure 10. Due to the fact that these mass profiles match reasonably well, we will use

$$\Upsilon = 50 \text{ yr}^{-1} \tag{29}$$

Journal of Physics: Conference Series **615** (2015) 012001 doi:10.1088/1742-6596/615/1/012001

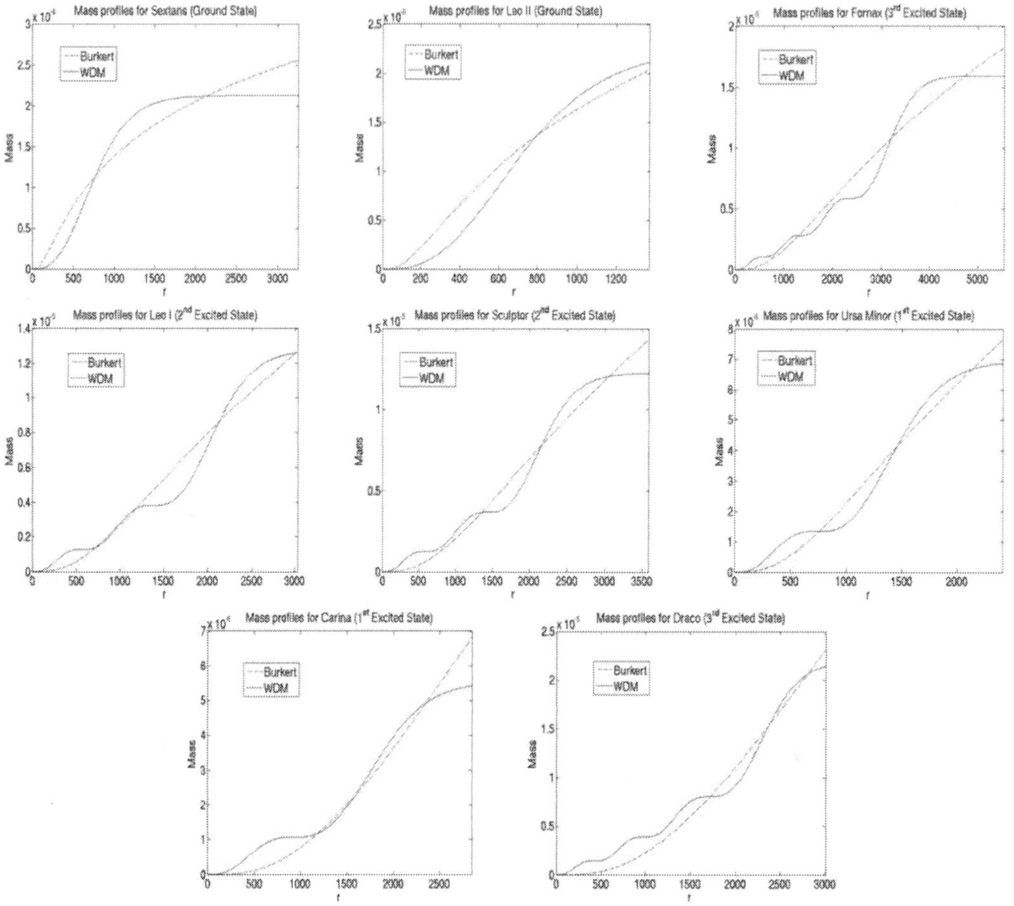

Figure 10. Static state mass profiles for $\Upsilon = 50$ which are each a best fit to the Burkert profiles of the corresponding dwarf spheroidal galaxy. For $\Upsilon = 50$, we picked an nth excited state whose best fit profile matched the Burkert profile qualitatively well. This shows that $\Upsilon = 50$ is a reasonable working value of Υ. However, it does not imply that the actual value of Υ is 50 or that these galaxies are correctly modeled by the presented n^{th} excited state. The units on Υ are yr^{-1}.

as a working value of Υ in our future work with wave dark matter until we have the capability to make a more accurate approximation or precise measurement of this value.

While we have chosen $\Upsilon = 50\ \text{yr}^{-1}$ as a working value of Υ since it corresponds to wave dark matter models compatible with other well fitting models, we note that the above does not constitute a precise measurement of the value of Υ. As said before, this is merely an example that a single value of Υ is sufficient to produce dark matter mass profiles that are qualitatively similar to the profiles found by Salucci et al.

4. Utilized Approximations

Now that we have presented our results, we list here the important approximations made in this paper which led to these results and explain briefly why we make them.

Approximation 1: Dark matter is correctly described by the wave dark matter model.

Approximation 2: Dark matter halos around dwarf spheroidal galaxies are spherically symmetric.

Approximation 3: The dwarf spheroidal galaxies used in this paper are in a state of dynamical equilibrium.

Approximation 4: The spacetime metrics describing these dwarf spheroidal galaxies are static.

Approximation 5: Wave dark matter predicts outcomes qualitatively similar to those of spherically symmetric static state solutions to the Einstein-Klein-Gordon equations.

Approximation 6: The Burkert mass profiles computed by Salucci et al. [6] fit the observational data very well.

Approximation 7: The spacetime is in the low field limit, that is, $M \ll r$.

Approximation 8: The spacetime is asymptotically Schwarzschild.

Approximation 1 is used because we are testing the wave dark matter model against observations. Approximation 2 is a common approximation for dwarf spheroidal galaxies and is also necessary because we are comparing the wave dark matter model to the spherically symmetric Burkert mass profile. Approximation 3 seems to be consistent with observations of dwarf spheroidal galaxies at least out to large radii [6, 38]. Approximation 6 is reasonable given Figure 1. Approximations 7 and 8 are standard when modeling galaxies.

Approximations 4 and 5 are used to simplify the types of solutions to the Einstein-Klein-Gordon equations we consider. We note here, however, that there is a question of the stability of the spherically symmetric static state solutions. It is known that, if the corresponding total mass is not too large, the ground state is stable under perturbations [22, 39] but that, on their own, the excited states are not [23] regardless of their mass. However, it has also been shown that a coupling of an excited state with a ground state can produce a stable configuration [21]. We hypothesize that luminous matter distributions coupled with combinations of static states will produce a stabilizing effect allowing for more dynamically interesting systems to be physically plausible.

5. Conclusions

To summarize the results of this paper, we have drawn effectively two conclusions, which we list here.

Conclusion 5.1 *Given Approximations 1 through 8 and Rejection Criteria 3.1, if the dark matter halos of all of the eight classical dwarf spheroidal galaxies are correctly modeled by* 20th *excited states or less, then*

$$\Upsilon < 1000\ yr^{-1}.$$

Conclusion 5.2 *Given Approximations 1 through 8, a value of* Υ *which yields one or more spherically symmetric static state mass profiles which match well the best fit Burkert mass profiles computed by Salucci et al. [6] for each of the eight classical dwarf spheroidal galaxies is*

$$\Upsilon = 50\ yr^{-1}.$$

We note here that the main result of this paper is more to describe a procedure of computing a working value and upper bound of Υ rather than stating that the values that appear in the above conclusions are the best ones. If one wished to alter the hypotheses of these conclusions, the corresponding values of Υ might differ from what we presented here. However, if one remains in the realm of using static states to model the dark matter, then the procedure presented in this paper for finding an upper bound would likely still apply and could be employed to get an upper bound under the new assumptions. As such, the conclusions above should be taken as an example of the procedure applied to a set of assumptions we currently find reasonable or useful and not as final precise estimates.

For the interested reader, the Matlab code used for this paper to generate the spherically symmetric static states and to compute the best fits to a Burkert profile can be found on Bray's Wave Dark Matter Web Page at `http://www.math.duke.edu/~bray/darkmatter/darkmatter.html` .

6. Acknowledgements
The authors gratefully acknowledge the support of National Science Foundation Grant # DMS-1007063.

References
[1] Oort J H 1932 *Bulletin of the Astronomical Institutes of the Netherlands* **6** 249
[2] Zwicky F 1933 *Helvetica Physica Acta* **6** 110–127
[3] Begeman K G 1989 *Astronomy and Astrophysics* **223**(1-2) 47–60
[4] Bosma A 1981 *The Astrophysical Journal* **86** 1791–1846 URL http://adsabs.harvard.edu/abs/1981AJ.....86.1791B
[5] Walker M G, Mateo M, Olszewski E W, Gnedin O Y, Wang X, Sen B and Woodroofe M 2007 *The Astrophysical Journal Letters* **667** L53 URL http://stacks.iop.org/1538-4357/667/i=1/a=L53
[6] Salucci P, Wilkinson M I, Walker M G, Gilmore G F, Grebel E K, Koch A, Frigerio Martins C and Wyse R F G 2012 *Monthly Notices of the Royal Astronomical Society* **420** 2034–2041 ISSN 1365-2966 URL http://dx.doi.org/10.1111/j.1365-2966.2011.20144.x
[7] Walker M G, Mateo M, Olszewski E W, Peñarrubia J, Evans N W and Gilmore G 2009 *The Astrophysical Journal* **704** 1274 URL http://stacks.iop.org/0004-637X/704/i=2/a=1274
[8] Walker M G, Mateo M, Olszewski E W, Peñarrubia J, Evans N W and Gilmore G 2010 *The Astrophysical Journal* **710** 886 URL http://stacks.iop.org/0004-637X/710/i=1/a=886
[9] Dahle H 2007 *[arXiv:astro-ph/0701598]* URL http://arxiv.org/abs/astro-ph/0701598v1
[10] Salucci P, Martins C F and Lapi A 2010 DMAW 2010 LEGACY the Presentation Review: Dark Matter in Galaxies with its Explanatory Notes URL http://www.sissa.it/ap/dmg/dmaw_presentation.html
[11] Hooper D and Baltz E A 2008 *Ann.Rev.Nucl.Part.Sci.* **58** 293–314
[12] Bertone G, Hooper D and Silk J 2005 *Physics Reports* **405** 279–390 ISSN 0370-1573 URL http://www.sciencedirect.com/science/article/pii/S0370157304003515
[13] Ostriker J 1993 *Annual Review of Astronomy and Astrophysics* **31** 689–716
[14] Trimble V 1987 *Annual Review of Astronomy and Astrophysics* **25** 425–472 URL http://www.annualreviews.org/doi/abs/10.1146/annurev.aa.25.090187.002233
[15] Binney J and Merrifield M 1998 *Galactic Astronomy* Princeton Series in Astrophysics (Princeton University Press)
[16] Binney J and Tremaine S 2008 *Galactic Dynamics* Princeton Series in Astrophysics (Princeton University Press)
[17] Bray H 2013 *AMS Contemporary Mathematics* **599** 1–64 URL http://arxiv.org/abs/1004.4016
[18] Bray H 2012 *[arXiv:1212.5745 [physics.gen-ph]]* URL http://arxiv.org/abs/1212.5745
[19] Lee J W 2009 *Journal of the Korean Physical Society* **54** 2622 URL http://arxiv.org/abs/0801.1442
[20] Matos T, Vázquez-González A and Magaña J 2009 *Monthly Notices of the Royal Astronomical Society* **393** 1359–1369 ISSN 1365-2966 URL http://dx.doi.org/10.1111/j.1365-2966.2008.13957.x
[21] Bernal A, Barranco J, Alic D and Palenzuela C 2010 *Phys. Rev. D* **81** 044031 URL http://arxiv.org/abs/0908.2435
[22] Seidel E and Suen W M 1990 *Phys. Rev. D* **42**(2) 384–403 URL http://link.aps.org/doi/10.1103/PhysRevD.42.384
[23] Balakrishna J, Seidel E and Suen W M 1998 *Phys. Rev. D* **58**(10) 104004 URL http://link.aps.org/doi/10.1103/PhysRevD.58.104004
[24] Bernal A, Matos T and Núñez D 2008 *Revista Mexicana de Astronomía y Astrofísica* **44** 149–160 URL http://arxiv.org/abs/astro-ph/0303455v3
[25] Sin S J 1994 *Phys. Rev. D* **50**(6) 3650–3654 URL http://link.aps.org/doi/10.1103/PhysRevD.50.3650
[26] Schunck F and Mielke E 2003 *Classical and Quantum Gravity* **20** R301–R356 URL http://stacks.iop.org/0264-9381/20/i=20/a=201

[27] Sharma R, Karmakar S and Mukherjee S 2008 *[arXiv:0812.3470 [gr-qc]]* URL http://arxiv.org/abs/0812.3470
[28] Ji S U and Sin S J 1994 *Phys. Rev. D* **50**(6) 3655–3659 URL http://link.aps.org/doi/10.1103/PhysRevD.50.3655
[29] Lee J W and Koh I G 1992 *Abstracts, Bulletin of the Korean Physical Society* **10**
[30] Lee J W and Koh I G 1996 *Phys. Rev. D* **53**(4) 2236–2239 URL http://link.aps.org/doi/10.1103/PhysRevD.53.2236
[31] Guzmán F S, Matos T and Villegas-Brena H 2001 *Rev. Mex. Astron. Astrofis.* **37** 63–72 URL http://arxiv.org/abs/astro-ph/9811143
[32] Parry A R 2012 *[arXiv:1212.6426 [gr-qc]]* URL http://arxiv.org/abs/1212.6426
[33] Burkert A 1995 *The Astrophysical Journal Letters* **447** L25 URL http://stacks.iop.org/1538-4357/447/i=1/a=L25
[34] Parry A R 2014 *Analysis and Mathematical Physics* **4** 333–375 ISSN 1664-2368 URL http://dx.doi.org/10.1007/s13324-014-0085-x
[35] Parry A R 2013 *Wave Dark Matter and Dwarf Spheroidal Galaxies* Ph.D. thesis Duke University URL http://arxiv.org/abs/1311.6087
[36] Bizoń P and Wasserman A 2000 *Communications in Mathematical Physics* **215** 357–373 ISSN 0010-3616 URL http://dx.doi.org/10.1007/s002200000307
[37] Kleyna J, Wilkinson M I, Evans N W, Gilmore G and Frayn C 2002 *Monthly Notices of the Royal Astronomical Society* **330** 792–806 ISSN 1365-2966 URL http://dx.doi.org/10.1046/j.1365-8711.2002.05155.x
[38] Côté P, Mateo M, Olszewski E W and Cook K H 1999 *The Astrophysical Journal* **526** 147 URL http://stacks.iop.org/0004-637X/526/i=1/a=147
[39] Lai C W and Choptuik M W 2007 *[arXiv:0709.0324 [gr-qc]]* URL http://arxiv.org/abs/0709.0324

Rotation curve for the Milky Way galaxy in conformal gravity

James G. O'Brien[1] **and Robert J. Moss**[2]

1 Department of Sciences, [2] Department of Computer Science and Networking, Wentworth Institute of Technology; 550 Huntington Ave. Boston, Massachusetts 02115

E-mail: obrienj10@wit.edu[1], mossr@wit.edu[2]

Abstract. Galactic rotation curves have proven to be the testing ground for dark matter bounds in galaxies, and our own Milky Way is one of many large spiral galaxies that must follow the same models. Over the last decade, the rotation of the Milky Way galaxy has been studied and extended by many authors. Since the work of conformal gravity has now successfully fit the rotation curves of almost 140 galaxies, we present here the fit to our own Milky Way. However, the Milky Way is not just an ordinary galaxy to append to our list, but instead provides a robust test of a fundamental difference of conformal gravity rotation curves versus standard cold dark matter models. It was shown by Mannheim and O'Brien that in conformal gravity, the presence of a quadratic potential causes the rotation curve to eventually fall off after its flat portion. This effect can currently be seen in only a select few galaxies whose rotation curve is studied well beyond a few multiples of the optical galactic scale length. Due to the recent work of Sofue et al and Kundu et al, the rotation curve of the Milky Way has now been studied to a degree where we can test the predicted fall off in the conformal gravity rotation curve. We find that – like the other galaxies already studied in conformal gravity – we obtain amazing agreement with rotational data and the prediction includes the eventual fall off at large distances from the galactic center.

1. Introduction

Since its inception over seventy five years ago, the concept of dark matter has been widely accepted as the explanation for the missing mass problem in spiral galaxies. Many current large scale collaborations have been created to search for dark matter with little notable success. In the current paradigm, the parameter space of cold dark matter is being constrained to higher and higher degrees, but physical searches have turned up with null or conflicting results. Most notably, the recent Cryogenic Dark Matter Search (CDMS) [1] faced the same unfortunate conclusions as many of its predecessors. As the community further pushes technology to probe deeper into the possible parameter space of dark matter, the lack of pure observable evidence and reliance on the rotation curves of galaxies has opened the door for many alternative theories to standard Einstein Gravity. Many of these alternative theories attempt to solve the rotation curve problem without the need for invoking dark matter. Although many of these theories have had success in fitting the rotation curves of spiral galaxies in the past, such as Modified Newtonian Dynamics (MOND) [2], Scalar Tensor Vector Gravity (STVG) [3], and more recently the Luminous Convolution Model (LCM) [4], still many of these theories struggle with the universality of galactic rotational dynamics. However, recently one of the authors has fit a diverse set of over 130 rotation curves of various galaxies using Conformal Gravity (CG) as an alternative gravitational model [5]. Mannheim and O'Brien have successfully fit the most recent data of the THINGS survey [6] as well as the Ursa Major galaxies of Verheijen et al [7]. Their research continued with less studied low surface brightness galaxies of Kim et al [8] and a very recent survey of dwarf galaxies by Swaters [9]. To further diversify the studies performed by Mannheim and O'Brien, they have fit conformal gravity to the three Tidal Dwarf Galaxies (TDG) of NGC 5291 with an astonishing degree of success [10]. We note that the TDG galaxies are a unique testing ground of gravitational physics in the sense that they offer an observable rotation curve [2] despite the fact that they should not be in existence due to the parameter space defined by ΛCDM.

With these surveys fully populating the spectrum of studied galactic rotation curves, Conformal Gravity emerges as an alternative gravitational theory that can universally explain the rotation curve problem, and unlike theories such as MOND the creation of CG was not formulated to address the rotation curve problem alone. For the discussion of this paper, the outstanding difference between conformal gravity and the other alternative gravitational theories is the presence of a quadratic potential, which neither forces rotation curves to forever remain flat or allows rotation curves to infinitely rise. Instead as shown in [11] the presence of the quadratic term begins to compete at large distance scales and eventually forces the stable orbits of particles in the presence of the gravitational potential of the galaxy to terminate. Hence, extremely large galaxies, viz. ones whose outermost data points are well beyond the optical scale lengths, serve as a perfect testing ground for this prediction of the conformal theory. In this paper, we discuss how the latest rotation curve data of the Milky Way galaxy provides us with a great comparison of various rotation curve models, with conformal gravity being able to once again explain the rotation observed without invoking copious dark matter. Moreover, due to the shape of the rotation curve of the Milky Way galaxy, conformal gravity's predication of an eventual fall off of the rotational velocity can be illuminated and discussed.

2. The Milky Way Galaxy Data

The Milky Way has long been a galaxy studied by astronomers. For many years the rotation curve of the Milky Way has been known, but was not of particular interest to the missing mass problem due to its relatively low density of data points, and its early termination of optical data. The standard rotation curve can be obtained as seen in [12] where the observable parameters are: the distance to the galactic center, R, the adopted distance from the Sun to the galactic center, R_{S0}, the relative velocity of the Local Standard of Rest, V_{LSR}, and the angular direction of the tracer stars, (l, b). The rotation velocity, V_{ROT}, is then explained by the formula

$$V_{ROT} = \frac{R}{R_{S0}} \left(\frac{V_{LSR}}{\sin l \cdot \cos l} - 220 \right). \tag{1}$$

With the advent of recent observational techniques, many new types of radio tracers can be used to increase the validity and density of the Milky Way rotation curve. In 2009, Sofue [13] published a unified rotation curve of the Milky Way galaxy synthesizing data from over eight different tracer sources (including the aformentioned [12]) to build one of the densest rotation curves ever produced, consisting of 610 data points. The vast data points observed by Sofue highlight the structure of the Milky Way, including a bulge in the inner region, and extend the data to 20 kpc.

The work of Sofue allowed for the structure of the Milky Way for $R \sim 20$ kpc [13] to be well established. The high density of points for $R < 20$ kpc can easily be seen in Fig. 1(a). The need to extend the rotation velocity past the disk region is expressed throughout the current literature [12]. Xue et al was able to then extend the Milky Way rotation curve to approximately 60 kpc [14]. Recent work by Kundu et al [15] has extended the rotation curve from the local

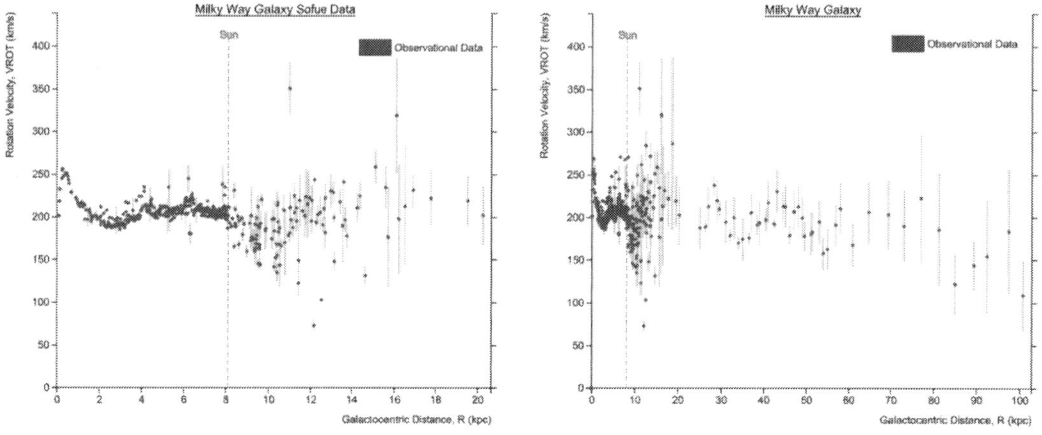

(a) (b)

Figure 1. (a) The dense number of observations taken by Sofue. Note that the uncertainty increases when tracing objects away from the galactic center. (b) The full synthesized data set of the rotation curve.

disk region to the outer regions of the galaxy, continuing far past the galactic disk using Blue Horizontal Branch tracers (BHB) to an astonishing distance. They note that the rotation curve is slowly declining as R increases. Here we synthesized the data sets from [13], [14] and [15] which accumulate to a total of 636 observed data points for the rotation curve as shown in Fig. 1(b). The falloff of the observed rotation velocity serves as a testing ground for the predictions of competing theories, and separates the conformal theory in its unique prediction of eventual falloff.

3. Models of the Milky Way rotation curve

3.1. General Relativity Prediction

Similar to all other observed rotation curves, the Milky Way suffers from the same missing mass problem. The prediction set forth by general relativity (GR) can be found by starting with a single point mass solution to the Einstein field equations, and then modeling the galaxy as a collection of point masses arranged in a disk in superposition. We assume for simplicity that the disk is infinitely thin, and the distribution of mass falls exponentially as

$$\Sigma(R) = \Sigma_0 e^{-\frac{R}{R_0}} \tag{2}$$

where R_0 is the luminous scale length, and Σ_0 is the central density. Upon integrating over the disk in cylindrical coordinates, one arrives at the familiar

$$v_{GR}(R) = \sqrt{\frac{N^*\beta^* c^2 R^2}{2R_0^3}\left[I_0\left(\frac{R}{2R_0}\right)K_0\left(\frac{R}{2R_0}\right) - I_1\left(\frac{R}{2R_0}\right)K_1\left(\frac{R}{2R_0}\right)\right]}, \tag{3}$$

where I_0, I_1, K_0, and K_1 are Bessel functions. This is the well established Freeman curve, and is assumed that each parameter is fixed. The only free parameter in this equation then is the overall number of stars,

$$N^* = \frac{M_{disk}}{M_\odot}. \tag{4}$$

Although this is a free parameter for fitting purposes, it is physically bounded by the preservation of the mass to light ratio to be on the order of unity.

3.2. Lambda Cold Dark Matter (ΛCDM) Model

To solve the missing mass problem, one can assume that since the velocity is a function of R and M, then more mass at the appropriate locations from the galactic center could rectify the GR prediction to match the data and hence cold dark matter can be introduced. In order to make the data match the prediction of equation (3), we can assume the total rotational velocity would be given as

$$v_{total}(R) = \sqrt{v_{GR}^2 + v_{dark}^2}. \tag{5}$$

The question then arises as how to put the dark matter into the galaxy as to match the prediction, but not force the inner region to then overshoot in turn. Following the prescription described in [6] the dark matter contribution can take the form

$$v_{dark}(R) = \sqrt{4\pi\beta^* c^2 \sigma_0 \left[1 - \frac{r_0}{R}\arctan\left(\frac{R}{r_0}\right)\right]} \tag{6}$$

where σ_0 is the dark matter density and r_0 is the dark matter halo radius. Since we have not yet physically observed dark matter, these two parameters have to be fit to the data using a χ^2 test against v_{total}. The two free parameters effectively shape the modeled rotation curve, and when coupled with the free parameter of the total luminous mass, M_{disk}, the theory has three total free parameters. It should be noted that although this is only two extra free parameters for the Milky Way, this process must be arbitrarily done for any studied galaxy, making the number of fitting parameters for a given sample twice the number of galaxies studied. Due to the nature of equation (6), the dark matter contribution for a given galaxy can cause the rotation curves to become asymptotically flat, hence can never predict an overall fall off. As described by the work in [11], we believe it is this very observation that can lead to overall departures from dark matter fitting in the largest of the studied galaxies. Since the Milky Way data is now the largest diameter galaxy we have studied, it makes our own galaxy an ideal test of conformal gravity against dark matter.

3.3. Conformal Gravity Prediction

The conformal gravity theory originally derived from Weyl, was later re-studied by Mannheim and Kazanas [5]. Conformal gravity retains a completely covariant metric theory of gravity but also includes the feature of local conformal invariance, where the action is unchanged due to local transformations $g_{\mu\nu}(x) \rightarrow e^{2\alpha(x)} g_{\mu\nu}(x)$ with local phase $\alpha(x)$. It should be noted that although it is an alternative theory of gravity, since it is still a metric theory of gravity, many of the familiar properties of General Relativity such as curvature of space and time, and the coupling of electromagnetic fields to gravity (bending of light) are retained. Moreover, the conformal theory was not originally studied to solve the rotation curve problem, but instead one may use the rotation curve problem as a testing ground for the overall theory. Conformal gravity assumes a scalar action which is described by

$$I_W = -\alpha_g \int d^4x(-g)^{1/2} C_{\lambda\mu\nu\kappa} C^{\lambda\mu\nu\kappa} = -2\alpha_g \int d^4x(-g)^{1/2} \left[R_{\mu\kappa}R^{\mu\kappa} - (1/3)(R^\alpha{}_\alpha)^2\right] \tag{7}$$

where

$$C_{\lambda\mu\nu\kappa} = R_{\lambda\mu\nu\kappa} - \frac{1}{2}\left(g_{\lambda\nu}R_{\mu\kappa} - g_{\lambda\kappa}R_{\mu\nu} - g_{\mu\nu}R_{\lambda\kappa} + g_{\mu\kappa}R_{\lambda\nu}\right) + \frac{1}{6}R^\alpha{}_\alpha\left(g_{\lambda\nu}g_{\mu\kappa} - g_{\lambda\kappa}g_{\mu\nu}\right), \tag{8}$$

$$4\alpha_g W^{\mu\nu} = 4\alpha_g \left[2C^{\mu\lambda\nu\kappa}{}_{;\lambda;\kappa} - C^{\mu\lambda\nu\kappa}R_{\lambda\kappa}\right] = 4\alpha_g \left[W^{\mu\nu}_{(2)} - \frac{1}{3}W^{\mu\nu}_{(1)}\right] = T^{\mu\nu}, \tag{9}$$

and

$$\begin{aligned} W^{\mu\nu}_{(1)} &= 2g^{\mu\nu}(R^{\alpha}{}_{\alpha})^{;\beta}{}_{;\beta} - 2(R^{\alpha}{}_{\alpha})^{;\mu;\nu} - 2R^{\alpha}{}_{\alpha}R^{\mu\nu} + \frac{1}{2}g^{\mu\nu}(R^{\alpha}{}_{\alpha})^2, \\ W^{\mu\nu}_{(2)} &= \frac{1}{2}g^{\mu\nu}(R^{\alpha}{}_{\alpha})^{;\beta}{}_{;\beta} + R^{\mu\nu;\beta}{}_{;\beta} - R^{\mu\beta;\nu}{}_{;\beta} - R^{\nu\beta;\mu}{}_{;\beta} - 2R^{\mu\beta}R^{\nu}{}_{\beta} + \frac{1}{2}g^{\mu\nu}R_{\alpha\beta}R^{\alpha\beta}. \end{aligned} \tag{10}$$

Since the goal is to construct a galaxy in the same procedure as done in the GR prediction, we note that $W^{\mu\nu}$ vanishes when $R^{\mu\nu}$ vanishes. However, Mannheim and Kazanas noted that the vanishing solutions may not be the only solutions to the fourth order theory. Hence they solved the metric outside a static, spherically symmetric source of radius a. They found, as described in [5], that in the conformal theory the exact line element is given by $ds^2 = -B(r)dt^2 + B(r)^{-1}dr^2 + r^2 d\Omega_2$ where the exterior metric coefficient $B(r > a)$ is given by

$$B(r > a) = 1 - \frac{2\beta}{r} + \gamma r - kr^2. \tag{11}$$

We can immediately see the departures from the GR prediction, where as the two new factors γ and k arise due to the fact that the conformal theory is fourth order and thus must contain two additional terms. It should also be noted that when γ and k are small, we return the exact Schwarzchild solution. For a more rigorous treatment of the derivation see [6]. Now that we have the "Schwarzchild like" solution for conformal gravity, we can effectively follow the procedure above by noting that a galaxy is a disk with exponential density falloff in the radial direction. The only other issue that conformal gravity needs to account for is local vs. global effects. Since the theory is fourth order in construction, we no long possess the power of a global guess law. Hence, the integration must be made both locally and globally [6] which gives rise to the total rotational prediction of the galaxy as

$$v_{CG}(R) = \sqrt{v_{GR}^2 + \frac{N^*\gamma^*c^2R^2}{2R_0} I_1\left(\frac{R}{2R_0}\right) K_1\left(\frac{R}{2R_0}\right) + \frac{\gamma_0 c^2 R}{2} - \kappa c^2 R} \tag{12}$$

where the integration constants are set as

$$\begin{aligned} \gamma^* &= 5.42\text{e}^{-41}\text{cm}^{-1}, \\ \gamma_0 &= 3.06\text{e}^{-30}\text{cm}^{-1}, \\ \kappa &= 9.54\text{e}^{-54}\ \text{cm}^{-2}. \end{aligned}$$

As can be seen in equation (11) and (12), the presence of the linear and quadratic potential terms are negligible on solar system scales, but would begin to dominate at galactic scales. The key feature of the solution is the requirement that the k term be negative [11] which forces the quadratic term to compete and eventually dominate over the linear term. The result is the termination of stable orbits in the galaxy and the ultimate fall off of a galactic rotation curve very far from the center. It is this feature that was tested in a sample of the fourteen largest studies galaxies [11], and due to the recent Milky Way data described above, can now be tested in our own galaxy.

4. Milky Way as a test of Conformal Gravity

4.1. Input Parameters for the Milky Way

For the Milky Way galaxy, we use the same set of input parameters as per a normal rotation curve as discussed in [6]. The gas mass of the Milky Way is adopted from McGaugh 2008 [16] as $M_g = 1.18 * 10^{10} M_\odot$ which is inclusive of the HI as well as helium. For the scale length, as will be shown in the figures, we adopt the range of scale lengths from Porcel 1997 [17] as $R_0 = 2.1 \pm 0.3\ kpc$. To be able to best compare our fits with the MOND fits of [16], we will plot the Milky Way galaxy for the range of scale lengths from $2.0 \leq R_0 \leq 2.5$. The luminosity is adopted from Spruit [18] as $L = 1.62 * 10^{10} L_\odot$ and will be used to compute the mass to light ratios. For the gas scale length we adopt the same fitting procedure as in [6] where we use a gas scale length four times the optical scale length. Lastly, we implement a bulge contribution for the noted galactic bulge as in [6] such that,

$$v_{bulge}(R) = \sqrt{\frac{2N_b^* \beta^* c^2}{\pi R} \int_0^{R/t} dz\, z^2 K_0(z)}. \tag{13}$$

The total contribution of rotation for the Milky Way galaxy in the conformal theory is then given by:

$$v_{tCG}(R) = \sqrt{v_{CG}^2 + v_{gas}^2 + v_{bulge}^2} \tag{14}$$

It should be noted that for comparison, the gas contribution and bulge contribution are also added to the other respective theories in the final fits that will be presented.

4.2. Fits

Here we present the fits for equations (3), (6), and (14). To produce the fits we have made use of a new computational tool called the Rotation Curve Modeler (RoCM) which was developed recently by the two authors [19]. The specific input and output parameters are listed in Table 1. We see that due to the synthesized rotation curve shown in Fig. 2, the outermost points are the ones where the predictions of conformal gravity can truly be tested. As seen in all of the plots, the essence of the inner most points including the bulge are well described by general relativity, ΛCDM, and conformal gravity alike. In all of the fits, general relativity falls far too quickly to account for the data as described in the missing mass problem. However, due to the nature of the dark matter equation (6), the outermost points of the rotation curve prediction will continue to remain flat even as the data begins to show a falloff. The falloff for each point of the Milky Way is predicted and captured by the conformal theory.

Plot	R_0 (kpc)	M_{disk} $(10^{10} M_\odot)$	M/L $(M_\odot/L_\odot)$
Fig. 2(a)	2.0	5.4738	3.38
Fig. 2(b)	2.1	5.189	3.2
Fig. 2(c)	2.2	5.1738	3.19
Fig. 2(d)	2.3	5.2831	3.26
Fig. 2(e)	2.4	5.3262	3.29
Fig. 2(f)	2.5	5.2122	3.22

Table 1. List of parameters used to fit the rotation curve with varying scale length, R_0, and disk mass, M_{disk}, to satisfy an acceptable mass to light ratio, M/L.

As another subject of comparison, we include in each figure, the dotted vertical line which is representative of where our sun lies in the Milky Way galaxy. For the dark matter theory,

Journal of Physics: Conference Series **615** (2015) 012002 doi:10.1088/1742-6596/615/1/012002

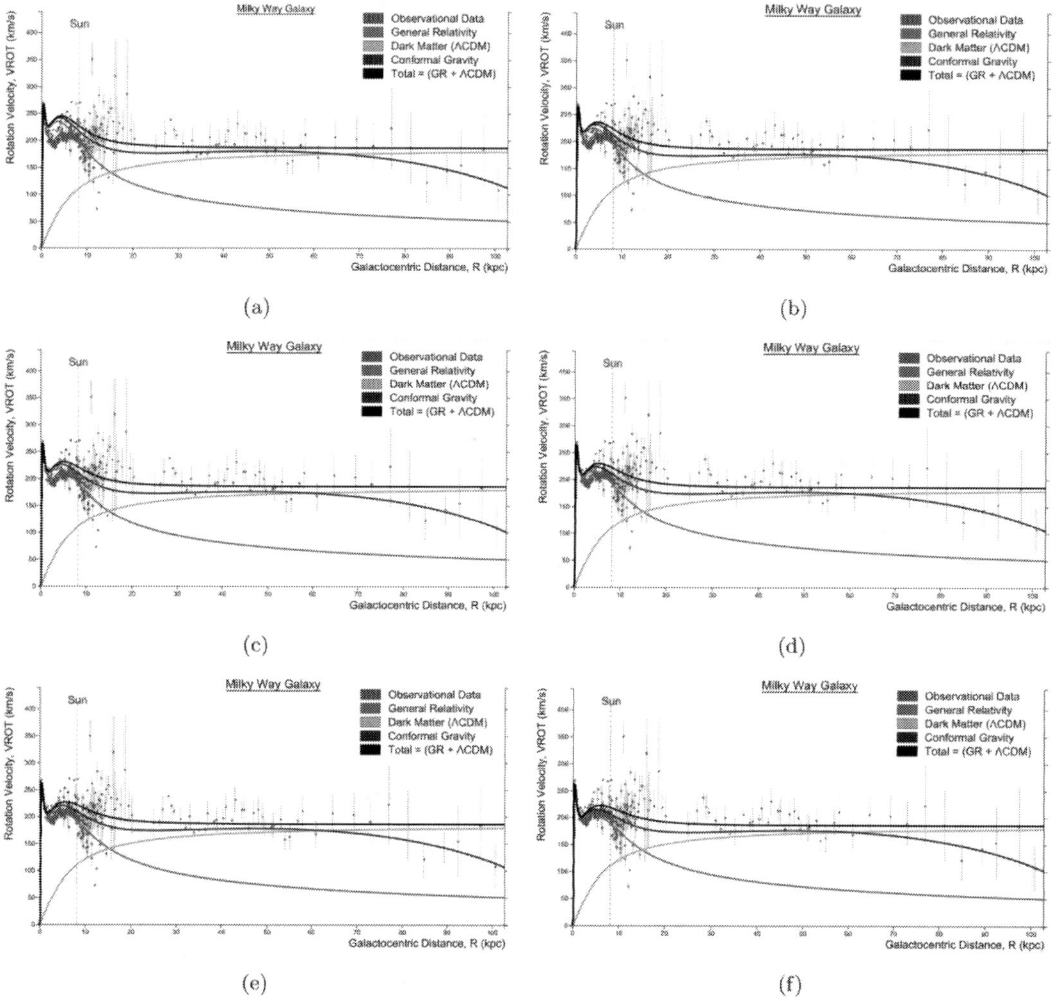

Figure 2. The scale length ranging from 2.0-2.5 kpc. These fits were produced using the RoCM software [19], and the colored predictions are described in the legend

we see that the fits are roughly generated by a $40 - 50$ percent dark matter contribution. This poses a problem for nearby neighborhood dark matter searches which have returned null results, such as the recent work illustrated in [20]. The conformal theory once again can establish the predicted rotation even at the location of our sun with only baryonic matter. It should also be noted that we chose to plot the six figures of various scale length since MOND [16] seems to prefer only a smaller scale length for the Milky Way galaxy. In our fits, it's shown that the conformal theory is much less sensitive to the scale length. Instead, conformal gravity provides a model of the galaxy at any of the accepted scale lengths without the need for copious amounts of dark matter while still preserving a physically acceptable mass to light ratio.

5. Analysis

In order to better capture the overall power of the conformal gravity fit versus the standard theory or ΛCDM we produce a density plot of the velocity discrepancy. Here we take a point

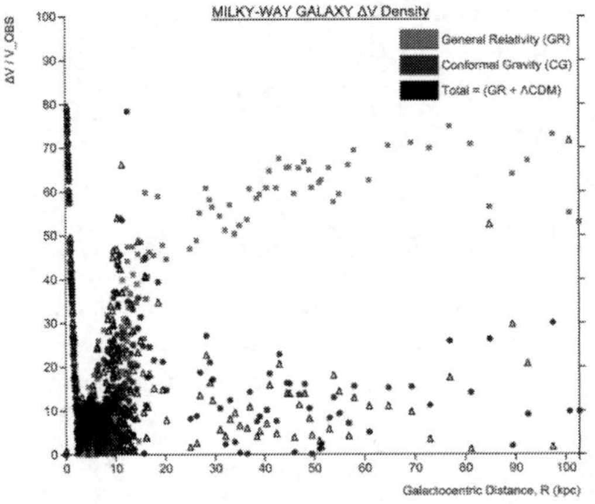

(a)

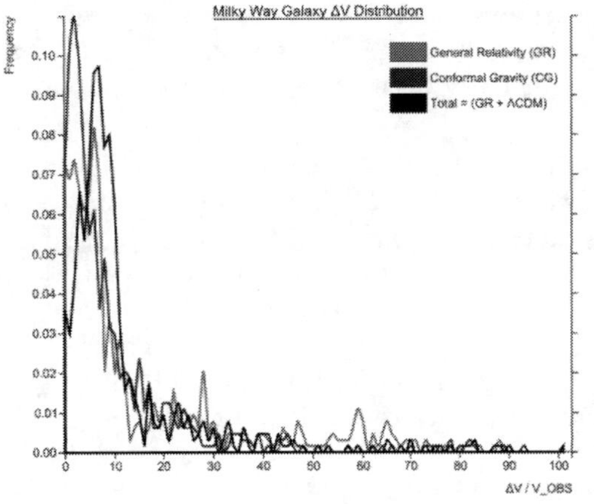

(b)

Figure 3. (a) The density plot of the velocity discrepancy relative to the galactocentric distance. (b) The frequency distribution of the velocity discrepancy which illustrates the aggregate comparison of each model to the observational data without the dependence of the galactocentric distance. Note that the vertical axis represents the frequency of the binned $\Delta V/V_{obs}$.

by point analysis of the 600+ points surveyed for the Milky Way and for each point compute $\Delta V/V_{obs}$ for each of the theories, where $\Delta V = |V_{obs} - V_{theory}|$. This gives us a more concrete comparison than a pure eyeball argument as to which prediction fits the data more precisely. The statistical fit is given in Fig. 3(a) and is shown for the Milky Way at scale length of $R_0 = 2.1$. This particular R_0 was chosen since it is the central value given in the literature. We see that much like the eyeball comparison of the fits, conformal gravity yields similar desirable statistics without the need for invoking dark matter.

To eliminate the dependency of the galactocentric distance from our analysis, we've generated a distribution of the velocity discrepancy for the Milky Way galaxy in Fig. 3(b). To clarify the comparison of each theory relative to the observations, the distribution can be used as an overall measure for the quality of the models. When comparing the distribution of conformal gravity vs. ΛCDM, it's straightforward to understand that conformal gravity predicts the observations to a higher degree. Each plot in Fig. 3 was generated by the RoCM tool [19] and can be produced for any arbitrary galaxy, and its data bin can be output. Future analysis will focus on comparing sets of galaxies with the density and distribution of the velocity discrepancy in order to achieve an aggregate analysis of each model. This can be accomplished using the power of RoCM and will result in a direct comparison of various models over entire sets of data.

6. Conclusion

In this work we presented the various rotation models of the synthesized data of the Milky Way galaxy. We leveraged the new tool developed by Moss and O'Brien [19] to simultaneously fit and analyze the rotation curve data for the Milky Way in the standard theory, conformal gravity, and the standard dark matter theory. We find that the Milky Way can now be added to the large galaxy survey of [11] as one of the largest galactic rotation curves currently in the literature. We find that like the other large galaxies in [11] the Milky Way can be fit by conformal to a high degree of precision without the need for dark matter. Further analysis will be to revisit the work of [11] with the current modeling power of [19] in order to build an overall velocity discrepancy analysis of an unbiased survey of galaxies without reference to particular distances as in Fig. 3(b). The authors are pleased to note that the tool [19] used in analyzing the data has currently been made open to the public domain at www.wit.edu/rotationcurve.

Acknowledgments
The authors would like to thank the International Association of Relativistic Dynamics (IARD) for the chance to present this work at the biannual IARD 2014 conference in Storrs, CT. J. G. O'Brien would also like to thank Martin Land for his help and support in the organization of IARD 2014. J. G. O'Brien would also like to thank Philip Mannheim for his continued efforts in both IARD local organization as well as contributions to conformal gravity. R. J. Moss would like to thank Patrick McGee, David Miller, and Alex Clement for their contribution to [19] which made the analysis of this work seamless.

References

[1] Agnese R, Ahmed Z, Anderson A J, Arrenberg S, Balakishiyeva D, Basu Thakur R, Bauer D A, Billard J, Borgland A, Brandt D and et al (CDMS Collaboration) 2013 *Phys. Rev. Lett.* **111**(25) 251301

[2] Famaey B and McGaugh S S 2012 *Living Reviews in Relativity* **15** 10 (*Preprint* 1112.3960)

[3] Brownstein J R and Moffat J W 2006 *A. J.* **636**

[4] Cisneros S, Oblath N S, Formaggio J A, Ott R A, Chester D, Battaglia D J, Ashley A, Robinson R and Rodriguez A 2014 *ArXiv e-prints* (*Preprint* 1407.7583)

[5] Mannheim P D and Kazanas D 1989 *ApJ* **342** 635–638

[6] Mannheim P D and O'Brien J G 2012 *Physical Review D* **85** 124020 (*Preprint* 1011.3495)

[7] Verheijen M A W and de Blok W J G 1999 *As. Sp. Sc.* **269-270** 673

[8] Kim J H 2007 Ph. d. dissertation University of Maryland

[9] Swaters R A, van Albada T S, van der Hulst J M and Sancisi R 2002 *A. A.* **390** 829

[10] Mannheim P D and O'Brien J G 2013 *Journal of Physics: Conference Series* **437**

[11] Mannheim P D and O'Brien J G 2011 *Phys. Rev. Lett.* **106**(12) 121101

[12] Demers S and Battinelli P 2007 *A. A.* **473** 143–148

[13] Sofue Y, Honma M and Omodaka T 2009 *Publications of the ASJ* **61** 227– (*Preprint* 0811.0859)

[14] Xue X X, Rix H W, Zhao G, Fiorentin P R, Naab T, Steinmetz M, van den Bosch F C, Beers T C, Lee Y S, Bell E F, Rockosi C, Yanny B, Newberg H, Wilhelm R, Kang X, Smith M C and Schneider D P 2008 *The Astrophysical Journal* **684** 1143 URL http://stacks.iop.org/0004-637X/684/i=2/a=1143

[15] Bhattacharjee P, Chaudhury S and Kundu S 2014 *Ap. J.* **785** 63

[16] McGaugh S 2008 *ApJ* **683** 137–148 (*Preprint* 0804.1314)

[17] Porcel C, Garzon F, Jimenez-Vicente J and Battaner E 1998 *aap* **330** 136–138 (*Preprint* astro-ph/9710197)

[18] Abbott B, Abbott R, Adhikari R, Agresti J, Ajith P, Allen B, Amin R, Anderson S B, Anderson W G, Arain M and et al 2008 *prd* **77** 062002 (*Preprint* 0704.3368)

[19] Moss R J and O'Brien J G 2014 *Journal of Undergrad Research in Physics* **In Review**

[20] Kafle P R, Sharma S, Lewis G F and Bland-Hawthorn J 2014 *The Astrophysical Journal* **794** 59

Static solutions of Einstein's field equations for compact stellar objects

Omair Zubairi[1,2], Alexis Romero[2], and Fridolin Weber[2,3]

[1] Computational Science Research Center, San Diego State University, 5500 Campanile Drive, San Diego, CA 92182, USA

[2] Department of Physics, San Diego State University, 5500 Campanile Drive, San Diego, CA 92182, USA

[3] Center for Astrophysics and Space Sciences, University of California, San Diego, La Jolla, CA 92093, USA

E-mail: zubairi@rohan.sdsu.edu[1,2], romero14@rohan.sdsu[2], fweber@mail.sdsu.edu[2,3]

Abstract. In this work, we present two solutions of Einstein's field equations for compact stellar objects such as quark or neutron stars. Due to their unique stellar properties, these compact objects pose as excellent laboratories to study matter in the most extreme conditions. In part one of this paper, we solve Einstein's field equations modified for a finite value for the cosmological constant for spherically symmetric mass distributions. This solution has been presented in the literature before, but with inconsistent results. In part two, we examine the structure of deformed (non-spherical) compact objects. The stellar structure equations are derived and solved for in the limiting case of isotropic pressure and energy-density. We calculate stellar properties such as masses and radii along with pressure and density profiles for these deformed objects and investigate changes from the standard spherical models.

1. The Cosmological Constant and Compact Stars

In recent years, scientists have discovered that the expansion rate of the Universe is increasing rather than decreasing [1, 2]. This cosmic acceleration leads to an additional term in Einstein's field equation which is known as the cosmological constant. Albert Einstein originally introduced this term in 1916 as a modification of his field equation to achieve a stationary Universe, but quickly abandoned the concept after Hubble's discovery of the Universe expanding. This work explores the consequences of a non-vanishing cosmological constant for spherically symmetric mass distributions of compact stars such as quark or neutron stars.

In order to study compact stars, we must first investigate the stellar structure of these objects. The stellar properties of any compact object are dependent on the star's internal structure, which is described by the equation of state (EoS). Using Einstein's field equations, J. R. Oppenheimer and G. M. Volkoff [3] derived a differential equation using the work of R. C. Tolman [4] that describes the stellar structure of a compact object of isotropic material in hydrostatic equilibrium. This first order differential equation known as the Tolman-Oppenheimer-Volkoff (TOV) equation (Eq. (1)) is

$$\frac{dP}{dr} = -\frac{\varepsilon\left(1+\frac{P}{\varepsilon}\right) m \left(1+\frac{4\pi P r^3}{m}\right)}{r^2\left(1-\frac{2m}{r}\right)} , \tag{1}$$

where the natural units of $G = c = 1$. In Eq. (1) P is the pressure, ε is the energy-density, and m is the mass inside a spherical mass shell of radius r.

It has been suggested by a few authors [5, 6, 7] what form Eq. (1) should have with a modification of a cosmological constant. In previous work conducted by Reddy and Silbar,they suggest that only the numerator of Eq. (1) would change [7]; however, upon further inspection we find that this is not the only modification, but another term must be in the denominator. Papers from Ghezzi [6] and Vinayaraj and Kuriakose [5] show a modified TOV equation that includes a cosmological constant; however, we find that there are some inconsistencies with some of the terms in their TOV equation. In Section 1.1, we present the key equations and perform a derivation of the TOV equation from Einstein's field equation for a non-vanishing cosmological constant. For the entire step by step derivation, see [8]. In Section 1.3, solutions for compact stellar mass distributions (such as mass, radius, pressure and density profiles) are then obtained by the use of a simple relativistic quark equation of state (EoS).

1.1. Derivation of Modified TOV Equation

We start our calculations with the line element for a spherically symmetric mass distribution which has the form [9, 10]

$$ds^2 = \mathrm{e}^{\nu}dt^2 - \mathrm{e}^{\lambda}dr^2 - r^2d\theta^2 - r^2\sin^2(\theta)d\phi^2 \ , \tag{2}$$

where $\nu = \nu(r)$ and $\lambda = \lambda(r)$ are unknown metric functions. The non-vanishing components of the metric tensor in covariant form are given by

$$\begin{aligned} g_{tt} &= \mathrm{e}^{\nu} \ , \\ g_{rr} &= -\mathrm{e}^{\lambda} \ , \\ g_{\theta\theta} &= -r^2 \ , \\ g_{\phi\phi} &= -r^2\sin^2(\theta) \ , \end{aligned} \tag{3}$$

so that Eq. (2) can be written as

$$ds^2 = g_{\mu\nu}dx^{\mu}dx^{\nu} \ . \tag{4}$$

Due to spherical symmetry and time independence, the metric functions only have radial dependence. We start our calculations from Einstein's field equations in mixed representation given by

$$G^{\mu}_{\lambda} \equiv R^{\mu}_{\lambda} - \frac{1}{2}\delta^{\mu}_{\lambda}R + \Lambda\delta^{\mu}_{\lambda} = -8\pi T^{\mu}_{\lambda} \ , \tag{5}$$

where R^{μ}_{λ} is the Ricci tensor, R is the Ricci scalar, Λ is the cosmological constant, and T^{μ}_{λ} is the energy-momentum tensor which is given in terms of pressure P and the energy-density ε of the stellar matter [9, 10]

$$T^{\mu}_{\lambda} = (\varepsilon + P)\frac{dx_{\lambda}}{ds}\frac{dx^{\mu}}{ds} - \delta^{\mu}_{\lambda}P \ . \tag{6}$$

The energy-momentum tensor, (Eq. (6)) is for a (ideal) fluid with no viscosity and heat conduction.

Starting with the Einstein tensor, we calculate the t, r, θ, and ϕ components of the Einstein tensor and obtain expressions that take the form

$$\begin{aligned} G_t^t &\equiv R_t^t - \tfrac{1}{2}R + \Lambda &= -T_t^t \,, \\ G_r^r &\equiv R_r^r - \tfrac{1}{2}R + \Lambda &= -T_r^r \,, \\ G_\theta^\theta &\equiv R_\theta^\theta - \tfrac{1}{2}R + \Lambda &= -T_\theta^\theta \,, \\ G_\phi^\phi &\equiv R_\phi^\phi - \tfrac{1}{2}R + \Lambda &= -T_\phi^\phi \,. \end{aligned} \tag{7}$$

Once we have determined the left hand side terms of Eq. (7), we will equate those terms with the right hand side respectively.

We calculate the components of the Ricci tensor to be [8]

$$R_t^t = \left[-\frac{\nu'^2}{4} + \frac{\nu'\lambda'}{4} - \frac{\nu''}{2} - \frac{\nu'}{2} \right] \mathrm{e}^{-\lambda} \,, \tag{8}$$

and

$$R_r^r = -\left[\frac{\nu'^2}{4} - \frac{\lambda'\nu'}{4} - \frac{\lambda'}{r} - \frac{\nu''}{2} \right] \mathrm{e}^{-\lambda} \,, \tag{9}$$

along with the Ricci scalar given by

$$\begin{aligned} R &= g^{\mu\nu} R_{\mu\nu} \\ &= \mathrm{e}^{-\lambda} \left[-\frac{\nu'^2}{2} - \nu'' + \frac{\nu'\lambda'}{2} - \frac{2\nu'}{r} + \frac{2\lambda'}{r} - \frac{2}{r^2} \right] + \frac{2}{r^2} \,. \end{aligned} \tag{10}$$

Substituting Eqs. (8), (9), and (10) into Eq. (7), we calculate the components of the Einstein tensor to be [8]

$$\begin{aligned} G_t^t &= \mathrm{e}^{-\lambda} \left(-\frac{\lambda'}{r} + \frac{1}{r^2} \right) - \frac{1}{r^2} + \Lambda \,, \\ G_r^r &= \mathrm{e}^{-\lambda} \left(\frac{\nu'}{r} + \frac{1}{r^2} \right) - \frac{1}{r^2} + \Lambda \,, \\ G_\theta^\theta &= G_\phi^\phi \,. \end{aligned} \tag{11}$$

Equating Eq. (6) to Eq. (11), we obtain

$$\begin{aligned} 8\pi\varepsilon &= \mathrm{e}^{-\lambda} \left(\frac{\lambda'}{r} - \frac{1}{r^2} \right) + \frac{1}{r^2} - \Lambda \,, \\ 8\pi P &= \mathrm{e}^{-\lambda} \left(\frac{\nu'}{r} + \frac{1}{r^2} \right) - \frac{1}{r^2} + \Lambda \,. \end{aligned} \tag{12}$$

Using the gravitational mass,

$$m(r) = 4\pi \int_0^r \varepsilon(r') r'^2 \, dr' \,, \tag{13}$$

and other mathematical means [8], we calculate the $r-$component of Eq. (2) to be

$$e^{-\lambda(r)} = 1 - \frac{2m(r)}{r} - \frac{\Lambda r^2}{3} \; . \tag{14}$$

Adding the expressions given in Eq. (12) and using Eq. (14), along with other mathematical methods [8], our modified TOV equation with a non-vanishing cosmological constant is

$$\frac{dP}{dr} = -\frac{\varepsilon\left(1+\frac{P}{\varepsilon}\right) m \left(1+\frac{4\pi P r^3}{m} - \frac{\Lambda r^3}{3m}\right)}{r^2\left(1-\frac{2m}{r}-\frac{\Lambda r^2}{3}\right)} \; , \tag{15}$$

where the units of Λ are given in MeV/fm^3. From Eq. (15), one can easily see if Λ is small ($\approx 10^{-41}$) [11], we will get no contributions for the bulk properties. Λ would have to be on the order of nuclear density scales (140 MeV/fm^3) to see any significant changes in mass and radius.

We can solve Eq. (15) for a given equation of state (EoS) model and produce stellar properties such as masses, radii, pressure and density profiles, along with the gravitational redshift. Section 1.2 briefly describes the EoS model used in this work.

1.2. Equation of State

For simplicity, we use the well known EoS of a relativistic gas of de-confined quarks also known as the MIT bag model [12, 13, 14]. This EoS model will allow us to mimic stellar properties such as masses and radii of neutron stars. In this model the pressure P^f, (where f are the quark flavors $f = u, d, s$) of the individual quarks are contained in a bag which is counterbalanced by the total external bag pressure, B, and will have the form

$$P + B = \sum_f P^f \; . \tag{16}$$

The total energy-density ε of the quarks inside the bag is is given by

$$\varepsilon = \sum_f \varepsilon^f + B \; , \tag{17}$$

where $\varepsilon^f = 3P^f$ denote the contributions that the individual quarks have on the total energy density. The term B is the known as the bag constant. The EoS is then obtained which has the form [12, 13, 15, 16]

$$P = \frac{\varepsilon - 4B}{3} \; . \tag{18}$$

One can see from Eq. (18) that the pressure becomes zero when $\varepsilon = 4B$. The mass contained in a spherical shell of radius r is given by Eq. (13). The total mass M of a stellar configuration is therefore given by $M = m(R)$, where R denotes the radius of the stellar mass distribution, which is defined by $P(r = R) = 0$. By using Eqs. (13) and (18), we solve Eq. (15) and produce stellar properties such as mass, radius, pressure and density profiles.

1.3. Results

Solving Eq. (15) in combination with Eq. (18) and inputting different values of Λ, we calculate the masses and radii of stellar configurations made of a relativistic quark gas (quark stars) [8]. Figures 1 and 2 show the mass-radius relationships for these objects.

Figures 1 and 2 show the mass-radius relationships of stellar sequences for negative and positive values of Λ respectively. We see that if $\Lambda = 0$ the maximum-mass is approximately $2M_\odot$, where $M_\odot$ denotes the mass of the sun (2.0×10^{30} kg). This is in agreement with known

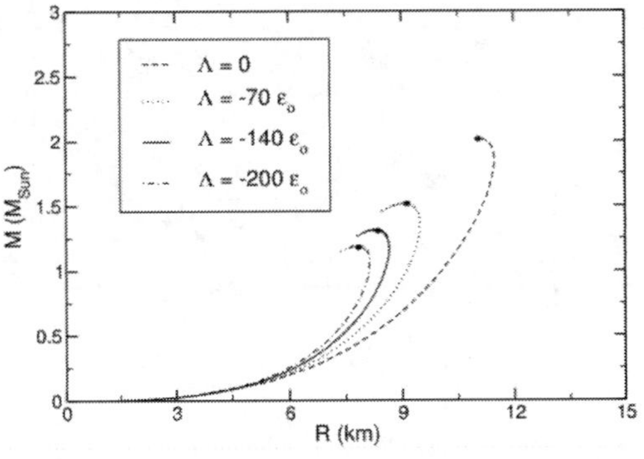

Figure 1. Mass-radius relationship for negative values of the cosmological constant Λ. (ε_0 denotes the energy-density of nuclear matter (140 MeV/fm^3).) The solid dots on each curve represent the maximum-mass star for each stellar sequence.

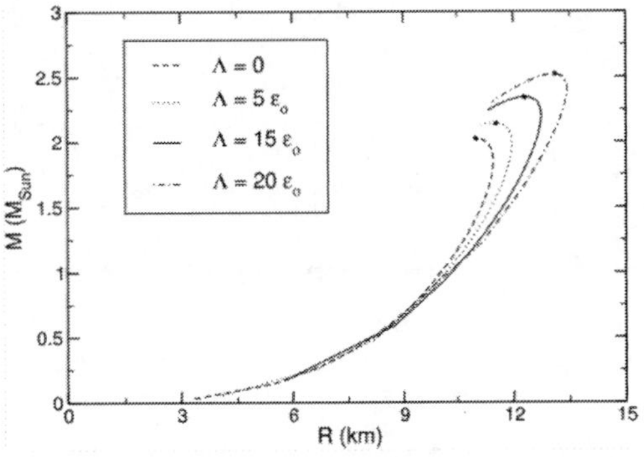

Figure 2. Mass-radius relationship for positive values of the cosmological constant Λ. (ε_0 denotes the energy-density of nuclear matter (140 MeV/fm^3).) The solid dots on each curve represent the maximum-mass star for each stellar sequence.

results from the literature [14, 15, 16, 17, 18, 19, 20]. If Λ becomes negative, we see that the star's mass and radius decrease. By inputting positive values for Λ, we get the opposite effect as shown in figure. 1.

Next, we calculate the pressure and density profiles of the maximum-mass objects shown in figures 1 and 2 [8]. The pressure profiles for each maximum mass stars are shown in figures 3 and 4. As it is shown in figures 3 and 4, the pressure becomes zero at the surface, which is in agreement with our boundary conditions.

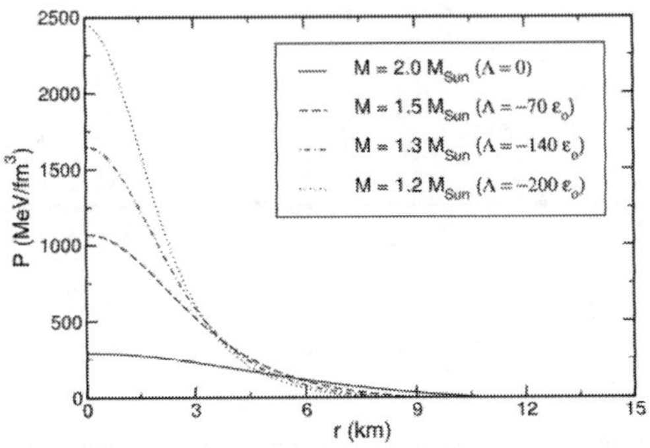

Figure 3. Pressure profiles for the maximum masses shown in figure 1. These profiles are for negative Λ values.

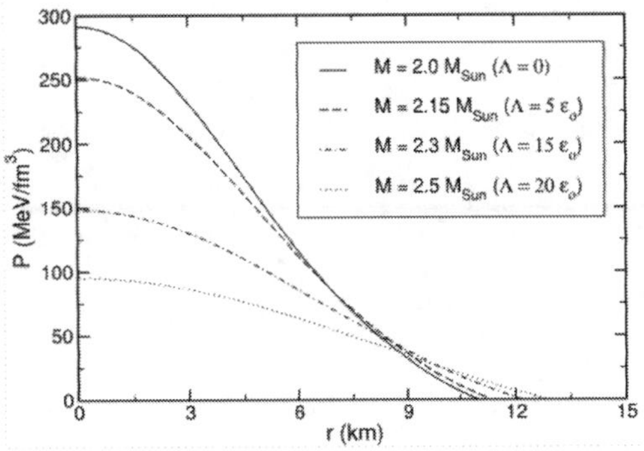

Figure 4. Pressure profiles for the maximum masses shown in figure 2. These profiles are for positive Λ values.

Next, we show the density profiles for each maximum mass in figures 1 and 2. The results are shown in figures 5 and 6. As one can see, the energy-density approaches to $4B$ as dictated by our EoS (Eq. (18)) for each maximum mass star.

From our results shown in figures 1 and 2, we have calculated the masses and radii of compact stars using a quark EoS. Since we now have numerical values for the total mass and radius, we

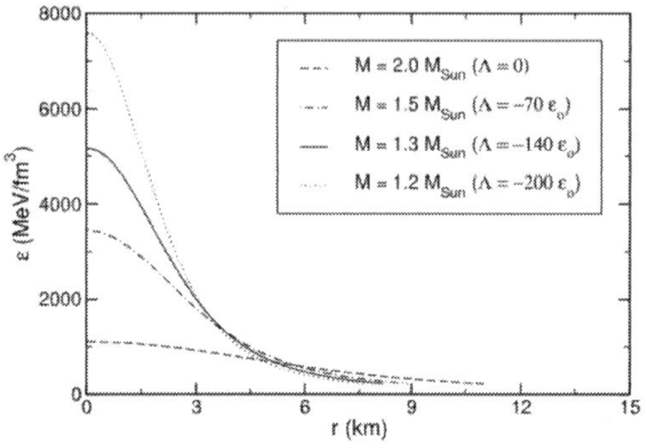

Figure 5. Energy-Density profiles for the maximum mass objects shown in figure 1.

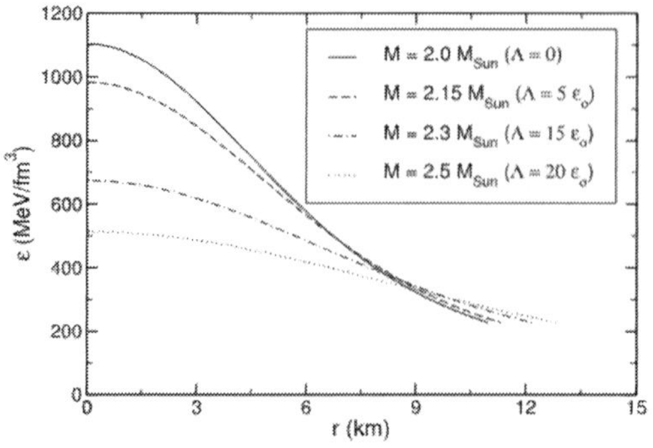

Figure 6. Energy-Density profiles for the maximum mass objects shown in figure 2.

can also calculate the gravitational redshift of these objects. The gravitational redshift is given by [9]

$$z = \frac{1}{\sqrt{1 - \frac{2M}{R}}} - 1 \,, \tag{19}$$

where, M and R are the mass and radius of a given stellar object. By using masses and radii that are associated with specific cosmological constant values and by Eq. (19), we calculate the gravitational redshift. The results are given in figure 7.

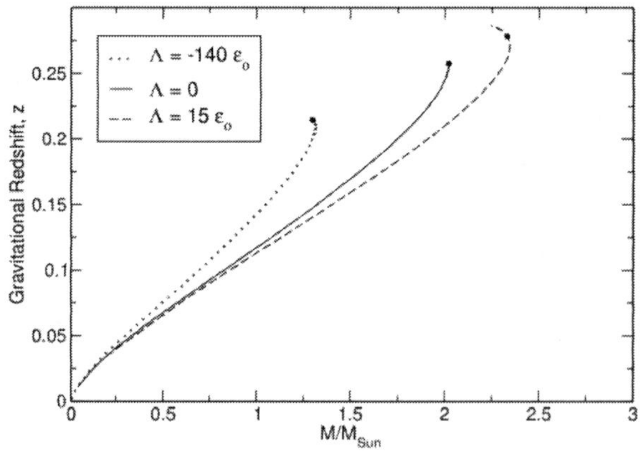

Figure 7. Gravitational redshift as a function of stellar mass for selected Λ values.

From Eq. (19), the limit of the ratio of mass and radius is less then 8/9 as described in [9] for any compact object satisfying the TOV equation. This limit will then yield a gravitational redshift of $z < 2$.

1.4. Conclusions

This work on the cosmological constant and compact stars consists in performing a derivation of the Tolman-Oppenheimer-Volkoff (TOV) equation modified for the cosmological constant Λ.

The form of this equation and solution has been stated in the literature by a few authors [5, 6, 7] already, however with disagreeing results and conclusions.

In order to investigate the role of the cosmological constant for compact objects, we have solved the modified TOV equation for mass distributions consisting of a relativistic quark gas [8]. We have obtained stellar properties such as mass, radius, pressure and density profiles along with gravitational redshifts [8] for a range of different Λ values. It follows from figures 1 through 6 that the stellar properties of compact objects in our Universe, where Λ is very small ($\Lambda \approx 10^{-41}\mathrm{MeV/fm^3}$) [9, 11, 21] do not depend on the cosmological constant, in contrast to claims expressed in the literature [5].

2. Non Spherical Stellar Models of Compact Stars

Conventionally, the structure of compact stellar objects such as neutron or quark stars are modeled with the assumption that they are perfect spheres. However, due to high magnetic fields up around 10^{18} G, certain classes of compact stars [22] (such as magnetars and neutron stars containing cores of color-superconducting quark matter) are expected to be deformed (non-spherical) making them ob-longed spheroids. In this part of our paper, we seek to investigate the stellar properties of deformed compact stars in the framework of general relativity. Using a metric that describes a non-spherical mass distribution, we derive the stellar structure equations of these non-spherical objects. We then calculate stellar properties such as mass and radii along with density and pressure profiles and investigate any changes from the standard spherical models.

2.1. Non-Spherical Symmetry

In order to understand how the deformation will impact the stellar configurations, we must first take a look at the symmetry of these objects. We begin with a standard spherically symmetric object of radius r.

We introduce a deformation constant γ and normalize it when $\gamma = 1$ to be a perfect sphere. The deformation of the stellar objects can have two outcomes: (1) Equatorial deformation (oblate spheroid) (2) polar deformation (prolate spheroid). In terms of γ the deformation can be defined as $\gamma < 1$, for equatorial deformation or $\gamma > 1$, for polar deformation.

For both cases, we are assuming axial symmetry (see figures 8 and 9)–meaning the axis of symmetry will be along the polar axis. The equatorial radii lie in the x-y plane of the star, while the polar radius z is orthogonal to the x-y plane. Hence, we only have to consider two distinct radii. Having our symmetry laid out in this fashion allows us now to derive the stellar structure equations that will govern the stellar properties of deformed compact stars.

2.2. Stellar Structure Equations

Since we are dealing with non-spherical symmetry as mentioned in section 2.1, we begin our calculations with an axial-symmetric metric in non-spherical coordinates given by [23, 25]

$$ds^2 = \mathrm{e}^{2\Phi}dt^2 - \mathrm{e}^{-2\Phi}\left[\mathrm{e}^{2\Lambda}\left(dr^2 + dz^2\right) + r^2 d\phi^2\right] \ , \tag{20}$$

where t is the time component, and r, z, and ϕ are the spatial components. Having this extra spatial dz component implies our calculations to be in 2-Dimensions. To simplify this scenario and as a first step towards this work, we parameterize the z-component in terms of γ and the radial component. This will allow two important facets:

(i) Use an EoS in the *limiting* case of isotropic pressure and density.

(ii) Still maintain deformation structure.

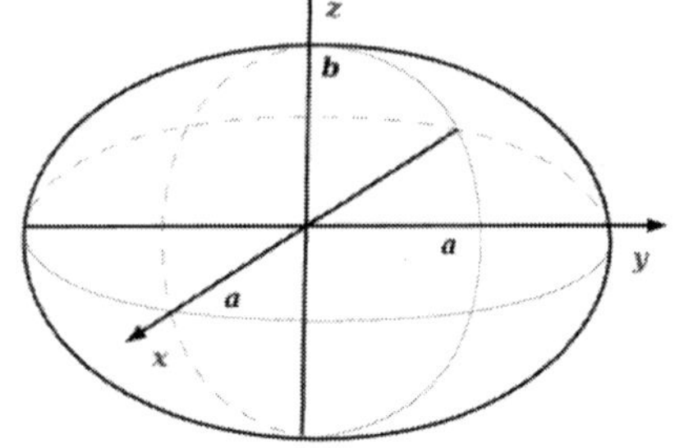

Figure 8. Oblate spheroid ($\gamma < 1$) with radii a and b.

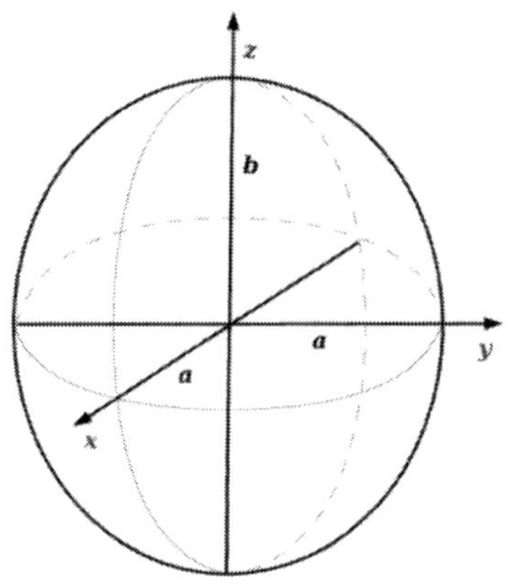

Figure 9. Prolate spheroid ($\gamma > 1$) with radii a and b.

Maintaining the deformation structure will allow us to examine how the mass and radii change due to the deformity. We begin with the examination of the components of the metric, specifically the components of this metric given in terms of [25]

$$\begin{aligned} e^{2\Phi} &= \left[\frac{k_1 + k_2 - 2m}{k_1 + k_2 + 2m}\right]^{\gamma} , \\ e^{2\Lambda} &= \left[\frac{(k_1 + k_2 + 2m)\,(k_1 + k_2 - 2m)}{4k_1 k_2}\right]^{\gamma^2} , \end{aligned} \tag{21}$$

where

$$\begin{aligned} k_1 &= \sqrt{r^2 + (z - m)^2} , \\ k_2 &= \sqrt{r^2 + (z + m)^2} . \end{aligned} \tag{22}$$

Hence, the components of this metric are given in terms of equatorial radius r, polar radius z, along with the mass m inside a mass shell.

In the case where $\gamma = 1$, we obtain the metric for a spherically symmetric object of the form [25, 26]

$$ds^2 = Adt^2 - A^{-1}\left[Bdr^2 + Cd\theta^2 + \left(r^2 - 2mr\right)\sin^2(\theta)d\phi^2\right] = g_{\mu\nu}dx^{\mu}dx^{\nu} , \tag{23}$$

where

$$A = \left(1 - \frac{2m}{r}\right)^{\gamma} ,$$
$$B = \left(\frac{r^2 - 2mr}{r^2 - 2mr + m^2 \sin^2(\theta)}\right)^{\gamma^2 - 1} ,$$
$$C = \frac{\left(r^2 - 2mr\right)^{\gamma^2}}{\left(r^2 - 2mr + m^2 \sin^2(\theta)\right)^{\gamma^2 - 1}} . \tag{24}$$

Hence, Eq. (23) is now given in terms of our deformation constant and is parameterized. Using Eq. (23) along with other mathematical means [24] we solve Einstein's equations for $\Lambda = 0$, (Eqs. (5) and (6)) and arrive at the stellar structure equation of hydrostatic equilibrium for a deformed compact star

$$\frac{dP}{dr} = -\frac{(\epsilon + P)\left[\frac{1}{2}r + 4\pi r^3 P - \frac{1}{2}r\left(1 - \frac{2m}{r}\right)^{\gamma}\right]}{\left[r^2\left(1 - \frac{2m}{r}\right)^{\gamma}\right]} . \tag{25}$$

Eq. (25) describes how the pressure behaves inside a deformed compact star. In the limiting case when $\gamma = 1$, we will obtain Eq. (1) for hydrostatic equilibrium.

The total mass of a deformed compact star is given by $M = m(R)$, where R denotes the radius of the star given by $P(r = R) = 0$.

Since Eq. (25) is governed by a deformation constant (γ), the stellar mass must also be governed by the same constant and will have the form [25, 26]

$$M = \gamma m . \tag{26}$$

Hence, we utilize a parameterization for Eq. (26). Once we have our mass parameterized in terms of γ, we can then use an EoS model and calculate stellar properties of a deformed compact star.

2.3. Parameterization

We begin our parameterization by considering the volume of a sphere given by

$$V = \frac{4}{3}\pi r^3 , \tag{27}$$

where r is the radius of the sphere. However, in our case, we have two distinct radii as shown in figures 8 and 9 and hence, the volume of these oblate or prolate spheroids is given by

$$V = \frac{4}{3}\pi r^2 z , \tag{28}$$

where r is the equatorial radius and z is the polar radius. We can parameterize the polar radius by $z = \gamma r$ to give us

$$V = \frac{4}{3}\pi r^3 \gamma , \tag{29}$$

and therefore, our mass will now become

$$\frac{dm}{dr} = 4\epsilon\pi r^2 \gamma . \tag{30}$$

Having Eqs. (25) and (30) parameterized in terms of γ, we now solve these two coupled equations with the use of an EoS of relativistic quark gas as described in section 1.2.

2.4. Results

We first calculate the mass and radii of deformed objects which are parameterized by the deformation constant γ. The results are shown in figures 10 and 11.

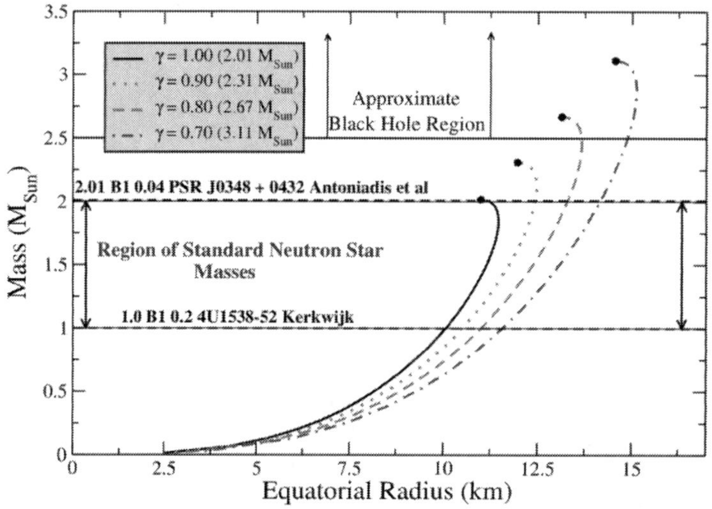

Figure 10. (Color online) Mass vs. equatorial radius for values of $\gamma < 1$. The dots on each curve represent the maximum mass star for each stellar sequence.

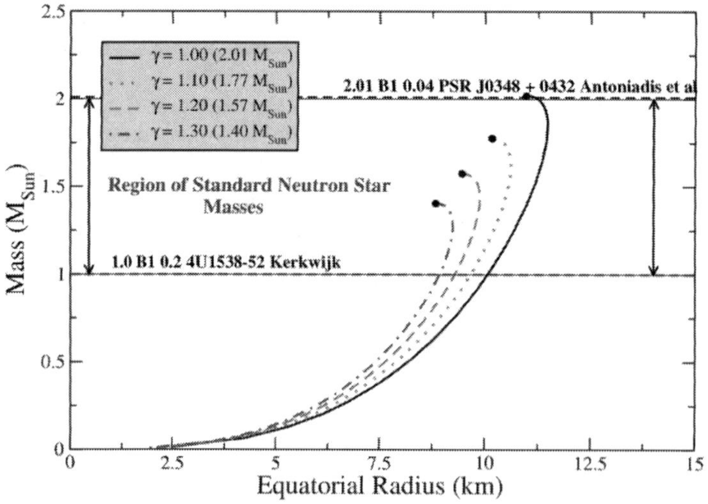

Figure 11. (Color online) Mass vs. equatorial radius for values of $\gamma > 1$. The dots on each curve represent the maximum mass star for each stellar sequence.

From figure 10, we see for the spherical case ($\gamma = 1$), we get a mass on the order of about 2 $M_{\odot}$ and an equatorial radius of about 10 km. However, if decrease γ by a mere 10 %, we get a 15 % increase in mass and an increase in equatorial radius by a few kilometers. A greater equatorial radius means a smaller polar radius, leading to an oblate compact star. If we continue to change γ, we see that the mass keeps increasing. However, this is not the case for when $\gamma > 1$. As shown in figure 11, if we increase γ by 10 % we get a 12 % decrease in mass and a decrease in equatorial radius. A smaller equatorial radii will result in a larger polar radius and thus we will have deformation in the polar direction leading to a prolate spheroid.

The change in mass due to the geometry of the deformation is interesting. From our results in figures 10 and 11, we can conclude that the deformation does change the mass of the compact star by a significant amount. The images in figure 12 show the total deformation for the maximum masses in figure 10.

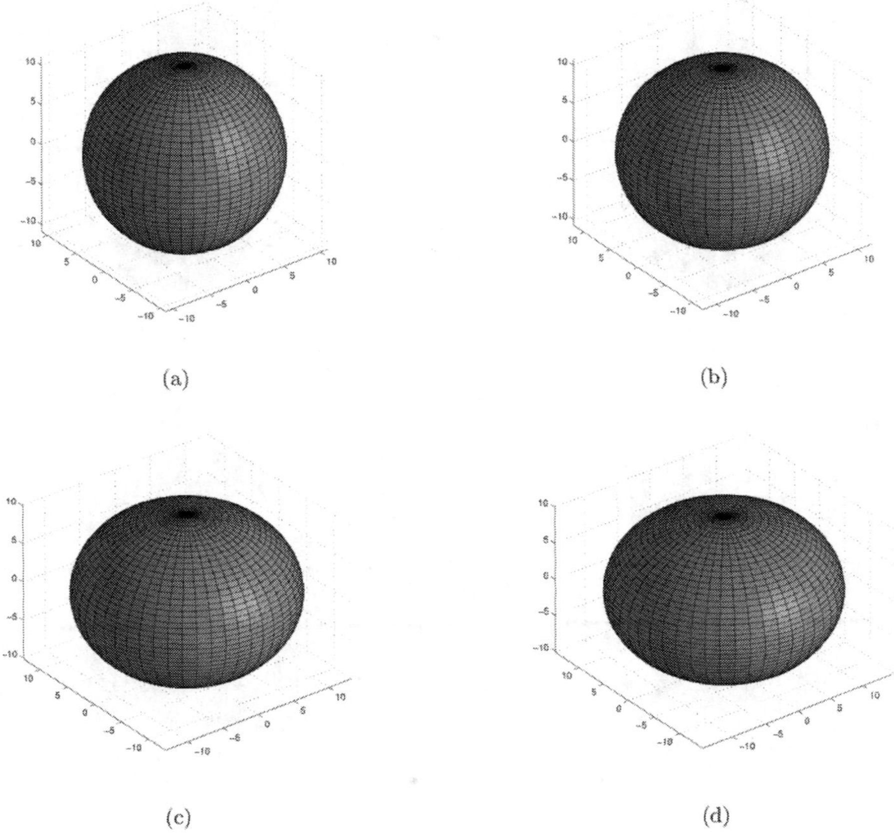

Figure 12. Shapes of the maximum mass stars given in figure 10. (a) $\gamma = 1.00$, M=2.01 $M_{\odot}$, (b) $\gamma = 0.90$, M=2.32 $M_{\odot}$, (c) $\gamma = 0.80$, M=2.67 $M_{\odot}$, (d) $\gamma = 0.70$, M=3.01 $M_{\odot}$

We next look at the prolate case for various values of our deformation constant γ. The images in figure 13 show the total deformation for the maximum masses in figure 11.

Since we have obtained stellar properties such as mass and radius the next step is to look at the pressure and density profiles associated with the maximum masses shown in figures 10 and 11. We calculate the pressure and density profiles in both the equatorial and polar radii for

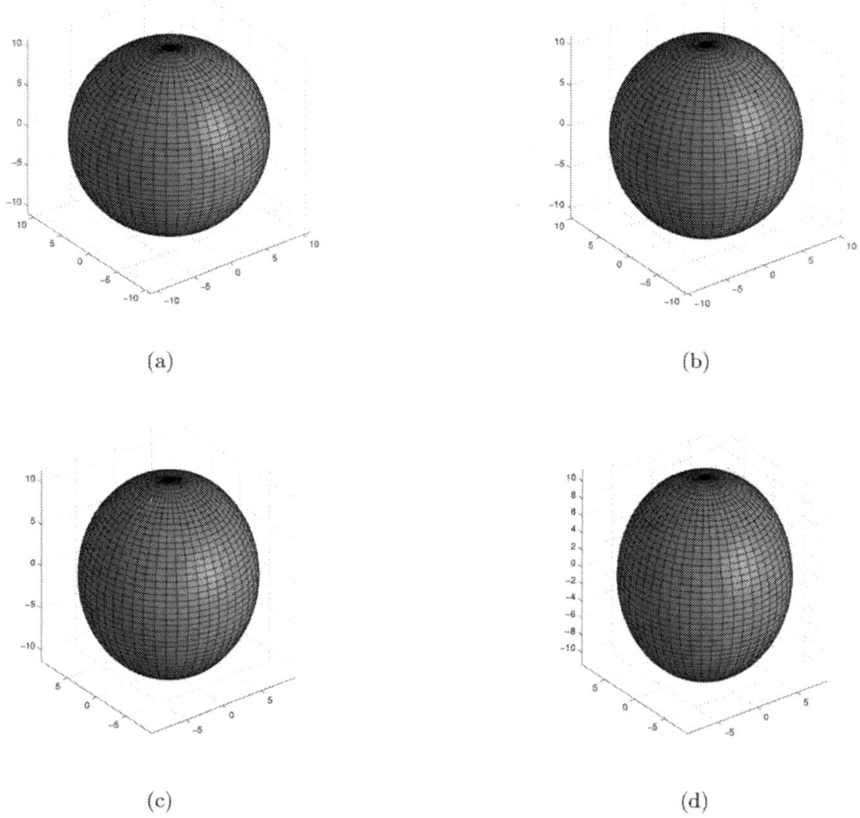

Figure 13. Shapes of the maximum mass stars given in figure 11. (a) $\gamma = 1.00$, M=2.01 $M_\odot$, (b) $\gamma = 1.10$, M=1.77 $M_\odot$, (c) $\gamma = 1.20$, M=1.57 $M_\odot$, (d) $\gamma = 1.30$, M=1.40 $M_\odot$

these maximum masses. The outcome is shown in figures 14 and 15, respectively.

From our results from figures 10 and 11, we see the that mass does change as we deformed the object either in the equatorial or polar directions. Since we have numerical values for the mass and radii, we can also investigate the gravitational redshift of these objects. However, since our metric is given in terms of γ, we must derive an expression of gravitational redshift which will relate stellar properties such as mass and radii in terms of γ.

We begin our redshift calculation with an examination of the components of the metric described by Eq. (23), in particular the $t-$component $g_{tt} = g_{00}$ is given by

$$g_{00} = \left(1 - \frac{2M}{R}\right)^{\gamma} , \tag{31}$$

where M and R represent the total mass and equatorial radii, along with our deformation constant γ.

Consider an idealized situation in which a photon is emitted from the surface of a star and travels through a gravitational field towards an observer located at infinity. The time interval dx^0 for this photon leaving the surface of the star is then given by [9]

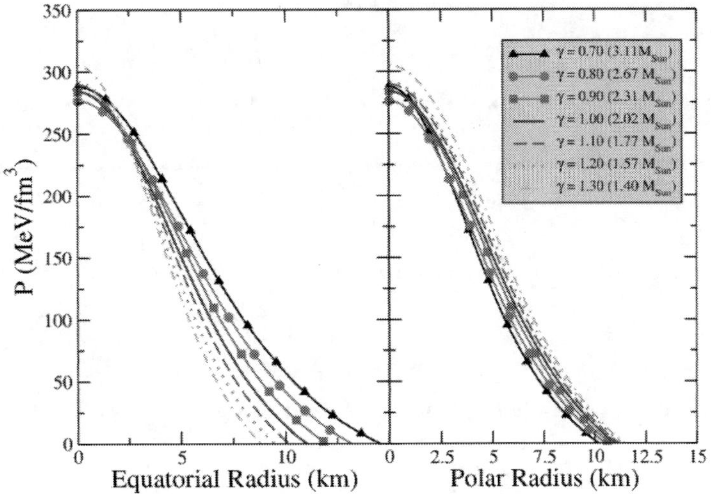

Figure 14. (Color online) Pressure profiles for the maximum mass stars shown in figures 10 and 11. All pressure profiles becomes zero for both the equatorial and polar radii as dictated by our boundary conditions.

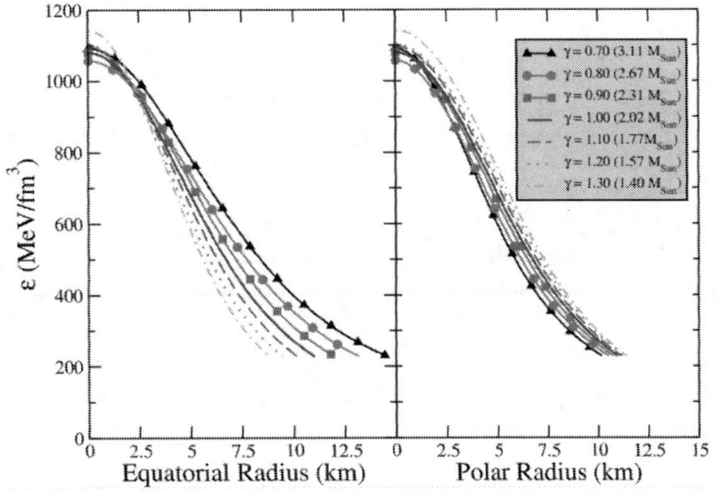

Figure 15. (Color online) Density profiles for the maximum mass stars shown in figures 10 and 11. As one can see, the density approaches to $4B$ as dictated by our EoS for the equatorial and polar radii.

$$d\tau_a = \sqrt{g_{00}(R)}dx^0 \ . \tag{32}$$

If we assume that light propagates in a radial direction in a static field, the time interval $d\tau^2$ is then described by

$$d\tau^2 = g_{00}(r)dt^2 - g_{11}(r)dr^2 = 0 \ . \tag{33}$$

Integrating Eq. (33), we find the constant time interval for each wave crest to travel from R to infinity to be

$$\Delta t = t_\infty - t_R = \int_R^\infty \left(\frac{g_{11}(r)}{g_{00}(r)}\right)^{1/2} dr \ , \tag{34}$$

where an observer located at ∞ will measure the proper time between the crests arriving from the star to be

$$d\tau_{a'} = \sqrt{g_{00}(\infty)}dx^0 \ . \tag{35}$$

If we consider an emission of a photon from ∞ toward the star, and we let the event be denoted by x'^μ. Then the equation that governs the coordinate time interval dx'^0 will be given by [9]

$$d\tau_a = \sqrt{g_{00}(\infty)}dx'^0 \ . \tag{36}$$

Using Eqs. (35) and (36), we can compare the observed frequency ω_o from the surface of the star with that of the photon emitted from ∞ to the star, ω_e.

The proper time between crests (multiplied by c) gives the wavelength of the light or the reciprocal of the frequency. Therefore, using Eqs. (35) and (36), respectively, we obtain the following relationship [9]

$$\frac{\omega_o}{\omega_e} = \frac{d\tau_a}{d\tau_{a'}} = \frac{dx'^0}{dx^0} = \left(\frac{g_{00}(R)}{g_{00}(\infty)}\right)^{1/2} \ . \tag{37}$$

As measured by the observer at infinity, the photon coming from the surface of the star will have a frequency of

$$\omega_o = e^{2\nu}\omega_e = \left(1 - \frac{2M}{R}\right)^\gamma \omega_e \ , \tag{38}$$

where in geometric normalized units, $G = 1$. The ratio ω_o/ω_e is given by

$$\frac{\omega_o}{\omega_e} = \left(1 - \frac{2M}{R}\right)^{\gamma/2} \ . \tag{39}$$

From Eq. (38), one can see that the photon arriving from the surface of the star to ∞ has a lower frequency than the photon arriving from ∞ to the surface of the star. The light is therefore red-shifted.

Conventionally, the gravitational redshift is defined by the fractional change between the observed and emitted wavelengths divided by the emitted wavelength [9]

$$z = \frac{\lambda_o - \lambda_e}{\lambda_e} = \frac{\lambda_o}{\lambda_e} - 1 = \frac{\omega_e}{\omega_o} - 1 \ . \tag{40}$$

Hence, using Eq. (40), we find the gravitational redshift to be

$$z = \frac{1}{\left(1 - \frac{2M}{R}\right)^{\gamma/2}} - 1 \ . \tag{41}$$

If $\gamma = 1$ then Eq. (41) will simplify for the spherical case as described by Eq. (19). By using the the mass and radii results shown in figures 10 and 11, we calculate the gravitational redshift z (Eq. (41)). The results are shown in figure 16.

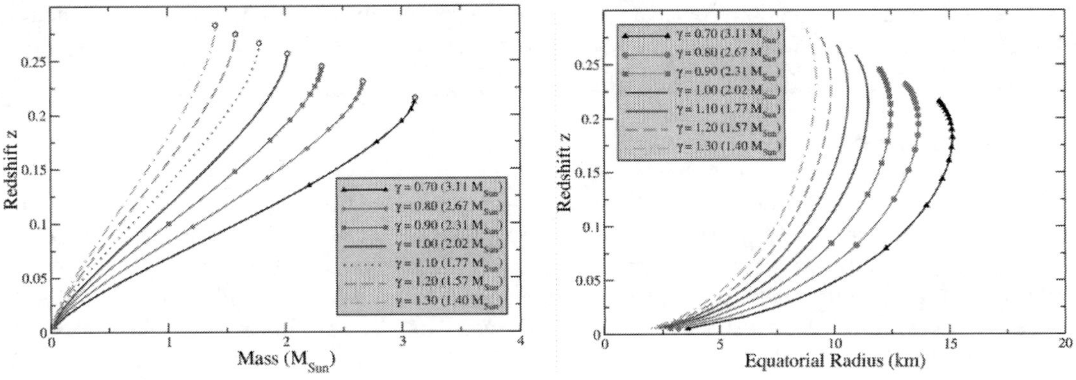

Figure 16. (Color online) Calculated redshift values as function of mass and equatorial radii for various values of our deformation constant γ. (a) Redshift z as function of mass M. The circular dots represent the maximum mass from figures 10 and 11. (b) Redshift z as function of equatorial radii R.

Using the results in figure 16, one can directly calculate either mass or radii if the redshift z and either the mass or radii is known. Hence, we can use this numerical model for observational findings.

2.5. Conclusions

The goal of part 2 of this paper was to investigate the structure of deformed compact stars in the framework of general relativity and investigate how the deformation impacts the stellar properties of such objects. We have derived the stellar structure equations that govern the deformation. We have introduced a γ factor which we used to parameterize the deformation.

Having this parameterization allowed us to use a EoS in the limiting case of isotropy but still maintain deformation structure. From our results given in figures. 10 and 11, we see a slight change in deformation results in significant changes in mass and radii, and therefore we can conclude that the deformation does not need to be high to see these changes.

These promising results leads us to continue this work in the case of anisotropic EoS. However, for this case, we can not simply use a 1-D model governed by some deformation constant. We must have distinct hydro-static equations in both the radial and polar directions (i.e. dP/dr and dP/dz). This implies that we need to construct a full 2-D model of the global structure of our compact objects in this framework. Having a 2-D model will result in a more accurate representation of the deformation and therefore obtain more realistic calculations for stellar properties of these deformed compact stars.

3. Summary and Discussion

In this paper, we investigated two different solutions for compact stars in the framework of general relativity. In the first part of this paper, we solved Einstein's field equations for a non-vanishing cosmological and obtained a stellar structure equation in this framework. This structure equation has been suggested by a few authors already, but with inconsistent results.

From our results from figures 1 to 6, we see that in a Universe where the cosmological constant is very small, there is no significant contributions to the bulk properties such as mass and radius.

In the second part of this paper, we examined the deformation of compact stars. We introduced a deformation constant γ and a parameterization of the polar coordinate to obtain a 1-D description of the stellar structure of these deformed compact objects. We can conclude from figures 10 and 11 that slight change in deformation results in appreciable masses, and hence, the deformation does not need to be high to see significant mass changes.

Acknowledgments

This work is supported through the National Science Foundation under grants PHY-1411708 and DUE-1259951. Additional computing resources are provided by the Computational Science Research Center and the Department of Physics at San Diego State University. The research reported in this publication is also supported by the National Institute of General Medical Sciences of the National Institutes of Health under award number 5T34GM008303-25.

The authors would like to thank Vivian de la Incera and Efrain Ferrer from The University of Texas at El Paso, for their insightful discussions and initial motivation on this work. The authors would also like to thank Dr. James O'Brien for the invitation to speak at the biannual 2014 IARD conference.

The content is solely the responsibility of the authors and does not necessarily represent the official views of the National Institutes of Health.

References

[1] Perlmutter S et al 1999 *Astrophys. J.* **517** 565-86.
[2] Riess A G et al 1988 *Astron. J.* **116** 1009-38.
[3] Oppenheimer J R and Volkoff G M 1939 *Phys. Rev.* **55** 374-81.
[4] Tolman R C 1939 *Phys. Rev.* **55** 364-73.
[5] Vinayaraj O K and Kuriakose V C 2008 *arXiv: 0802.1155v2* [astro-ph].
[6] Ghezzi C R 2011 *Ap & SS* **333** 437-47.
[7] Silbar R and Reddy S 2004 *Am. J. Phys.* **72** 892-905.
[8] Zubairi O and Weber F 2014 *Astron. Nachr.* **335** 593-98.
[9] Glendenning N K 1997 *Compact Stars: Nuclear Physics, Particle Physics, and General Relativity* (New York: Springer-Verlag) P 7-123.
[10] Weber F 1999 *Pulsars as Astrophysical Laboratories for Nuclear and Particle Physics* (London: IoP Publishing) P 311-357.
[11] Carroll S M 2001 *Living Rev. Relativity* **3** 1.
[12] Chodos A, Jaffe R L, Johnson K, Thorn C B, and Weisskopf V F 1974a *Phys. Rev. D.* **9** 3471-95.
[13] Chodos A, Jaffe R L, Johnson K, and Thorn C B, 1974b *Phys. Rev. D.* **10** 2599-2604.
[14] Farhi E and Jaffe R L 1984 *Phys. Rev. D.* **30** 2379.
[15] Weber F 2005 *Progress in Particle and Nuclear Physics* **54** 193-288.
[16] Weber F 2003 *Nucl. Phys.* **721** 1032c-35c.
[17] Alcock C, Farhi E, and Olinto A 1986 *Astrophys. J.* **310** 261-72.
[18] Alcock C and Olinto A 1988 *Ann. Rev. Nucl. Part. Sci.* **38** 161-84.
[19] Witten E 1984 *Phys. Rev. D.* **30** 272-85.
[20] Glendenning N K, Kettner Ch., and Weber F. 1995 *Astrophys. J.* **450** 253-61.
[21] Tegmark M et al 2004 *Phys. Rev. D.* **69** 103501.
[22] Ferrer E, Incera V, Keith J, Portillo I, and Springsteen P 2010 *Phys. Rev. C.* **82** 6.
[23] Weyl H. 1918 *Ann. Phys.* (Leipzig), **54** 117.
[24] Zubairi O, Spinella W, Mellinger R, Weber F, Osaria M, and Contrera G 2014. In preparation
[25] Herrera L Filipe M, and Santos N O, 1999 *J. Math Phys.* **40** 8.
[26] Esposito F and Witten L 1975 *Phys. Lett.* **58B** 357.

Torsion, magnetic monopoles and Faraday's law via a variational principle

Philip D. Mannheim
Department of Physics, University of Connecticut, Storrs, CT 06269, USA
E-mail: philip.mannheim@uconn.edu

Abstract. Even though Faraday's Law is a dynamical law that describes how changing **E** and **B** fields influence each other, by introducing a vector potential A_μ according to $F_{\mu\nu} = \partial_\mu A_\nu - \partial_\nu A_\mu$ Faraday's Law is satisfied kinematically, with the relation $(-g)^{-1/2}\epsilon^{\mu\nu\sigma\tau}\nabla_\nu F_{\sigma\tau} = 0$ holding on every path in a variational procedure or path integral. In a space with torsion $Q_{\alpha\beta\gamma}$ the axial vector $S^\mu = (-g)^{1/2}\epsilon^{\mu\alpha\beta\gamma}Q_{\alpha\beta\gamma}$ serves as a chiral analog of A_μ, and via variation with respect to S_μ one can derive Faraday's Law dynamically as a stationarity condition. With S_μ serving as an axial potential one is able to introduce magnetic monopoles without S_μ needing to be singular or have a non-trivial topology. Our analysis permits torsion and magnetic monopoles to be intrinsically Grassmann, which could explain why they have never been detected. Our procedure permits us to both construct a Weyl geometry in which A_μ is metricated and then convert it into a standard Riemannian geometry.

1. Introduction

The Maxwell equations of electromagnetism in flat space break up into two groups, the Maxwell-Ampere and Electric Gauss Laws

$$\boldsymbol{\nabla} \times \boldsymbol{B} - \frac{\partial \boldsymbol{E}}{\partial t} = \boldsymbol{J}_e, \qquad \boldsymbol{\nabla} \cdot \boldsymbol{E} = \rho_e, \tag{1}$$

and the Faraday and Magnetic Gauss Laws

$$\boldsymbol{\nabla} \times \boldsymbol{E} + \frac{\partial \boldsymbol{B}}{\partial t} = 0, \qquad \boldsymbol{\nabla} \cdot \boldsymbol{B} = 0. \tag{2}$$

Since can one derive second-order wave equations for the propagation of the **E** and **B** field strengths in a source-free region only when all of the above equations are taken in conjunction, as such all of these equations should be regarded as being on an equal dynamical footing. And if they are to be on an equal dynamical footing, then each one of these equations should, like all dynamical equations, be derivable via stationary variation of an action.

However, the standard treatment of electrodynamics is not formulated in this way, as it does not in fact derive all of these equations from a variational procedure. Rather, in order to develop the variational procedure that it does use, it relies on the fact that the Faraday-Magnetic Gauss equations immediately admit of an exact solution

$$\boldsymbol{E} = -\frac{\partial \boldsymbol{A}}{\partial t} - \boldsymbol{\nabla}\phi, \qquad \boldsymbol{B} = \boldsymbol{\nabla} \times \boldsymbol{A}, \tag{3}$$

Journal of Physics: Conference Series **615** (2015) 012004 doi:10.1088/1742-6596/615/1/012004

a solution that is unique up to gauge transformations of the form $\boldsymbol{A} \to \boldsymbol{A} + \boldsymbol{\nabla}\chi$, $\phi \to \phi - \partial\chi/\partial t$. The introduction of $\mathbf{A}$ and ϕ serves two purposes. When inserted into Eq. (1) they enable one to solve for the $\mathbf{E}$ and $\mathbf{B}$ fields once $\mathbf{J}_e$ and ρ_e are specified. And in addition they allow one to develop a variational procedure.

To discus the variational procedure it is more convenient to first write the Maxwell equations covariantly in a curved space where they generalize to

$$\nabla_\nu F^{\nu\mu} = J^\mu, \tag{4}$$

$$(-g)^{-1/2}\epsilon^{\mu\nu\sigma\tau}\nabla_\nu F_{\sigma\tau} = 0. \tag{5}$$

In Eq. (5) the antisymmetric rank two tensor $F^{\mu\nu}$ is the field strength with components $F^{01} = -E_x$, $F^{12} = -B_z$ etc., and $J^\mu = (\rho_e, \mathbf{J}_e)$. Using $-(-g)^{-1}\epsilon_{\mu\nu\sigma\tau}\epsilon^{\mu\alpha\beta\gamma} = \delta^\alpha_\nu\delta^\beta_\sigma\delta^\gamma_\tau + \delta^\alpha_\tau\delta^\beta_\nu\delta^\gamma_\sigma + \delta^\alpha_\sigma\delta^\beta_\tau\delta^\gamma_\nu - \delta^\alpha_\nu\delta^\beta_\tau\delta^\gamma_\sigma - \delta^\alpha_\sigma\delta^\beta_\nu\delta^\gamma_\tau - \delta^\alpha_\tau\delta^\beta_\sigma\delta^\gamma_\nu$, Eq. (5) can also be written in the form

$$\nabla_\nu F_{\sigma\tau} + \nabla_\tau F_{\nu\sigma} + \nabla_\sigma F_{\tau\nu} = 0. \tag{6}$$

For brevity we shall refer to Eq. (5) as Faraday's Law even as it encompass Gauss' Law of Magnetism as well. With Eq. (5) possessing an exact solution of the form $F_{\mu\nu} = \nabla_\mu A_\nu - \nabla_\nu A_\mu$, one introduces the Maxwell action

$$I_{\rm MAX} = \int d^4x(-g)^{1/2}\left[-\frac{1}{4}F_{\mu\nu}F^{\mu\nu} - A_\mu J^\mu\right], \tag{7}$$

with its stationary variation with respect to A_μ immediately leading to Eq. (4).

Since this variation is a variation in which Eq. (5) is not varied, Eq. (5) is required to hold on every variational path. Thus even though Faraday's Law is a dynamical equation, the variation that is done is a constrained one in which Faraday's Law is imposed on every variational path, even on those that are not stationary, with the quantum path integral $\int DA_\mu \exp(iI_{\rm MAX})$ then being constrained this way as well. We shall thus seek to construct a variational procedure in which Faraday's Law is to only hold at the stationary minimum.

2. Setting up the Variational Procedure

If we do not want Faraday's Law to hold on non-stationary paths, we cannot set $F_{\mu\nu} = \nabla_\mu A_\nu - \nabla_\nu A_\mu$, since this would immediately cause $(-g)^{-1/2}\epsilon^{\mu\nu\sigma\tau}\nabla_\nu F_{\sigma\tau}$ to vanish [1]. If however, we wish to recover Faraday's Law at the stationary minimum, then with 8 equations being embodied in Eqs. (4) and (5), we need not one but two 4-vector potentials, one of course being the standard vector potential A_μ and the other needing to be some as yet to be identified axial vector S_μ. Moreover, without regard to variational issues, we note that in the event of magnetic monopoles one would ordinarily (though not quite in fact as we show below) modify Eq. (5) to

$$(-g)^{-1/2}\epsilon^{\mu\nu\sigma\tau}\nabla_\nu F_{\sigma\tau} = K^\mu, \tag{8}$$

with there then being both vector and axial vector current sources, for a total of 8 components. In the same way as we couple A_μ to J^μ via $A_\mu J^\mu$ we should equally anticipate a coupling $S_\mu K^\mu$ in the axial current sector, a coupling that is parity conserving if S_μ is an axial vector. The issue of constructing a variational principle for Faraday's Law is thus related to the coupling of electromagnetism to magnetic currents, and our objective will be to set up a variational principle with respect to A_μ and S_μ that would recover Eqs. (4) and (8) at the stationary minimum, with Eq. (5) then following in the limit in which we could set the monopole current to zero.

Recalling the two-potential study [2, 3] of the monopole problem [4], it is very convenient to introduce

$$X^{\mu\nu} = \nabla^\mu A^\nu - \nabla^\nu A^\mu - \frac{1}{2}(-g)^{-1/2}\epsilon^{\mu\nu\sigma\tau}(\nabla_\sigma S_\tau - \nabla_\tau S_\sigma) \tag{9}$$

as a generalized $F^{\mu\nu}$. On setting $S^{\mu\nu} = \nabla^\mu S^\nu - \nabla^\nu S^\mu$, we can rewrite $X^{\mu\nu}$ in terms of $F^{\mu\nu}$ and the dual $\hat{S}^{\mu\nu} = (1/2)(-g)^{-1/2}\epsilon^{\mu\nu\sigma\tau}S_{\sigma\tau}$ of $S^{\mu\nu}$ according to:

$$X^{\mu\nu} = F^{\mu\nu} - \hat{S}^{\mu\nu}, \quad \hat{X}^{\mu\nu} = \hat{F}^{\mu\nu} + S^{\mu\nu}. \tag{10}$$

(If $\epsilon^{0123} = +1$, $\epsilon_{0123} = -1$.) Given this $X^{\mu\nu}$, Eqs. (4) and (8) are to be replaced by

$$\begin{aligned} \nabla_\nu X^{\nu\mu} &= \nabla_\nu F^{\nu\mu} = J^\mu, \qquad \nabla_\nu \hat{X}^{\nu\mu} = \nabla_\nu S^{\nu\mu} = K^\mu, \\ \nabla_\nu \hat{F}^{\nu\mu} &= 0, \qquad \nabla_\nu \hat{S}^{\nu\mu} = 0, \end{aligned} \tag{11}$$

with it now being $\nabla_\nu \hat{X}^{\nu\mu} = K^\mu$ and not in fact $\nabla_\nu \hat{F}^{\nu\mu} = K^\mu$ that is to describe the monopole. If we introduce a second set of field strengths $S^{01} = -B'_x$, $S^{12} = +E'_z$, $\hat{S}^{01} = E'_x$, $\hat{S}^{12} = B'_z$ etc., on setting $K^\mu = (\rho_m, -\mathbf{J}_m)$, we find that in flat space Eq. (11) breaks up into two sectors, namely Eqs. (1) and (2) and the analog

$$\begin{aligned} \nabla \times \boldsymbol{B}' - \frac{\partial \boldsymbol{E}'}{\partial t} = 0, &\qquad \nabla \cdot \boldsymbol{E}' = 0, \\ \nabla \times \boldsymbol{E}' + \frac{\partial \boldsymbol{B}'}{\partial t} = \boldsymbol{J}_m, &\qquad \nabla \cdot \boldsymbol{B}' = \rho_m. \end{aligned} \tag{12}$$

Moreover, if we define $\mathbf{E}_{\rm TOT} = \mathbf{E} + \mathbf{E}'$, $\mathbf{B}_{\rm TOT} = \mathbf{B} + \mathbf{B}'$, we can combine Eqs. (1), (2), and (12) into

$$\begin{aligned} \nabla \times \boldsymbol{B}_{\rm TOT} - \frac{\partial \boldsymbol{E}_{\rm TOT}}{\partial t} = \boldsymbol{J}_e, &\qquad \nabla \cdot \boldsymbol{E}_{\rm TOT} = \rho_e, \\ \nabla \times \boldsymbol{E}_{\rm TOT} + \frac{\partial \boldsymbol{B}_{\rm TOT}}{\partial t} = \boldsymbol{J}_m, &\qquad \nabla \cdot \boldsymbol{B}_{\rm TOT} = \rho_m. \end{aligned} \tag{13}$$

Thus even if $\mathbf{J}_m$ and ρ_m can be neglected, it is $\mathbf{E}_{\rm TOT}$ and $\mathbf{B}_{\rm TOT}$ that are measured in electromagnetic experiments.

On introducing the action

$$I = \int d^4x(-g)^{1/2}\left[-\frac{1}{4}X_{\mu\nu}X^{\mu\nu} - A_\mu J^\mu - S_\mu K^\mu\right], \tag{14}$$

we find that stationary variation with respect to A_μ and S_μ then immediately leads to Eq. (11), just as we want. Moreover, up to surface terms this action decomposes into two sectors according to

$$I = \int d^4x(-g)^{1/2}\left[-\frac{1}{4}F_{\mu\nu}F^{\mu\nu} - A_\mu J^\mu - \frac{1}{4}S_{\mu\nu}S^{\mu\nu} - S_\mu K^\mu\right]. \tag{15}$$

Thus with the introduction of a magnetic current sector we can formulate a variational principle for Faraday's Law and for theories that involve magnetic monopoles, and can do so without the use of singular potentials or non-trivial topologies [5]. However, we still need to ascribe a physical meaning to S_μ, and to this end we turn to torsion. This will lead us directly to the action given in Eq. (15), and suggest a rationale for why the $S_{\mu\nu}$ sector has escaped detection and why a purely A_μ-based quantum electrodynamics works as well as it does.

3. Torsion

To construct covariant derivatives in a metric theory one introduces a connection $\Gamma^{\lambda}{}_{\mu\nu}$. For a torsionless Riemann space one uses the Levi-Civita and spin connections

$$\begin{aligned} \Lambda^{\lambda}{}_{\mu\nu} &= \frac{1}{2}g^{\lambda\alpha}(\partial_{\mu}g_{\nu\alpha}+\partial_{\nu}g_{\mu\alpha}-\partial_{\alpha}g_{\nu\mu})=\Lambda^{\lambda}{}_{\nu\mu}, \\ -\omega^{ab}_{\mu} &= V^{b}_{\nu}\partial_{\mu}V^{a\nu}+V^{b}_{\lambda}\Lambda^{\lambda}{}_{\nu\mu}V^{a\nu}=\omega^{ba}_{\mu}, \end{aligned} \tag{16}$$

to construct covariant derivatives such as $\nabla_{\mu}g^{\lambda\nu}=\partial_{\mu}g^{\lambda\nu}+\Lambda^{\lambda}{}_{\alpha\mu}g^{\alpha\nu}+\Lambda^{\nu}{}_{\alpha\mu}g^{\lambda\alpha}$ and $D_{\mu}V^{a\lambda}=\partial_{\mu}V^{a\lambda}+\Lambda^{\lambda}{}_{\nu\mu}V^{a\nu}+\omega^{ab}_{\mu}V^{\lambda}_{b}$ that transform as tensors under local translations and local Lorentz transformations. In Eq. (16) we have introduced vierbeins V^{a}_{μ} that carry an index a associated with a fixed special-relativistic reference system, with the metric being writable as $g_{\mu\nu}=\eta_{ab}V^{a}_{\mu}V^{b}_{\nu}$. The covariant derivatives of $g_{\mu\nu}$ and $V^{\mu a}$ constructed with $\Lambda^{\lambda}{}_{\mu\nu}$ obey the metricity conditions $\nabla_{\mu}g^{\lambda\nu}=0$, $D_{\mu}V^{a\lambda}=0$. If one generalizes $\Lambda^{\lambda}{}_{\mu\nu}$ to $\tilde{\Gamma}^{\lambda}{}_{\mu\nu}$ by adding a rank-3 tensor to it, covariant derivatives constructed with $\tilde{\Gamma}^{\lambda}{}_{\mu\nu}$ will still transform as true tensors. However, they may not necessarily obey metricity conditions $\tilde{\nabla}_{\mu}g^{\lambda\nu}=0$, $\tilde{D}_{\mu}V^{a\lambda}=0$ with respect to $\tilde{\Gamma}^{\lambda}{}_{\mu\nu}$.

To extend the geometry to include torsion one takes the connection to no longer be symmetric on its two lower indices, and defines the Cartan torsion tensor $Q^{\lambda}{}_{\mu\nu}$

$$Q^{\lambda}{}_{\mu\nu}=\Gamma^{\lambda}{}_{\mu\nu}-\Gamma^{\lambda}{}_{\nu\mu}. \tag{17}$$

To implement metricity one defines a contorsion tensor

$$K^{\lambda}{}_{\mu\nu}=\frac{1}{2}g^{\lambda\alpha}(Q_{\mu\nu\alpha}+Q_{\nu\mu\alpha}-Q_{\alpha\nu\mu}), \tag{18}$$

and with $K^{\lambda}{}_{\mu\nu}$ one constructs connections of the form

$$\begin{aligned} \tilde{\Gamma}^{\lambda}{}_{\mu\nu} &= \Lambda^{\lambda}{}_{\mu\nu}+K^{\lambda}{}_{\mu\nu}, \\ -\tilde{\omega}^{ab}_{\mu} &= -\omega^{ab}_{\mu}+V^{b}_{\lambda}K^{\lambda}{}_{\nu\mu}V^{a\nu}=\tilde{\omega}^{ba}_{\mu}. \end{aligned} \tag{19}$$

To couple spinors to gravity in a Riemannian space without torsion one uses the covariantized Dirac action $I_{\rm D}=(1/2)\int d^4x(-g)^{1/2}i\bar{\psi}\gamma^{a}V^{\mu}_{a}(\partial_{\mu}+\Sigma_{bc}\omega^{bc}_{\mu})\psi+H.c.$, where $\Sigma_{ab}=(1/8)(\gamma_a\gamma_b-\gamma_b\gamma_a)$. To generalize this action to include torsion one replaces ω^{bc}_{μ} by $\tilde{\omega}^{bc}_{\mu}$ and obtains

$$\tilde{I}_{\rm D}=\frac{1}{2}\int d^4x(-g)^{1/2}i\bar{\psi}\gamma^{a}V^{\mu}_{a}(\partial_{\mu}+\Sigma_{bc}\tilde{\omega}^{bc}_{\mu})\psi+H.c. \tag{20}$$

Integration parts, use of properties of the Dirac gamma matrices, and introduction of a coupling to A_{μ} yields [6]

$$\tilde{I}_{\rm D} = \int d^4x(-g)^{1/2}i\bar{\psi}\gamma^{a}V^{\mu}_{a}(\partial_{\mu}+\Sigma_{bc}\omega^{bc}_{\mu}-iA_{\mu}-i\gamma^{5}S_{\mu})\psi, \tag{21}$$

where

$$\begin{aligned} &S^{\mu}=\frac{1}{8}(-g)^{-1/2}\epsilon^{\mu\alpha\beta\gamma}Q_{\alpha\beta\gamma}, \\ &-(-g)^{-1/2}\epsilon_{\mu\alpha\beta\gamma}S^{\mu}=\frac{1}{4}[Q_{\alpha\beta\gamma}+Q_{\gamma\alpha\beta}+Q_{\beta\gamma\alpha}]. \end{aligned} \tag{22}$$

In the action $\tilde{I}_{\rm D}$ we note that even though the torsion is only antisymmetric on two of its indices, the only components of the torsion that appear in its torsion-dependent S^μ term are the four that constitute that part of the torsion that is antisymmetric on all three of its indices. As well as being locally gauge invariant under $\psi \to e^{i\alpha(x)}\psi$, $A_\mu \to A_\mu + \partial_\mu \alpha(x)$, $\tilde{I}_{\rm D}$ is also locally chiral invariant [6] under $\psi \to e^{i\gamma^5 \beta(x)}\psi$, $S_\mu \to S_\mu + \partial_\mu \beta(x)$. Additionally, as noted in [7], $\tilde{I}_{\rm D}$ is locally conformal invariant under $V^a_\mu(x) \to \Omega(x) V^a_\mu(x)$, $\psi(x) \to \Omega^{-3/2}(x)\psi(x)$ since, just like the vector potential A_μ, the equally minimally coupled S_μ also has zero conformal weight [8]. The $\tilde{I}_{\rm D}$ action thus has a remarkably rich local invariance structure, as it is invariant under local translations, local Lorentz transformations, local gauge transformations, local axial gauge transformations, and local conformal transformations.

With S^μ having a structure identical to the Faraday Law structure given in Eqs. (5) and (6), and with S^μ precisely being an axial 4-vector, S^μ is thus the natural quantity to act as the second potential that appears in $X_{\mu\nu}$ [9], and thus the natural axial vector needed to set up a variational procedure for Faraday's Law of electromagnetism [10]. However, in order to set up a variational procedure we will need to construct a kinetic energy term for it. To generate such a kinetic energy term we appeal to the Dirac action. Specifically, we recall [11], [6] that when one does a path integration $\int D\bar{\psi} D\psi \exp(i\tilde{I}_{\rm D})$ over the fermions (equivalent to a one fermion loop Feynman graph) one generates an effective action of the form [12]

$$I_{\rm EFF} = \int d^4x(-g)^{1/2} C\left[\frac{1}{20}\left[R_{\mu\nu}R^{\mu\nu} - \frac{1}{3}(R^\alpha{}_\alpha)^2\right] + \frac{1}{3}F_{\mu\nu}F^{\mu\nu} + \frac{1}{3}S_{\mu\nu}S^{\mu\nu}\right], \tag{23}$$

where C is a log divergent constant and $R_{\mu\nu}$ is the standard (torsionless) Ricci tensor. The action $I_{\rm EFF}$ possesses all the local symmetries possessed by $\tilde{I}_{\rm D}$, with the appearance of the $R_{\mu\nu}R^{\mu\nu} - (1/3)(R^\alpha{}_\alpha)^2$ term being characteristic of a gravity theory that is locally conformal invariant (see e.g. [13, 14]). Also, we take note of the fact that path integration over the fermions has converted terms that are linear in A_μ and S_μ in $\tilde{I}_{\rm D}$ into terms that are quadratic in A_μ and S_μ in $I_{\rm EFF}$. Comparing now with Eq. (15), we see that the action $I_{\rm EFF}$ contains precisely the kinetic energy term we seek. Thus not only does torsion provide a natural origin for the second potential needed for $X_{\mu\nu}$, up to renormalization constants it also provides precisely the correct action whose variation, on adding appropriately coupled sources, leads to Eq. (11) and a derivation of Faraday's Law via a variational principle. S_μ thus serves as an analog of the electromagnetic A_μ, an analog that is purely geometrical.

Given the geometrical structure of S_μ, we note that it is also possible to give A_μ an analogous such structure. Specifically, we recall that Weyl had suggested that one could metricate electromagnetism by introducing a B_μ-dependent connection for a real field B_μ of the form

$$W^\lambda{}_{\mu\nu} = -\frac{2}{3}(g^\lambda_\nu B_\mu + g^\lambda_\mu B_\nu - g_{\mu\nu}B^\lambda) = W^\lambda{}_{\nu\mu}, \tag{24}$$

as written here with a convenient 2/3 charge normalization. However, if we now use $\Lambda^\lambda{}_{\mu\nu} + K^\lambda{}_{\mu\nu} + W^\lambda{}_{\mu\nu}$ in the spin connection, as noted in [15] the B_μ term drops out of the Dirac action identically, with Weyl's B_μ not coupling to the Dirac spinor at all. The reason for this is that the Weyl connection generates individual non-Hermitian terms of the generic form $i(\partial_\mu + B_\mu)\psi$, and in the full Hermitian $\tilde{I}_{\rm D}$ such terms must cancel identically. However, given this, suppose we instead take B_μ to be anti-Hermitian and set $B_\mu = iA_\mu$ where A_μ is Hermitian. Now, not only is there now no cancellation, use of this anti-Hermitian connection is found to precisely lead to none other than the above $\tilde{I}_{\rm D}$ as given in Eq. (21). Thus starting from the torsionless $I_{\rm D}$ we can derive Eq. (21) in two distinct ways. If we demand local invariance of the action under $\psi \to e^{i\alpha(x)}\psi$ and $\psi \to e^{i\gamma^5\beta(x)}\psi$, we can introduce A_μ and S_μ by minimal coupling or by changing the geometry. The two potentials needed for electromagnetism can thus be put on a completely equal footing. Now a drawback in using a B_μ-dependent $W^{\lambda}{}_{\mu\nu}$ is that with it parallel transport is path dependent, with the geometry being a Weyl geometry rather than a Riemannian one. However, with iA_μ the geometry associated with $I_{\rm EFF}$ is a regular Riemannian one that uses only the connections given in Eq. (16). Thus by using iA_μ instead of B_μ we convert a Weyl geometry into a Riemannian one.

4. The Nature of Torsion

While we have seen that the axial 4-vector S_μ gives electromagnetism a chiral structure, we need to comment on the fact that experimentally there is no apparent sign of S_μ. Moreover, since S_μ is associated with torsion it is not simply a typical spacetime axial vector field. To underscore the special nature of torsion, we note that even if the standard (torsionless) Riemann tensor is zero, torsion is not obliged to vanish. Torsion could thus exist in a spacetime with no Riemann curvature at all. In a space that is flat as far as the geometry of its four spacetime x^μ coordinates is concerned, we note that since the Minkowski metric is independent of the x^μ, a non-zero torsion might not depend on the x^μ coordinates either. Given the antisymmetry of $Q_{\lambda\mu\nu}$, we can thus envisage that S_μ, and thus concomitantly its monopole source K^μ as well, might depend instead on a set of Grassmann coordinates, coordinates that anticommute with each other.

To realize this possibility, on comparing Eqs. (5) and (6) with Eq. (22), we can consider the possibility that the torsion can written as $Q_{\lambda\mu\nu} = \nabla_\lambda A_{\mu\nu}$, where $A_{\mu\nu}$ is an antisymmetric rank two tensor. In [10] it was suggested that this $A_{\mu\nu}$ could be the antisymmetric part of a 16-component metric tensor. To this end, we now note that if we introduce a set of Grassmann vierbeins ξ^a_μ, then the quantity $A_{\mu\nu} = \eta_{ab}\xi^a_\mu\xi^b_\nu$ will be antisymmetric since $\xi^a_\mu\xi^b_\nu + \xi^b_\mu\xi^a_\nu = 0$. Thus we can envisage spacetime being enlarged to encompass both ordinary coordinates and Grassmann coordinates, and spaces of this type were constructed in e.g. [16], where it was shown that a canonical quantization in which vanishing anticommutators were replaced by non-vanishing ones led to the Dirac equation. As also noted in [16], because of the Pauli principle finite degree of freedom Grassmann coordinates ξ^b_μ (as opposed to infinite degree of freedom Grassmann fields $\psi(x)$) could not be macroscopically occupied. Consequently, a Grassmann torsion could only be microscopic, with only the sector of electromagnetism that is based on A_μ ordinarily being observable in macroscopic systems.

Now we had found in Eq. (23) that at the classical level the A_μ and S_μ sectors were decoupled from each other. However, according to the Dirac action given in Eq. (21) both sectors couple to the fermions. Thus quantum mechanically one could have transitions between the two sectors mediated by fermion loops with both vector and axial vector insertions (axial analog of light on light scattering). This would be a small effect, and would also be microscopic, with a quantized Grassmann torsion not making any substantial modifications to QED. Thus torsion, and equally magnetic monopoles, might only be manifest microscopically, where they could potentially contribute to physics beyond the standard model [17]. Finally, if the torsion/monopole sector is only manifest microscopically, then macroscopically we can set $\mathbf{E}'$, $\mathbf{B}'$, ρ_m, and $\mathbf{J}_m$ to zero, with the standard sourceless Faraday Law then holding for macroscopic electrodynamics.

References

[1] If A_μ is singular (Dirac string) or has a non-trivial topology (grand unified monopoles) one could evade the vanishing of $(-g)^{-1/2}\epsilon^{\mu\nu\sigma\tau}\nabla_\nu(\nabla_\sigma A_\tau - \nabla_\tau A_\sigma)$ – see e.g. Preskill J 1984 *Ann. Rev. Nucl. Part. Sci.* **34** 461; Milton K A 2006 *Rep. Prog. Phys.* **69** 1637. However, the intent of this paper is to obtain a non-vanishing $(-g)^{-1/2}\epsilon^{\mu\nu\sigma\tau}F_{\sigma\tau}$ without singularities or non-trivial topologies.

[2] Shanmugadhasan S 1952 *Can. Jour. Phys.* **30** 218

[3] Cabibbo N and Ferrari E 1962 *Nuovo Cim.* **23** 1147

[4] While our work follows these earlier studies, it differs in three key aspects. First, our motivation is to address Faraday's Law and not magnetic monopoles per se, with it being Faraday's Law that forces us to two potentials and not monopoles. Second, we provide an explicit origin for the potential S_μ as being due torsion. And third, we consider a coupling of S_μ to fermions, and with it we are able to derive an action for its dynamics rather than having to postulate one. Some earlier work on relating torsion and magnetic monopoles may be found in Israelit M 1998 *Found. Phys.* **28** 205, and Hammond R T 1993 *Nuovo Cim.* B **108** 725

[5] This of course does not preclude their possible existence.

[6] Shapiro I L 2002 *Phys. Rept.* **357** 113

[7] Fabbri L and Mannheim P D 2014 *Phys. Rev.* D **90** 024042

[8] Under a local conformal transformation the torsion transforms as [6] ${Q^\lambda}_{\mu\nu} \to {Q^\lambda}_{\mu\nu} + q\Omega^{-1}(x)(\delta^\lambda_\mu\partial_\nu - \delta^\lambda_\nu\partial_\mu)\Omega(x)$ where q is its conformal weight. No matter what the value of q, under a conformal transformation S_μ transforms into itself with all derivatives of $\Omega(x)$ dropping out.

[9] Since $X_{\mu\nu}$ is invariant under $A_\mu \to A_\mu + \partial_\mu\alpha(x)$ and $S_\mu \to S_\mu + \partial_\mu\beta(x)$, $X_{\mu\nu}$ is recognized as the chiral generalization of $F_{\mu\nu}$. On allowing for these two transformations, A_μ and S_μ together contain 6 degrees of freedom, exactly the same number as $\mathbf{E}_{\rm TOT}$ and $\mathbf{B}_{\rm TOT}$ combined.

[10] Using torsion one can also develop a gravitational analog of Faraday's Law. See Mannheim P D and Poveromo J J 2014 *Gen. Rel. Gravit.* **46** 1795

[11] 't Hooft G 2010 Probing the small distance structure of canonical quantum gravity using the conformal group *Preprint* arXiv:1009.0669 [gr-qc]

[12] The Riemann and Maxwell sector terms are given in [11] and references therein. The $S_{\mu\nu}S^{\mu\nu}$ term is given in [6].

[13] Mannheim P D 2006 *Prog. Part. Nucl. Phys.* **56** 340

[14] Mannheim P D 2012 *Found. Phys.* **42** 388

[15] Hayashi K, Kasuya M and Shirafuji T 1977 *Prog. Theor. Phys.* **57** 431

[16] Mannheim P D 1985 *Phys. Rev.* D **32** 898

[17] As an alternative to Grassmann coordinates one could construct the antisymmetric $A_{\mu\nu}$ via $A_{\mu\nu} = C_{ab}V^a_\mu V^b_\nu$, where C_{ab} is the charge conjugation matrix. Torsion would then be macroscopic, but could still have escaped detection if the bridge between the A_μ and S_μ sectors is purely quantum-mechanical. As discussed in Mannheim P D 2014 *PT* Symmetry, Conformal Symmetry, and the Metrication of Electromagnetism *Preprint* ArXiv:1407.1820 [hep-th], another possible way for S_μ to escape detection is for it to acquire a sufficiently large Higgs mechanism-type mass.

The Coleman-Weinberg mechanism in a conformal (Weyl) invariant theory: application to a magnetic monopole

Ariel Edery[1], **Noah Graham**[2]

[1]Department of Physics, Bishop's University, 2600 College Street, Sherbrooke, Québec, Canada, J1M 1Z7

[2]Department of Physics, Middlebury College, Middlebury, VT 05753

E-mail: [1]`aedery@ubishops.ca`

E-mail: [2]`ngraham@middlebury.edu`

Abstract. We consider a massless conformally (Weyl) invariant classical action consisting of a magnetic monopole coupled to gravity in an anti-de Sitter background spacetime. We implement quantum corrections and this breaks the conformal (Weyl) symmetry, introduces a length scale via the process of renormalization and leads to the trace anomaly. We calculate the one-loop effective potential and determine from it the vacuum expectation value (VEV). Spontaneous symmetry breaking is radiatively induced *a la* Coleman-Weinberg and the scalar coupling constant is exchanged for the dimensionful VEV via dimensional transmutation. An important result is that the Ricci scalar of the AdS background spacetime is determined entirely by the value of the VEV.

1. Introduction

There has been a general interest in scale and conformal (Weyl) symmmetry as a possible fundamental principle in physics and cosmology, as pointed out for example in [1]. These authors point out that the classical action of the standard model is already consistent with global scale symmetry if the Higgs mass is dropped. Furthermore, on cosmic scales, the nearly-scale invariant spectrum of primordial fluctuations seems to demand an explanation based on a fundamental symmetry in nature. The authors therefore argue that scale and conformal symmetry could be the clue to fitting observations simultaneously on both small and large scales. In [1], authors discuss a model containing a Higgs field H, a dilaton field ϕ, standard model fields as well as gravity. However, their theory is classical and the authors mention that in a conformally (Weyl) invariant theory there is no mechanism at the classical level to set the scale of ϕ_0, the minimum of the dilaton field in their model. They then mention that adding quantum corrections could alleviate this problem via the Coleman-Weinberg mechanism [2]. This mechanism was recently implemented in the context of the standard model [3] where the Higgs mass was induced quantum mechanically starting with a scale invariant standard model. In our work, we implement the Coleman Weinberg mechanism starting with a classically Weyl invariant action containing a magnetic monopole. We provide here a review of the most salient features. The reader can find details in [4].

The classical Weyl invariant theory of a magnetic monopole coupled to gravity was studied in [5]. The gravity sector does not include an Einstein-Hilbert term because that term is not Weyl invariant; instead, one includes a Weyl squared term (which is Weyl invariant). A cosmological constant is also forbidden by Weyl invariance. In this massless model, the conformal coupling term $-R\phi^2/6$, where R is the Ricci scalar "replaces" the usual negative mass term $a^2\phi^2$ in the symmmetry breaking potential. The vacuum expectation value (VEV) is now expressed in terms of R. As pointed out in [1], there is no mechanism at the classical level that can introduce a length scale in a Weyl invariant theory; one must therefore choose a VEV value by hand (or alternatively specify the asymptotic value of the Ricci scalar). Magnetic monopole solutions in a background anti-de Sitter (AdS) space are then found numerically [5]. The static spherically symmetric solution at large distances (outside the monopole core) is Schwarzschild-AdS. It should be stressed that the solution in [5] is not a black hole (BH) solution since the core is non-singular and static. In contrast, a BH has a horizon and there exits a region in the interior which is non-stationary, where all Killing vectors are spacelike [6].

In this work, we introduce a length scale by considering the effects of quantum fluctuations of scalar fields at one loop in a classical gravitational and gauge field background. We derive the effective potential at one loop for a massless theory with a triplet of scalar fields and extract the vacuum expectation value (VEV) from it. The coupling constant λ is defined at the renormalization scale M and the dimensionless λ is traded for the dimensionful VEV of the theory, through dimensional transmutation [2]. The asymptotic (background) spacetime is AdS and an important result is that its Ricci scalar is completely determined by the VEV.

2. Schwinger-Dewitt coefficient and trace anomaly

The effects due to vacuum fluctuations of matter fields can be included via an effective action W which can be into two parts: a divergent part $W_{\rm div}$ and a renormalized finite part $W_{\rm ren} = W - W_{\rm div}$. $W_{\rm div}$ is due to high-frequency fluctuations and is therefore local and state-independent [8]. It is expressed as an integral over the Schwinger-Dewitt coefficient $a_2(x)$, which can be calculated via a set of "curvatures" [10]. $W_{\rm div}$ is incorporated into the original action S by adding the appropriate counterterms. We label this S_{ren}, the renormalized local part of the one loop effective action. $W_{\rm ren}$ is the finite state-dependent nonlocal part of the effective action due to long wavelength fluctuations that sample the entire geometry. For example, the finite part of the vacuum expectation value of the energy-momentum tensor, $\langle T_{\mu\nu}\rangle$, that will appear on the right hand side of the gravitational equations of motion, can be viewed as a Casimir effect which also arises from long wavelength quantum fluctuations [11].

The conformally invariant action that leads to a magnetic monopole coupled to gravity contains a metric $g_{\mu\nu}$, a triplet of scalar fields ϕ^a and nonabelian gauge fields A^a_μ. It is given by [5]

$$\begin{aligned} S &= \int d^4x\,\sqrt{-g}\,\Big(C_{\mu\nu\sigma\tau}C^{\mu\nu\sigma\tau} - \frac{1}{4e^2}\,F^a_{\mu\nu}\,F^{\mu\nu}_a + D_\mu\phi^a\,D^\mu\phi_a + \frac{1}{6}\,R\,\phi^a\phi_a - \frac{\lambda}{4!}(\phi^a\,\phi_a)^2\Big) \\ &= \int d^4x\,\sqrt{-g}\,\Big(C^2 - \frac{1}{4e^2}\,F^2 + (D\phi)^2 + \frac{1}{6}\,R\,\phi^2 - \frac{\lambda}{4!}\phi^4\Big) \end{aligned} \tag{1}$$

where $C_{\mu\nu\sigma\tau}$ is the Weyl tensor, $F^a_{\mu\nu}$ is the gauge field strength defined by $\nabla_\mu A^a_\nu - \nabla_\nu A^a_\mu + \varepsilon^a{}_{bc}\,A^b_\mu\,A^c_\nu$, and $D_\mu\phi^a$ is the covariant derivative defined by $\nabla_\mu\,\phi^a + \varepsilon^a{}_{bc}\,A^b_\mu\,\phi^c$. The square of the Weyl tensor is denoted as C^2 and $F^2 \equiv F^a_{\mu\nu} F^{\mu\nu}_a$, $(D\phi)^2 \equiv D_\mu\phi^a D^\mu\phi_a$ and $\phi^2 \equiv \phi^a\phi_a$. The Ricci-Levi-Civita three-dimensional coefficients ε_{abc} are totally antisymmetric. Raising or lowering of the internal (latin) indices does not change the sign so that $\phi_a = \phi^a$ and implicit summation is assumed throughout e.g. $\phi_a\phi^a = \phi_1^2 + \phi_2^2 + \phi_3^2$. The action (1) is invariant under the conformal transformations $g_{\mu\nu} \to \Omega^2(x) g_{\mu\nu}$ and $\phi^a \to \phi^a/\Omega(x)$ where $\Omega(x)$ is called the conformal factor, an arbitrary positive smooth function.

We consider the vacuum fluctuations of the triplet ϕ^a in a gravitational and gauge field classical background. The Euclidean effective action is given by $W = \frac{1}{2}\,\mathrm{tr}\,(\ln \hat{H}\,)$ where $\hat{H}$ is the hessian of the Euclidean version of the action (1). It takes the following form [10]

$$\hat{H} = g^{\mu\nu} D_\mu D_\nu + \hat{P} - \frac{1}{6} R\hat{1}\,. \tag{2}$$

The above operator is a 3×3 matrix in the internal vector space of the triplet ϕ^a and $\hat{1}$ is the identity matrix. Here $\hat{P}$ [1] arises from the self interaction term $\lambda(\phi^a\,\phi_a)^2/4!$, $\frac{1}{6}R\hat{1}$ arises from the conformally coupled scalar field term $\frac{1}{6}\,R\,\phi^a\phi_a$ and $g^{\mu\nu}D_\mu D_\nu$ arises from the covariant derivative term $D_\mu\phi^a\,D^\mu\phi_a$. The gravitational and nonabelian gauge fields are treated classically. One identifies three "curvatures" for the operator (2) that enter into the calculation of the effective action [10]. These are the Riemann curvature associated with $g_{\mu\nu}$, the commutator curvature $\hat{\mathscr{R}}_{\mu\nu}$ associated with the covariant derivative D_μ and defined by

$$[D_\mu, D_\nu]\,\phi^a = \mathscr{R}^a{}_{b\,\mu\,\nu}\,\phi^b \text{ with } \hat{\mathscr{R}}_{\mu\nu} \equiv \mathscr{R}^a{}_{b\,\mu\,\nu} \tag{4}$$

and the potential $\hat{P}$ which is its own curvature. The divergent part $W_{\rm div}$ of the Euclidean effective action can be expressed entirely in terms of these curvatures as [10]

$$W_{\rm div} = \frac{1}{n-4}\int d^4x\,\sqrt{g}\;\;\mathrm{tr}\,\hat{a}_2(x,x) \qquad (n \to 4) \tag{5}$$

$$\hat{a}_2(x,x) = \frac{1}{180}\Big(R_{\mu\nu\sigma\tau}R^{\mu\nu\sigma\tau} - R_{\mu\nu}R^{\mu\nu} + \Box R\Big)\hat{1} + \frac{1}{12}\hat{\mathscr{R}}_{\mu\nu}\hat{\mathscr{R}}^{\mu\nu} + \frac{1}{2}\hat{P}^2 + \frac{1}{6}\Box\hat{P}\,. \tag{6}$$

The curvatures $\hat{P}$ and $\hat{\mathscr{R}}_{\mu\nu}$ for the Euclidean version of the action (1) are calculated in [4]. Here we simply state the result, the quantity of interest that appears in the integrand of $W_{\rm div}$. This is

$$a_2(x) \equiv \mathrm{tr}\,\hat{a}_2(x,x) = \frac{1}{60}\Big(R_{\mu\nu\sigma\tau}R^{\mu\nu\sigma\tau} - R_{\mu\nu}R^{\mu\nu} + \Box R\Big) - \frac{1}{6}F^2 + \Big(\frac{11}{3}\Big)\frac{\lambda^2}{4!}\phi^4 - \frac{5}{36}\lambda\Box\phi^2 \tag{7}$$

where the factor of 11/3 in front of λ^2 stems from having a triplet of scalar fields (it is equal to 3 for a single scalar field). To cancel the divergent part $W_{\rm div}$, we simply add the appropriate counterterm to the original action (1). The renormalized local part of the action at one loop is simply

$$S_{\rm ren} = \int d^4x\,\sqrt{-g}\,\Big(\alpha\,R^2 + \beta\,R_{\mu\nu}R^{\mu\nu} - \frac{1}{4e^2}\,F^2 + (D\phi)^2 + \frac{1}{6}\,R\,\phi^2 - \frac{\lambda}{4!}\,\phi^4\Big)\,. \tag{8}$$

[1] The matrix elements of the potential $\hat{P}$ are

$$P_{ij} = -\frac{\partial}{\partial\phi_i}\frac{\partial}{\partial\phi_j}\frac{\lambda}{4!}(\phi_a\phi^a)^2 = -\frac{\lambda}{6}\,(\delta_{ij}\,\phi_a\phi^a + 2\,\phi_i\,\phi_j\,) \text{ where } i,j = 1,2,3\,. \tag{3}$$

Journal of Physics: Conference Series **615** (2015) 012005 doi:10.1088/1742-6596/615/1/012005

The total derivatives $\Box R$ and $\Box(\phi_a\,\phi^a)$ appearing in (7) are not included in the above action as they lead to boundary terms that have no bearing on the equations of motion. The terms $C_{\mu\nu\sigma\tau}C^{\mu\nu\sigma\tau}$ and $R_{\mu\nu\sigma\tau}R^{\mu\nu\sigma\tau}$ do not appear in (8) since they can be eliminated in favor of $R_{\mu\nu}R^{\mu\nu}$ and R^2 using the equality $C_{\mu\nu\sigma\tau}C^{\mu\nu\sigma\tau} = R_{\mu\nu\sigma\tau}R^{\mu\nu\sigma\tau} - 2\,R_{\mu\nu}R^{\mu\nu} + R^2/3$ and the fact that the integral of $R_{\mu\nu\sigma\tau}R^{\mu\nu\sigma\tau} - 4\,R_{\mu\nu}R^{\mu\nu} + R^2$, the Gauss-Bonnet integral, is a topological invariant (the Euler number) in four dimensions. The quantities e, λ, α, and β are renormalized constants. They are running coupling constants governed by a renormalization group equation which we do not need to state here for our purposes (for details see [12, 13]). At one loop, the constants α and β are related by $\alpha = -\beta/3$ and the sum of the first two terms in (8) are conformally invariant. At higher loops, non-conformally invariant terms like R^2 are generated [12, 13] and the two constants become independent. Our calculations will be at the one loop level, so we assume the relation $\alpha = -\beta/3$.

2.1. Trace Anomaly

To find the trace, we first note that under a conformal tranformation the inverse metric transforms according to $g^{\mu\nu}(x) \to e^{2\sigma(x)}g^{\mu\nu}(x)$ and the scalar field transforms according to $\phi^a(x) \to e^{\sigma(x)}\phi^a(x)$ where $\sigma(x)$ is introduced for convenience and is related to the conformal factor $\Omega(x)$ (previously introduced) by $e^{-\sigma(x)} \equiv \Omega(x)$. This yields the functional relation

$$\frac{\delta}{\delta\sigma(x)} = 2g^{\mu\nu}(x)\frac{\delta}{\delta g^{\mu\nu}(x)} + \phi^a(x)\frac{\delta}{\delta\phi^a(x)}\,. \tag{9}$$

The functional variation of the one-loop effective action with respect to $\sigma(x)$ is proportional to the Scwinger-Dewitt coefficient $a_2(x)$ [10]

$$\frac{\delta\,W_{\rm ren}}{\delta\sigma} = \frac{\sqrt{-g}\,a_2(x)}{16\,\pi^2} \tag{10}$$

and using (9) this yields

$$\frac{1}{\sqrt{-g}}(2\,g^{\mu\nu}\frac{\delta W_{\rm ren}}{\delta g^{\mu\nu}} + \phi^a\frac{\delta W_{\rm ren}}{\delta\phi^a}) = \frac{a_2(x)}{16\pi^2}\,. \tag{11}$$

The vacuum expectation value of the energy-momentum tensor is defined as

$$\langle T_{\mu\nu}\rangle = \frac{2}{\sqrt{-g}}\frac{\delta W_{\rm ren}}{\delta g^{\mu\nu}}\,. \tag{12}$$

¿From (11), its trace is given by

$$\langle T^\mu_\mu\rangle = \frac{a_2(x)}{16\pi^2} - [E] \tag{13}$$

where

$$[E] \equiv \frac{\phi^a}{\sqrt{-g}}\frac{\delta W_{\rm ren}}{\delta\phi^a} \tag{14}$$

and $a_2(x)$ is given by (7). The renormalized composite operator $[E]$ can be obtained readily once the effective potential has been calculated.

2.1.1. AdS space As already discussed, the magnetic monopole solution is obtained when the spacetime is asymptotically anti-de Sitter space. This is a maximally symmetric spacetime with constant Ricci scalar R (positive in our notation). It is the submanifold obtained by embedding a hyperboloid in a flat five-dimensional spacetime of signature(+,+,-,-,-). The universal covering space of AdS space has the topology of R^4 and can be represented by the following metric:

$$ds^2 = (1 + k\,r^2)\,dt^2 - \frac{dr^2}{1 + k\,r^2} - r^2(d\theta^2 + \sin^2\theta\,d\phi^2) \tag{15}$$

where (r, θ, ϕ) cover the usual range of spherical coordinates $(r \geq 0\,,\ 0 \leq \theta \leq \pi\,,\ 0 \leq \phi < 2\,\pi)$ and $-\infty < t < \infty$. The constant k is positive with the Ricci scalar given by $R = 12k$.

The symmetry group (isometry group) of AdS space is the ten parameter group SO(2,3). The only maximally form invariant rank two tensor under this group is the metric $g_{\mu\nu}$ (times a constant) so that the expectation value of the energy-momentum tensor in AdS space can be expressed in terms of its trace:

$$\langle T_{\mu\nu}\rangle_0 = \frac{1}{4}\,g_{\mu\nu}\langle T^{\mu}_{\ \mu}\rangle_0 \tag{16}$$

where the zero subscript means that the quantity is evaluated asymptotically in AdS space, the vacuum spacetime.

3. The effective potential

In this section we obtain the one loop effective potential by summing all the one loop one-particle irreducible (1PI) Feynman diagrams in the presence of $\lambda\phi^4/24$ and $-R\phi^2/6$ interactions. The coupling constant λ is defined at a renormalization scale M. We choose the renormalization scale M to coincide with the VEV, the minimum of the effective potential. This in turn fixes the value of λ; the dimensionless constant λ is traded for the dimensionful VEV through dimensional transmutation. The expectation value of the composite operator $[E]$, which appears both in the trace anomaly and in the equations of motion for the scalar field, can be readily obtained from the effective potential. Note that at one loop the effective potential and the expectation value of composite operators do not generate new geometrical curvature terms; they are generated starting only at the two loop level [12, 13]. In particular, for the calculation of the effective potential, the $\sqrt{-g}$ factor plays no role at one loop. The calculation proceeds in the same manner as in flat space, though of course the Ricci scalar R is non-zero and acts as a vertex for the $R\phi^2$ interaction.

The (classical) potential is given by

$$U = \frac{\lambda}{4!}\,\phi^4 - \frac{1}{6}R\,\phi^2\,. \tag{17}$$

We have a triplet of scalar fields and without loss of generality we take the vacuum expectation value to lie along the third component ϕ_3. For ϕ_3 loops, U generates two vertices: $-R/3$ and $\lambda\phi_3^2/2$. They can be combined into a single vertex given by the second derivative $U''(\phi_3) = -R/3 + \lambda\phi_3^2/2$. The vacuum expectation value of ϕ_1 and ϕ_2 are zero but they can still fluctuate in loops. The vertex for both is $U''(\phi_1) = U''(\phi_2) = -R/3 + \lambda\phi_3^2/6$ which is equivalent to replacing the coupling constant λ by $\lambda/3$ in the previous case. There are three sets of one loop 1PI Feynman diagrams; these are depicted in Fig.4.1. Let the classical field $\phi_c(x)$ be defined as the vacuum expectation value of ϕ_3 in the presence of some external source $J(x)$

$$\phi_c(x) = \left.\frac{\langle 0|\phi_3(x)|0\rangle}{\langle 0|0\rangle}\right|_J\,. \tag{18}$$

where J appears in the action in the usual fashion via the source term $J\phi$. The effective potential is obtained by summing all the diagrams in Fig.4.1. Note that the propagator is massless. For the first set of diagrams, the one-loop contribution yields

$$\begin{aligned} V &= i\int \frac{d^4k}{(2\pi)^4}\sum_{n=1}^{\infty}\frac{1}{2n}\left[\frac{-R/3+\lambda\phi_c^2/2}{k^2+i\epsilon}\right]^n \\ &= \frac{1}{16\pi^2}\int_0^{\Lambda} k_E^3 \ln\left[1+\frac{-R/3+\lambda\phi_c^2/2}{k_E^2}\right] \end{aligned} \tag{19}$$

where k_E is the Euclidean momenta and Λ is a momentum cut-off. The integral in (19) can be readily evaluated but we do not write it out explicitly here. The other two sets of Feynman diagrams can be evaluated by simply replacing λ by $\lambda/3$ in (19). The one loop contribution to the potential is then

$$V_1 = V + 2\,V[\lambda \to \lambda/3]\,. \tag{20}$$

As it stands, the expression V_1 is divergent in the infinite Λ limit. This is handled in the usual fashion by adding the necessary counterterms and then imposing the appropriate renormalization conditions. The total potential is given by

$$V_{tot} = \frac{\lambda}{4!}\,\phi_c^4 - \frac{1}{6}R\,\phi_c^2 + V_1 + A\phi_c^2 + B\phi_c^4 \tag{21}$$

where the last two terms are the counterterms. The constants A and B are determined via the renormalization conditions

$$\lambda = \left.\frac{d^4V_{tot}}{d\phi_c^4}\right|_{\phi_c=M} \quad \text{and} \quad -\frac{R}{3} = \left.\frac{d^2V_{tot}}{d\phi_c^2}\right|_{\phi_c=0}\,. \tag{22}$$

The renormalization scale M sets the scale for the theory. Substituting A and B back into (21), taking the infinite Λ limit and then collecting terms into compact expressions, we obtain the one loop effective potential

$$\begin{aligned} V_{\text{eff}} = \frac{\lambda}{4!}\,\phi_c^4 - \frac{1}{6}R\,\phi_c^2 + \frac{1}{1152\pi^2}\Bigg[& 18\Big(-\frac{R}{3}+\frac{\lambda\phi_c^2}{2}\Big)^2 \ln\Big[\frac{2R-3\lambda\phi_c^2}{2R}\Big] \\ & + 36\Big(-\frac{R}{3}+\frac{\lambda\phi_c^2}{6}\Big)^2 \ln\Big[\frac{2R-\lambda\phi_c^2}{2R}\Big] + 5R\,\lambda\,\phi_c^2 \\ & + \lambda^2\phi_c^4\Big(\frac{9}{2}\ln\Big[\frac{2R}{2R-3\lambda M^2}\Big] + \ln\Big[\frac{2R}{2R-\lambda M^2}\Big]\Big)\Bigg] \\ & + \Delta \end{aligned} \tag{23}$$

where

$$\Delta = \lambda^2\phi_c^4\,\frac{(-1584R^4 + 11904M^2R^3\lambda - 20360M^4R^2\lambda^2 + 12480M^6R\lambda^3 - 2475M^8\lambda^4)}{13824\,\pi^2(-2R+M^2\lambda)^2(-2R+3M^2\lambda)^2}\,. \tag{24}$$

Let $\phi_c = v$ be the vacuum expectation value (VEV). It takes on this value in the asymptotic (background) spacetime, which is AdS space. We will see later, in section 4.1, that solving the equations of motion asymptotically yields the relation (42) between R_0, the Ricci scalar of AdS space, and the VEV: $R_0 = \sqrt{110}\,\lambda\, v^2$. The VEV occurs at the minimum of the effective potential in the AdS background spacetime, where

$$\frac{dV_{\rm eff}}{d\phi_c}\Bigg|_{\substack{\phi_c = v \\ R=\sqrt{110}\lambda\, v^2}} = 0\,. \tag{25}$$

We set the arbitrary scale M to be equal to the VEV i.e. $\phi_c = v = M$. Equation (25) then yields a numerical value of $\lambda = C/D = 2519.926$, where the exact expressions for C and D are

$$\begin{aligned} C &= 288\pi^2(197355230\sqrt{110} - 1144879587) \text{ and} \\ D &= 299795296\sqrt{110} - 3734763307 + \\ &\quad (12\ln(2\sqrt{110}-1) + 18\ln(2\sqrt{110}-3) - 15\ln(440))(-96294467\sqrt{110} + 524292560)\,. \end{aligned} \tag{26}$$

The ratio of the one loop correction to the tree (classical) result for the potential can be readily calculated to be -0.504. Such ratios are typical of one loop corrections in massless theories (e.g. in massless $\lambda\phi^4$ theory in flat space with a single scalar field the ratio is close to -1 [14]). We discuss in the conclusions how adding gauge field fluctuations can effect this scenario.

We started with a classical massless theory, a conformally invariant theory with $\lambda\phi^4$ and $R\,\phi^2$ interactions. After including one loop quantum corrections, the dimensionelss parameter λ has been traded for the dimensionful VEV. An important result is that the Ricci scalar of AdS space is now completely determined by the value of the VEV.

3.1. Composite operator [E]

The composite operator [E(y)] is defined via (14). It appears in the trace but also as a quantum correction to the equations of motion for the scalar fields (see section 4 below). Inserting this operator into an n-point Green's function $\Gamma^n(x_1, ..., x_n)$ and integrating over all y yields n times the same Green's function [13]. The Feynman diagrams are therefore identical to those used to evaluate the effective potential $V_{\rm eff}$, namely those of Fig.4.1 (the only difference is that the symmetry factor is multiplied by n). The upshot is that $[E]$ can be obtained by taking the negative of the derivative of the one loop part of the effective potential (23), $V_{\rm loop} = V_{\rm eff} - U$, and then multiplying it by ϕ_c,

$$\begin{aligned} [E] &= -\phi_c \frac{dV_{\rm loop}}{d\phi_c} \\ &= -\frac{1}{1152\pi^2}\Bigg[36\left(-\frac{R}{3}+\frac{\lambda\phi_c^2}{2}\right)\ln\left[\frac{2R-3\lambda\phi_c^2}{2R}\right]\lambda\,\phi_c^2 - 108\left(-\frac{R}{3}+\frac{\lambda\phi_c^2}{2}\right)^2\frac{\lambda\phi_c^2}{2R-3\lambda\phi_c^2} \\ &\quad + 24\left(-\frac{R}{3}+\frac{\lambda\phi_c^2}{6}\right)\ln\left[\frac{2R-\lambda\phi_c^2}{2R}\right]\lambda\phi_c^2 - 72\left(-\frac{R}{3}+\frac{\lambda\phi_c^2}{6}\right)^2\frac{\lambda\phi_c^2}{2R-\lambda\phi_c^2} \\ &\quad + 10R\,\lambda\,\phi_c^2 + \lambda^2\phi_c^4\left(18\ln\left[\frac{2R}{2R-3\lambda M^2}\right] + 4\ln\left[\frac{2R}{2R-\lambda M^2}\right]\right)\Bigg] \\ &\quad - 4\Delta\,. \end{aligned} \tag{27}$$

4. Equations of motion

For the magnetic monopole, we seek static spherically symmetric solutions where the spatial symmetry (isometry) and gauge symmetry are both SO(3). These can be viewed as the lowest energy or ground state solution [5]. The metric, scalar triplet, and non-abelian gauge fields take on the following spherically symmetric form [5]:

$$\text{metric: } ds^2 = N(r)\,dt^2 - \psi(r)dr^2 - r^2\left(d\theta^2 + \sin^2(\theta)\,d\phi^2\right) \tag{28}$$

$$\text{scalar: } \phi^a(r) = f(r)\,\frac{r^a}{r} = f(r)[\sin\theta\sin\varphi, \sin\theta\cos\varphi, \cos\theta] \tag{29}$$

$$\text{gauge: } A^\mu_a = q(r)\,\xi^\mu_a \text{ where } \xi^\mu_a \text{ are the Killing vectors for SO(3), namely}$$
$$\xi^\mu_1 = [0,0,\cos\varphi,-\sin\varphi\cot\theta], \xi^\mu_2 = [0,0,-\sin\varphi,-\cos\varphi\cot\theta] \text{ and } \xi^\mu_3 = [0,0,0,1]\,. \tag{30}$$

It will be convenient to work with $a(r) \equiv 1 + r^2\,q(r)$ instead of $q(r)$. There are four functions of r to determine: the "metric" fields $N(r)$ and $\psi(r)$, the "gauge field" $a(r)$ and the "scalar" field $f(r)$. It is convenient to obtain the equations of motion by direct variation of these functions. The Lagrangian corresponding to S_{ren} is given by

$$\begin{aligned} L &= 4\pi \int_0^\infty dr\sqrt{N\,\psi}\,r^2\Big(\alpha\,R^2 + \beta\,R_{\mu\nu}R^{\mu\nu} - \frac{1}{4e^2}\,F^a_{\mu\nu}\,F^{\mu\nu}_a + D_\mu\phi^a\,D^\mu\phi_a + \frac{1}{6}\,R\,\phi^a\phi_a - \frac{\lambda}{4!}(\phi^a\phi_a)^2\Big) \\ &= 4\pi\int_0^\infty \mathscr{L}\,dr\,. \end{aligned} \tag{31}$$

The quantities that appear in (31) are evaluated using Eqs. 28-30 and are given in [4].

The equation of motion for N is given by

$$\begin{aligned} &\frac{\sqrt{N\psi}\,r^2}{2N}(\alpha\,R^2 + \beta\widetilde{R}_j\widetilde{R}^j) + \sqrt{N\psi}\,r^2\Big(2\,\alpha\,R\frac{\partial R}{\partial N} + 2\,\beta\,\widetilde{R}_j\frac{\partial\widetilde{R}^j}{\partial N}\Big) \\ &- \Big[\sqrt{N\psi}\,r^2\big(2\,\alpha\,R\frac{\partial R}{\partial N'} + 2\,\beta\,\widetilde{R}_j\frac{\partial\widetilde{R}^j}{\partial N'}\big)\Big]' + \Big[\sqrt{N\psi}\,r^2\,\big(2\,\alpha\,R\frac{\partial R}{\partial N''} + 2\,\beta\,\widetilde{R}_j\frac{\partial\widetilde{R}^j}{\partial N''}\big)\Big]'' \\ &+ \frac{1}{6\sqrt{N\psi}}\big(f^2 - f^2\psi - f^2 r\frac{\psi'}{\psi} - r^2 f f'\frac{\psi'}{\psi} + 4r f f' + 2r^2 f f''\big) - \frac{\lambda\,r^2 f^4\psi}{48\sqrt{N\psi}} \\ &- \frac{\big((a^2-1)^2\psi + 2r^2 a'^2\big)}{4\,e^2\,r^2\sqrt{N\psi}} - \frac{a^2 f^2\psi}{\sqrt{N\psi}} - \frac{r^2 f'^2}{6\sqrt{N\psi}} = \frac{\sqrt{N\,\psi}\,r^2}{2}\langle T^{tt}\rangle \end{aligned} \tag{32}$$

where the explicit expressions for the Ricci scalar R and for $\widetilde{R}_j = \widetilde{R}^j$ are found in [4]. Implicit summation over $j = 1,2,3,4$ is assumed. The equation of motion for ψ is given by

$$\begin{aligned} &\frac{\sqrt{N\psi}\,r^2}{2\psi}(\alpha\,R^2 + \beta\widetilde{R}_j\widetilde{R}^j) + \sqrt{N\psi}\,r^2\Big(2\,\alpha\,R\frac{\partial R}{\partial\psi} + 2\,\beta\,\widetilde{R}_j\frac{\partial\widetilde{R}^j}{\partial\psi}\Big) \\ &- \Big[\sqrt{N\psi}\,r^2\big(2\,\alpha\,R\frac{\partial R}{\partial\psi'} + 2\,\beta\,\widetilde{R}_j\frac{\partial\widetilde{R}^j}{\partial\psi'}\big)\Big]' + \Big[\sqrt{N\psi}\,r^2\,\big(2\,\alpha\,R\frac{\partial R}{\partial\psi''} + 2\,\beta\,\widetilde{R}_j\frac{\partial\widetilde{R}^j}{\partial\psi''}\big)\Big]'' \\ &- \frac{\lambda\,r^2 f^4 N}{48\sqrt{N\psi}} - \frac{N(a^2-1)^2}{4e_R^2\,r^2\sqrt{N\psi}} + \frac{N^2\,a'^2}{2\,e^2\,(N\psi)^{3/2}} + \frac{N^2 r^2 f'^2}{2\,(N\psi)^{3/2}} \\ &+ \frac{r f f'\,(4N^2 + rN'N)}{6\,(N\psi)^{3/2}} + \frac{f^2\,(N^2 - N^2\psi - 6a^2N^2\psi + rNN')}{6\,(N\psi)^{3/2}} \\ &= -\frac{\sqrt{N\,\psi}\,r^2}{2}\langle T^{rr}\rangle\,. \end{aligned} \tag{33}$$

The equation of motion for a is

$$2\,a\,(a^2-1)+4\,a\,e^2\,f^2\,r^2-\frac{2}{\psi}\,a''\,r^2+\frac{a'\,r^2}{\psi}\Big(\frac{\psi'}{\psi}-\frac{N'}{N}\Big)=0 \tag{34}$$

and finally the equation of motion for f is given by

$$-\frac{4a^2f^2}{r^2}+\frac{f^2\,R}{3}-\frac{\lambda}{6}\,f^4+\frac{2f\,f''}{\psi}+\frac{f\,f'}{\psi}\Big(\frac{4}{r}+\frac{N'}{N}-\frac{\psi'}{\psi}\Big)=-[E] \tag{35}$$

where $[E]$ is given by (27).

The above equations of motion are for static field configurations. Therefore the higher derivative terms in the metric field equations (32) and (33) pose no issues as they are spatial not time derivatives. Higher spatial derivatives appear in many branches of physics e.g. in physical acoustics the wave equation is modified by a term with four spatial derivatives when the bending stiffness of a vibrating string is included.

4.1. Relation between Ricci scalar of AdS space and the VEV: asymptotic analytical solution

We now solve the equations of motion (32),(33), (34) and (35) analytically in the asymptotic region to show that the Ricci scalar of AdS space is determined entirely by the VEV. The vacuum expectation value of the energy momentum tensor in AdS space is given by (16) and the non-zero components are

$$\begin{aligned}
\langle T^{tt}\rangle &= \frac{1}{4}\frac{1}{1+k\,r^2}\langle T^{\mu}_{\ \mu}\rangle_0 \quad ; \quad & \langle T^{rr}\rangle &= -\frac{1}{4}(1+k\,r^2)\langle T^{\mu}_{\ \mu}\rangle_0\\
\langle T^{\theta\theta}\rangle &= -\frac{1}{4\,r^2}\langle T^{\mu}_{\ \mu}\rangle_0 \quad ; \quad & \langle T^{\phi\phi}\rangle &= -\frac{1}{4\,r^2\sin^2\theta}\langle T^{\mu}_{\ \mu}\rangle_0
\end{aligned} \tag{36}$$

where the AdS metric (15) was used. The trace $\langle T^{\mu}_{\ \mu}\rangle$ is given by

$$\begin{aligned}
\langle T^{\mu}_{\ \mu}\rangle &= \frac{a_2(x)}{16\pi^2}-[E]\\
&= \frac{1}{16\pi^2}\Big\{\frac{1}{60}\Big(R_{\mu\nu\sigma\tau}R^{\mu\nu\sigma\tau}-R_{\mu\nu}R^{\mu\nu}+\Box R\Big)-\frac{1}{6}\,F^2+\frac{11}{3}\frac{\lambda^2}{4!}\phi^4-\frac{5}{36}\,\lambda\Box\,[\phi^2]\Big\}-[E],
\end{aligned} \tag{37}$$

where (13) and (7) were used and $[E]$ is given by (27). We now evaluate (37) in AdS space, the asymptotic spacetime. Asymptotically, $F^a_{\mu\nu}\to 0$, $\Box\,[\phi^2]\to 0$, $R_{\rho\sigma\mu\nu}=k\big(g_{\rho\mu}\,g_{\sigma\nu}-g_{\rho\nu}\,g_{\sigma\mu}\big)$ and $R_{\mu\nu}=3\,k\,g_{\mu\nu}$ so that $R=12k$, $\Box\,R=0$ and $R_{\mu\nu\sigma\tau}R^{\mu\nu\sigma\tau}-R_{\mu\nu}R^{\mu\nu}=-12\,k^2$. Substituting these values into (37), we obtain

$$\langle T^{\mu}_{\ \mu}\rangle_0=-\frac{1}{80}\frac{k^2}{\pi^2}+\frac{11}{1152}\frac{\lambda^2}{\pi^2}\,v^4-[E]_0 \tag{38}$$

where $[E]_0$ is $[E]$ evaluated in asymptotic AdS space.

The boundary conditions for the magnetic monopole [5] are that asymptotically, as $r\to\infty$, the spacetime is AdS where $N\to 1+kr^2$, $\psi\to 1/(1+k\,r^2)$, $f\to v$, $r\,f'\to 0$ (f' drops off faster than $1/r$), $r^2\,f''\to 0$, $a\to 0$ and $a'\to 0$. Both v and k are positive constants. Substituting these boundary conditions into the gravity equation (32) (or (33)) and using (38) yields the following relation

$$\frac{v^2}{2}-\frac{\lambda v^4}{48\,k}=\frac{1}{8\,k}\Big(-\frac{1}{80}\frac{k^2}{\pi^2}+\frac{11}{1152}\frac{\lambda^2}{\pi^2}\,v^4-[E]_0\Big)\,. \tag{39}$$

Journal of Physics: Conference Series **615** (2015) 012005 doi:10.1088/1742-6596/615/1/012005

We can eliminate $[E]_0$ above by solving the scalar equation (35) asymptotically. This yields

$$4\,k\,v^2 - \frac{\lambda v^4}{6} = -[E]_0 \tag{40}$$

where we used that $R = R_0 = 12\,k$ in AdS space. Substituting the above into (39) yields the solution

$$k = \frac{\sqrt{110}}{12}\lambda\,v^2 \tag{41}$$

so that

$$R_0 = \sqrt{110}\,\lambda\,v^2\,. \tag{42}$$

The Ricci scalar of AdS space is therefore determined solely by the VEV since the value of λ is known (it is no longer a free parameter having been traded for the dimensionful VEV). Substituting (41) into (40), yields $[E]_0 = -\frac{\lambda v^4}{6}\,(2\sqrt{110} - 1)$. This agrees with expression (27) asymptotically i.e. after substituting $R = \sqrt{110}\,\lambda\,v^2$ and $\phi_c = v = M$. The remaining unbroken U(1) is associated with $F_3^{\mu\nu}$ and the magnetic field is defined via $F_3^{ij} = \epsilon^{ijk}B^k$. Asymptotically, $a(r) \to 0$ and the function $q(r)$ appearing in (30) approaches $-1/r^2$. It is easy to verify that one obtains a radial magnetic field that varies as $1/r^2$ at large distances, corresponding to a magnetic monopole.

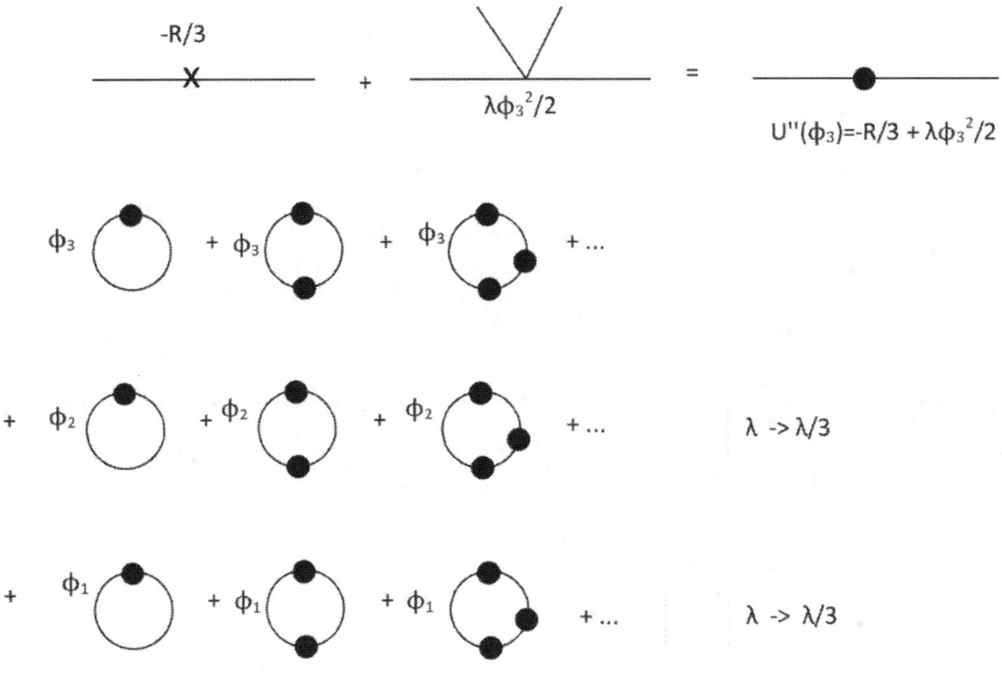

Figure 1: One loop Feynman diagrams for a triplet of scalar fields. The two vertices corresponding to $-R\phi^2/6$ and $\lambda\phi^4/24$ interactions, which are replaced by a single vertex (black dot). The last two sets of diagrams, where the components ϕ_2 and ϕ_1 run around the loop, have a vertex with 1/3 the coupling constant λ of the first set of diagrams. The propagator is massless.

5. Conclusions

In this work, we introduced a renormalization scale into a massless theory by considering quantum corrections. Symmetry breaking was radiatively induced *a la* la Coleman-Weinberg [2]. The dimensionless λ, defined at the renormalizaton scale M, was traded for the dimensionful VEV. An important result was that the Ricci scalar of the background AdS spacetime was determined entirely by the VEV. Our work can now be naturally extended in a few ways. First, one can add gauge field fluctuations in the calculation of the effective potential. Then the coupling contant λ can be expressed in terms the electromagnetic coupling constant e as in [2]. In that scenario, the two parameters in the theory become e and the dimensionful VEV. Second, we used the symmetry of AdS space to solve for $\langle T_{\mu\nu}\rangle_0$, the asymptotic value of $\langle T_{\mu\nu}\rangle$. This allowed us to solve the equations of motion analytically in the asymptotic regime and to obtain an expression relating the Ricci scalar of AdS space to the VEV. The interior spacetime obeys spherical symmetry but not the symmetry of AdS space. Therefore, the finite and nonlocal part of $\langle T_{\mu\nu}\rangle$ for the interior would require a more elaborate calculation. One could then obtain numerical solutions of the interior. It is of interest to see how these numerical solutions containing quantum corrections in an AdS background compare with those obtained in the classical context of General Relativity (GR) [17]. Third, we worked with a background AdS spacetime because the Ricci scalar of AdS space had the right sign for classical SSB [5]. However, the VEV here is obtained from the quantum-corrected effective potential. In the presence of quantum corrections, de Sitter (dS) space could well be a viable background spacetime. The Ricci scalar of the background AdS space was determined solely by the VEV, which should apply to dS space as well. Such solutions could be of greater cosmological interest.

Acknowledgments

We thank the organizers of IARD 2014 for a very interesting and well organized conference. AE acknowledges support from an NSERC discovery grant. He thanks the Kavli Institute of Theoretical Physics (KITP, Santa Barbara) for a four week stay during the summer of 2012 where part of this work was completed. This research was supported in part by the National Science Foundation under Grant No. NSF PHY11-25915. NG was supported in part by the National Science Foundation (NSF) through grant PHY-1213456.

References

[1] I. Bars, P. Steinhardt and N. Turok, Phys. Rev. D **89**, 043515 (2014) [arXiv:1307.1848].
[2] S. Coleman and E. Weinberg, Phys. Rev. D **7**, 1888 (1973).
[3] C.T. Hill, Phys.Rev. D bf 89, 073003 (2014) [arXiv:1401.4185].
[4] A. Edery and N. Graham, JHEP **11**, 109 (2013) [arXiv:1310.7878].
[5] A. Edery, L. Fabbri and M. B. Paranjape, Class. Quant. Grav. **23**, 6409 (2006) [hep-th/0603131].
[6] A. Edery and B. Constantineau, Class. Quant. Grav. **28**, 045003 (2011)[arXiv:1010.5844].
[7] E.C. Thomas, F.R. Urban and A. R. Zhitnitsky, JHEP **08**,043 (2009) [arXiv:0904.3779]
[8] V.F. Mukhanov and S. Winitzki, *Introduction to Quantum Effects in Gravity*, Cambridge University Press, (2007).
[9] N.D. Birrell and P.C. W. Davies, *Quantum Fields in Curved Space*, Cambridge University Press, (1982).
[10] A. O. Barvisnsky, Yu. V. Gusev, G.A. Vilkovisky and V.V. Zhytnikov, Nucl. Phys. B **439**, 561 (1995)[hep-th/9404187].
[11] V.M. Mostepanenko and N.N. Trunov, *The Casimir Effect And Its Applications*, Oxford University Press, (1997)
M. Bordag , U. Mohideen and V.M. Mostepanenko, Phys. Rep. **353**, 1 (2001)
A. Zee, *Quantum Field Theory in a Nutshell*, Princeton University Press, (2003)
A. Edery, J. Phys. A: Math. Gen. **39**, 685 (2006) [math-ph/0510056]

A. Edery, Phys. Rev. D **75**, 105012 (2007) [hep-th/0610173]
A. Edery, J. Stat. Mech. P06007 (2006) [hep-th/0510238].
[12] L.S. Brown and J.C. Collins, Ann. Phys. **130**, 215 (1980).
[13] S.J. Hathrell, Ann. Phys. **139**, 136 (1982).
[14] T. Cheng and L. Li, *Gauge Theory of Elementary Particle Physics*, Oxford University Press, (1984).
[15] M.E. Peskin and D.V. Schroeder,*An Introduction to Quantum Field Theory*, Westview Press, (1995).
[16] S. Caroll, *Spacetime and Geometry: An Introduction to General Relativity*, Benjamin Cummings, (2003).
[17] P. Breitenlohner, P. Forgács and D. Maison, Nucl. Phys. B **383**, 357 (1992)
K. Lee, V. P. Nair and E. J. Weinberg, Phys. Rev. D **45**, 2751 (1992)
K. Lee, V. P. Nair and E. J. Weinberg, Phys. Rev. Lett. **68**, 1100 (1992)
M.E. Ortiz, Phys. Rev. D **45**, 2586 (1992)
H. Hollmann, Phys. Lett. B **338**, 181 (1994).
[18] G. 't Hooft, Nucl. Phys. B **79**, 276, (1974).
[19] A.M. Polyakov, JETP Lett. **20**, 194 (1974).

Journal of Physics: Conference Series **615** (2015) 012006 doi:10.1088/1742-6596/615/1/012006

Boundary conditions for a composite model of leptons and quarks

G B Mainland

The Ohio State University at Newark, Newark, OH 43055, USA

E-mail: mainland.1@osu.edu

Abstract. Because the existence of families of elements and hadrons was ultimately understood by the realization that atoms and hadrons are composite, an obvious approach to explaining the existence of lepton and quark families is to assume that the particles in these families are also composite. The mass and spin spectra of leptons and quarks suggest that if these particles are composite, they are most likely bound states of a scalar and spin-1/2 fermion interacting via electrodynamics. However, if they are composite, the bound states must be highly relativistic since in each family the least massive member has a small mass compared with the others. Also, composite leptons and quarks must be extremely tightly bound since no internal structure has ever been conclusively detected. Highly relativistic, bound-state, Bethe-Salpeter solutions of a scalar and a spin-1/2 fermion bound by minimal electrodynamics are discussed. These specific solutions cannot describe leptons or quarks as bound states because the magnitude of the charges of the constituents are an order of magnitude larger than e. The boundary conditions, however, allow solutions when the constituents have charges with magnitudes on the order of e.

1. Introduction

Because the existence of families of elements and hadrons was ultimately understood by the realization that atoms and hadrons are composite, although speculative, as is any physics beyond the standard model, an obvious approach to explaining the existence of lepton and quark families is to assume that the particles in these families are composite. The theoretical high-energy physics community found this circumstantial evidence sufficiently compelling that in the 1970's many physicists devoted substantial effort to constructing a composite model. But by the early 1980's most physicists had quit working on the idea because a mechanism was never discovered that could tightly bind the constituents[1].

The existence of families, the circumstantial evidence that leptons and quarks are composite, is not conclusively corroborated by direct experimental evidence. There are indications that there might be discrepancies between the experimental value of the of the muon g-2 and the value calculated from the standard model[2]. Even if this disagreement between experiment and theory is confirmed, in addition to muon substructure, it could also be explained, for example, by substructure of the W-boson or by the existence of supersymmetric partners of existing particles.

If leptons and quarks are composite, several characteristics that any composite model must possess can be inferred from the mass and spin spectra of leptons and quarks. The masses of these particles satisfy the inequalities $m_{\text{electron}} \ll m_{\text{muon}} \ll m_{\tau}$, $m_{\text{up}} \ll m_{\text{charm}} \ll m_{\text{top}}$ and $m_{\text{down}} \ll m_{\text{strange}} \ll m_{\text{bottom}}$. Since no structure has been conclusively detected for leptons or quarks, any composite system must be very strongly bound. This precludes a composite model where, for example, the electron, muon and tau are successively much more massive because the constituents are successively much more massive. The large differences in the masses of the particles in each family is consistent with strong binding only if the bound system is highly relativistic. In each family the most tightly bound state must be one for which the binding energy is almost equal to the sum of the masses of the constituents. That is, the interaction must create a state that has almost zero energy. The development of supercomputers makes it possible to study such highly relativistic bound states.

Although the existence of families is explained by the atomic model of atoms, the quark model for mesons and baryons and, as discussed here, perhaps a preon model for leptons and quarks, each of the composite models is radically different. In the atomic model, aside from small mass defects, the mass of an atom is the sum of the masses of the constituents. For mesons and baryons most of the mass results from the kinetic energy of the quarks so the mass of a meson or baryon is substantially greater than the sum of the masses of the constituent quarks. If leptons and quarks are highly relativistic bound states, the mass of each lepton or quark is less - and for the least massive particle in each family, much less - than the sum of the masses of the constituents.

To construct a composite model of leptons and quarks, two questions must initially be answered: (a) What particles constituent the system? (b) What interaction binds the system? For simplicity the bound system would likely consist of two or three constituents. If the composite system were comprised of a spin-0 boson and a spin-1/2 fermion, all states would have total angular momentum or spin one-half if all states have zero orbital angular momentum. If the composite system were comprised of three or more constituents, it is difficult to imagine a mechanism that would prevent higher-spin bound states. Thus, the lepton and quark mass and spin spectra suggest that if the leptons and quarks are composite, they are likely relativistic bound states of a spin-0 boson and a spin-1/2 fermion.

Leptons interact gravitationally, weakly and electromagnetically. Among the three, the electromagnetic interaction is the only one that might be able to provide the requisite strong binding. The similarity of the charged-lepton and quark mass spectra and the fact that the two quark families and the charged lepton family each have three members suggests that the same mechanism might be responsible for binding in all four families, a possible scenario when the strong interactions can be neglected. This would be the situation if only the scalar quark constituent interacts strongly. A heretofore unknown interaction might, of course, be responsible for the binding, but assuming the existence of such a force would represent a more speculative and a much more difficult approach.

2. Possible composite model of leptons and quarks

If each of the lepton and quark families is a bound state of a scalar and a spin-1/2 fermion, the four families can be created from two constituent fermions with respective charges q_{f1}, q_{f2} and two constituent scalars with respective charges Q_{s1}, Q_{s2}. The four constituents can combine in four ways to create four bound states or four families. The four charges must satisfy the four constraints in Table 1. Note that for weak-interaction transitions between the two lepton families or between the two quark families, with the choice of charges in Table 1, the W-boson couples to the constituent fermions. In Table 1 there are three independent equations (constraints) and four unknown charges so there are an infinite number of solutions.

Possible charges of the constituents with magnitudes ≤ 2 that are multiples of $\pm 1/3$ are are given in in Table 2. For each solution the charges of both constituent fermions have the same sign and the charges of both constituent scalars have the opposite sign. Thus only a fermion and a scalar can form a bound state.

Table 1. Constraints on constituent charges.

Family	Constituent Fermion Charge	Constituent Boson Charge	Constraint
Electron	q_{f1}	Q_{s1}	$q_{f1} + Q_{s1} = -1$
Neutrino	q_{f2}	Q_{s1}	$q_{f2} + Q_{s1} = 0$
Neg. quarks	q_{f1}	Q_{s2}	$q_{f1} + Q_{s2} = -\frac{1}{3}$
Pos. quarks	q_{f2}	Q_{s2}	$q_{f2} + Q_{s2} = \frac{2}{3}$

Table 2. Possible charges of constituent fermions and scalars.

Constituent Fermion Charges		Constituent Scalar Charges	
q_{f1}	q_{f2}	Q_{s1}	Q_{s2}
1	2	-2	-4/3
2/3	5/3	-5/3	-1
1/3	4/3	-4/3	-2/3
-4/3	-1/3	1/3	1
-5/3	-2/3	2/3	4/3
-2	-1	1	5/3

3. Bethe-Salpeter equation

When a spin-0 field $\phi(x)$, representing a scalar with charge Q_s and mass m_s, interacts via minimal electrodynamics with a spin-1/2 field $\Psi(x)$, representing a fermion with charge q_f and mass m_f, the renormalizable Lagrangian is[3]

$$L =: [(-\imath\partial^\mu - Q_s A^\mu)\phi^\dagger][(\imath\partial_\mu - Q_s A_\mu)\phi] - m_s^2\phi^\dagger\phi$$
$$+ \bar{\Psi}\gamma_\mu(\imath\partial^\mu - q_f A^\mu)\Psi - m_f\bar{\Psi}\Psi - \frac{1}{4}F_{\mu\nu}F^{\mu\nu} - \frac{1}{2}(\partial^\mu A_\mu)^2 : , \quad (1)$$

where $F_{\mu\nu} = \partial_\nu A_\mu - \partial_\mu A_\nu$ and the final term is the gauge-fixing term for the Feynman gauge.

The Bethe-Salpeter equation is used to calculate the properties of the bound state. Since every Feynman diagram contributes to the exact equation, some approximation must be made. Usually only the effect of the lowest-order diagram shown to the right is included. Because of the structure of the Bethe-Salpeter equation, contributions occur from the diagram and iterations of the diagram, which form a ladder with the photon propagator forming the rungs. Thus the approximation is called the ladder approximation.

Following standard procedures[4] and making the ladder approximation, the Bethe-Salpeter equation describing a bound state of a minimally interacting scalar and spinor is[5]

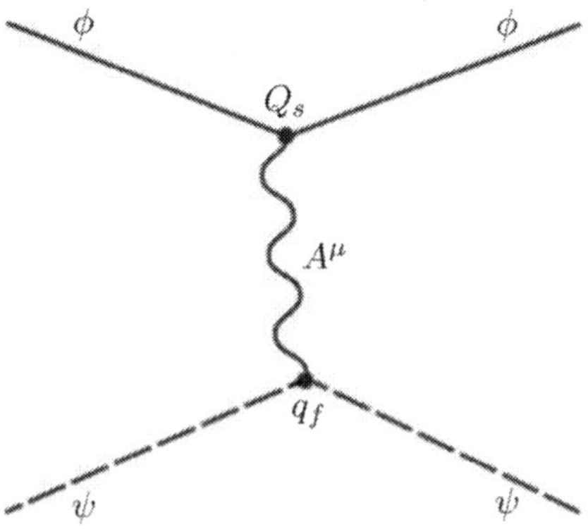

Figure 1. Feynman diagram for a spin-1/2 field ψ with charge q_f and a scalar field ϕ with a charge Q_s interacting with the electromagnetic field A^μ.

$$
\begin{aligned}
&[(p^0+\xi E)\gamma_0 + p^i\gamma_i - m_f]\{[p^0+(\xi-1)E]^2 - \mathbf{p}^2 - m_s^2\}\chi(p) \\
&= \frac{iq_fQ_s}{(2\pi)^4}\int_{-\infty}^{\infty}\frac{\mathrm{d}^4q}{(p-q)^2+i\varepsilon}\{\gamma^0[p_0+q_0+2(\xi-1)E]+\gamma^i(p_i+q_i)\}\chi(q)\,. \qquad (2)
\end{aligned}
$$

The equation has been written in the center-of-mass rest frame where the four-momentum K^μ takes the form $K^\mu = (E, 0, 0, 0)$.

In contrast, the non-relativistic, two-body, bound-state problem described by the Schrödinger equation can be written in the form

$$
H(\mathbf{x})\psi(\mathbf{x}) = E\psi(\mathbf{x})\,, \qquad (3)
$$

where H is the Hamiltonian of the system.

Significant differences between bound-state Bethe-Salpeter equation (2) and the bound-state Schrödinger equation (3) are as follows: (a) Since the energy E appears multiple times in the Bethe-Salpeter equation, there is no relativistic operator that has an eigenvalue E. That is, a Hamiltonian does not exist for the relativistic, bound-state problem. (b) The Bethe-Salpeter equation is covariant, implying that there is no action at a distance. If the fourth component of the photon propagator, which is under the integral, were neglected, the interaction would become instantaneous. The interaction would then depend on the distance r between the constituent particles and there would be action at a distance. (c) The Bethe-Salpeter equation is an integral equation so the boundary conditions are part of the equation itself and do not have to be specified independently. (d) Because the energy E appears multiple times in the Bethe-Salpeter equation and the coupling constant $q_fQ_s/(4\pi)$ appears only once, the equation is solved by specifying the energy and solving for the coupling constant as an eigenvalue. When solving the Schrödinger equation, the coupling constant is specified and the energy is calculated as an eigenvalue. Both procedures yield equivalent information that allows the coupling constant to be plotted as a function of the energy or vice versa. (e) The Bethe-Salpeter equation is always separable when energy $E=0$ because it is invariant under O(3,1) rotations. When the energy is nonzero, only one Bethe-Salpeter equation, the Wick-Cutkosky Model[6, 7], has been separated. (f) When the energy $E=0$, there are a few analytical solutions of the Bethe-Salpeter equation, but when the energy is nonzero, all solutions are numerical.

Zero-energy solutions, which are the zero-energy limit of finite-energy solutions, are discussed here. Such solutions are important because the structure of the generalized matrix eigenvalue equation that yields zero-energy solutions provides guidance[8] in constructing an appropriate discretized equation when the energy is nonzero.

To obtain solutions to (2) in the limit $E=0$, the singularity in the kernel is first removed and the equation is transformed from Minkowski to Euclidean space by making a Wick rotation[6], which is always possible in the ladder approximation and is sometimes possible when the effects of higher-order Feynman diagrams are included. The Wick-rotated Bethe-Salpeter equation is

$$
\begin{aligned}
&(\tilde{\gamma}\cdot p + m_f)(p\cdot p + m_s^2)\chi(ip^0,\mathbf{p}) \\
&= -\frac{q_fQ_s}{(2\pi)^4}\int_{-\infty}^{\infty}\frac{\mathrm{d}^4q}{(p-q)\cdot(p-q)}\,\tilde{\gamma}\cdot(p+q)\chi(iq^0,\mathbf{q})\,. \qquad (4)
\end{aligned}
$$

In the above equation the Euclidean scalar product $p\cdot p \equiv p^0p^0 + \mathbf{p}\cdot\mathbf{p}$ and $\tilde{\gamma}\cdot p \equiv \tilde{\gamma}^0p^0 + \tilde{\gamma}^ip^i$. The matrices $\tilde{\gamma}^\mu$ are given by $\tilde{\gamma}^0 \equiv -i\gamma^0, \tilde{\gamma}^i \equiv \gamma^i$.

Dimensionless variables are introduced by defining $m_f \equiv m(1-\Delta), m_s \equiv m(1+\Delta), p' \equiv p/m$ and $q' \equiv q/m$. Writing (4) in terms of dimensionless variables and then omitting primes since all momenta are now dimensionless,

$$[\tilde{\gamma}\cdot p + (1-\Delta)][p\cdot p + (1+\Delta)^2]\chi(ip^0,\mathbf{p}) \\ = -\frac{q_f Q_s}{(2\pi)^4}\int_{-\infty}^{\infty}\frac{\mathrm{d}^4 q}{(p-q)\cdot(p-q)}\tilde{\gamma}\cdot(p+q)\chi(iq^0,\mathbf{q})\,. \tag{5}$$

Spherical coordinates are introduced as follows:

$$p^0 = |p|\cos\theta_1 \qquad p_z = |p|\sin\theta_1\cos\theta_2 \\ p_x = |p|\sin\theta_1\sin\theta_2\sin\phi \qquad p_y = |p|\sin\theta_1\sin\theta_2\cos\phi \tag{6}$$

The four-vector q is written similarly in terms of primed angles. Then

$$\mathrm{d}^4 q = |q|^3\sin^2\theta_1'\sin\theta_2'\,\mathrm{d}\theta_1'\,\mathrm{d}\theta_2'\,\mathrm{d}\phi'\,\mathrm{d}|q| \equiv |q|^3\mathrm{d}|q|\mathrm{d}\Omega_{(3)}'\,, \tag{7}$$

and

$$(p-q)\cdot(p-q) = |p|^2 + |q|^2 - 2|p|\,|q|\cos\Theta\,, \tag{8}$$

where Θ is the angle between the vectors p and q.

In the zero-energy limit the angular dependence of the equation separates[9, 10] when solutions are written as products of the functions $\Psi_1^{(\pm)}(\theta_1,\theta_2,\phi)$ and $\Psi_2^{(\pm)}(\theta_1,\theta_2,\phi)$ that depend on hyperspherical harmonics in four-dimensional, Euclidean space-time and the functions $F^{(\pm)}(|p|)$ and $G^{(\pm)}(|p|)$ that depend only on the magnitude of the Euclidean four-momentum $|p|$,

$$\chi(ip^0,\mathbf{p}) = F^{(\pm)}(|p|)\Psi_1^{(\pm)}(\theta_1,\theta_2,\phi) + G^{(\pm)}(|p|)\Psi_2^{(\pm)}(\theta_1,\theta_2,\phi)\,. \tag{9}$$

The four-component column vectors $\Psi_1^{(\pm)}(\theta_1,\theta_2,\phi)$ and $\Psi_2^{(\pm)}(\theta_1,\theta_2,\phi)$, which are defined in Ref. 10, are related by

$$\Psi_1^{(-)}(\theta_1,\theta_2,\phi) = \gamma_5\Psi_1^{(+)}(\theta_1,\theta_2,\phi) \quad\text{and}\quad \Psi_2^{(-)}(\theta_1,\theta_2,\phi) = \gamma_5\Psi_2^{(+)}(\theta_1,\theta_2,\phi)\,, \tag{10}$$

and satisfy the relationships

$$\tilde{\gamma}\cdot p\,\Psi_1^{(\pm)}(\theta_1,\theta_2,\phi) = \mp|p|\Psi_2^{(\pm)}(\theta_1,\theta_2,\phi) \quad\text{and}\quad \tilde{\gamma}\cdot p\,\Psi_2^{(\pm)}(\theta_1,\theta_2,\phi) = \pm|p|\Psi_1^{(\pm)}(\theta_1,\theta_2,\phi)\,. \tag{11}$$

After substituting $\chi(ip^0,\mathbf{p})$ as given in (9) with upper superscripts into the Bethe-Salpeter equation (5), the equation can be simplified using (11), and the angular integrations on the right-hand side can be performed using Hecke's theorem[11]. (All formulas necessary for carrying out the angular integration are given in the appendix of Ref. 10.) The coefficients of $\Psi_1^{(+)}(\theta_1,\theta_2,\phi)$ and $\Psi_2^{(+)}(\theta_1,\theta_2,\phi)$ must vanish independently, yielding the following two separated, coupled equations that depend on the single variable $|p|$:

$$[|p|^2 + (1+\Delta)^2]\left[(1-\Delta)F^{(+)}(|p|) + |p|G^{(+)}(|p|)\right] \\ = -\frac{q_f Q_s}{(2\pi)^4}|p|\int_0^{\infty}\mathrm{d}|q|\,|q|^3\Lambda_{k_1+1/2}^{(2)}(|p|,|q|)G^{(+)}(|q|) \\ -\frac{q_f Q_s}{(2\pi)^4}\int_0^{\infty}\mathrm{d}|q|\,|q|^4\Lambda_{k_1-1/2}^{(2)}(|p|,|q|)G^{(+)}(|q|) \tag{12a}$$

$$\begin{aligned}
&\left[|p|^2 + (1+\Delta)^2\right]\left[-|p|F^{(+)}(|p|) + (1-\Delta)G^{(+)}(|p|)\right] \\
&\quad = \frac{q_f Q_s}{(2\pi)^4}|p| \int_0^\infty \mathrm{d}|q|\,|q|^3 \Lambda^{(2)}_{k_1-1/2}(|p|,|q|)F^{(+)}(|q|) \\
&\quad + \frac{q_f Q_s}{(2\pi)^4} \int_0^\infty \mathrm{d}|q|\,|q|^4 \Lambda^{(2)}_{k_1+1/2}(|p|,|q|)F^{(+)}(|q|) \qquad (12b)
\end{aligned}$$

In the above equation

$$\Lambda^{(2)}_n(|p|,|q|) = \frac{2\pi^2}{|p||q|}\,\frac{R(|p|,|q|)^{n+1}}{n+1}\,, \qquad (13)$$

where

$$R(|p|,|q|) = \begin{cases} \frac{|q|}{|p|} & \text{if} \quad |q| \le |p|\,, \\ \frac{|p|}{|q|} & \text{if} \quad |q| \ge |p|\,. \end{cases} \qquad (14)$$

The index $k_1 = 1/2, 3/2, \ldots$, and the angular momentum j of the bound state satisfies $j \le k_1$.

When $\chi(ip^0, \mathbf{p})$ with lower superscripts in (9) is substituted into the Bethe-Salpeter equation (5), the two equations that are obtained are identical to (12) after making the substitutions $F^{(-)}(|p|) \to F^{(+)}(|p|)$ and $G^{(-)}(|p|) \to -G^{(+)}(|p|)$. Since no additional eigenvalues are obtained by considering this second set of equations, attention is restricted to (12), and the superscripts $(+)$ are dropped for $F^{(+)}(|p|)$ and $G^{(+)}(|p|)$.

4. Analytical Boundary Conditions

The boundary conditions satisfied by the solutions as $|p|$ approaches zero and infinity must be determined before solutions can be calculated numerically. That is, the convergence parameters f_0, g_0, f_∞ and g_∞ must be determined that, respectively, satisfy the equations

$$F(|p|) \underset{|p|\to 0}{\longrightarrow} F_0|p|^{f_0}\,, \qquad G(|p|) \underset{|p|\to 0}{\longrightarrow} G_0|p|^{g_0}\,, \qquad (15)$$

$$F(|p|) \underset{|p|\to \infty}{\longrightarrow} F_\infty|p|^{-f_\infty}\,, \qquad G(|p|) \underset{|p|\to \infty}{\longrightarrow} G_\infty|p|^{-g_\infty}\,, \qquad (16)$$

where F_0, G_0, F_∞ and G_∞ are constants. The solutions for f_0 and g_0 were determined in Ref. 5 and are given by

$$f_0 = k_1 - \frac{1}{2}\,, \qquad g_0 \ge k_1 + \frac{1}{2}\,. \qquad (17)$$

At large momenta the Bethe-Salpeter equation can be solved analytically. Each solution yields (a) an equation for the convergence parameters f_∞ or g_∞ in terms of the coupling constant and (b) an algebraic equation that expresses one of the convergence parameters in terms of the other. But the analytical solutions at large momenta only determine allowed ranges for the values of the convergence parameters, not the specific values for which there are solutions (at all momenta).

Among the five possible classes of solutions found in Ref. 5, solutions were found only for "Boundary Solutions I" where f_∞ and g_∞ satisfy the following conditions:

$$f_\infty = -k_1 + \frac{5}{2} + \epsilon_f\,, \qquad 0 < \epsilon_f < 2k_1 + 1\,, \qquad (18a)$$

$$g_\infty = -k_1 + \frac{7}{2} + \epsilon_g\,, \qquad 0 < \epsilon_g < 2k_1 + 1\,, \qquad (18b)$$

$$\epsilon_g \le \epsilon_f \le \epsilon_g + 2\,. \qquad (18c)$$

The four integrals in (12) can be evaluated analytically at large momentum $|p|$[12]. In this limit (12a) and (12b) can then be written, respectively, as

$$\begin{aligned}
&(1-\Delta)F_\infty\,|p|^{k_1-\frac{1}{2}-\epsilon_f}+G_\infty\,|p|^{k_1-\frac{1}{2}-\epsilon_g}\\
&\underset{|p|>>1}{=}-\frac{q_fQ_s}{4\pi}\frac{1}{\pi}\left[\frac{2(1+\epsilon_g)\,G_\infty}{\epsilon_g(2k_1+1-\epsilon_g)(2+\epsilon_g)}|p|^{k_1-\frac{1}{2}-\epsilon_g}\right.\\
&\qquad\left.+\frac{1}{2}\left(\frac{c_{g1}}{k_1+\frac{3}{2}}+\frac{c_{g2}}{k_1+\frac{1}{2}}\right)|p|^{-k_1-\frac{3}{2}}\right], \qquad (19a)\\
&-F_\infty\,|p|^{k_1+\frac{1}{2}-\epsilon_f}+(1-\Delta)G_\infty|p|^{k_1-\frac{3}{2}-\epsilon_g)}\\
&\underset{|p|>>1}{=}\frac{q_fQ_s}{4\pi}\frac{1}{\pi}\left[\frac{2(2k_1+2-\epsilon_f)F_\infty}{\epsilon_f(2k_1+1-\epsilon_f)(2k_1+3-\epsilon_f)}|p|^{k_1+\frac{1}{2}-\epsilon_f}+\frac{c_{f1}}{2(k_1+\frac{1}{2})}|p|^{-k_1-\frac{1}{2}}\right], \qquad (19b)
\end{aligned}$$

where c_{g1}, c_{g2} and c_{f1} are constants.

From (18c) there are four possible relationships between ϵ_f and ϵ_g: either $\epsilon_g < \epsilon_f$ or $\epsilon_g = \epsilon_f$ and either $\epsilon_f < \epsilon_g+2$ or $\epsilon_f = \epsilon_g+2$. In this talk attention is restricted to the case where boundary conditions allow coupling constants with magnitudes on the order of the fine structure constant α only when the bound states have spin-1/2:

$$\epsilon_f > \epsilon_g\,, \qquad \epsilon_f < \epsilon_g + 2\,. \qquad (20)$$

At very large momenta (19a) simplifies by noting that (20) implies $|p|^{k_1-\frac{1}{2}-\epsilon_f} \ll |p|^{k_1-\frac{1}{2}-\epsilon_g}$ and (18b) implies $|p|^{-k_1-\frac{3}{2}} \ll |p|^{k_1-\frac{1}{2}-\epsilon_g}$. Thus the terms proportional to $|p|^{k_1-\frac{1}{2}-\epsilon_f}$ and $|p|^{-k_1-\frac{3}{2}}$ can be neglected yielding

$$G_\infty|p|^{k_1-\frac{1}{2}-\epsilon_g} = -\frac{q_fQ_s}{4\pi}\frac{1}{\pi}\left[\frac{2(1+\epsilon_g)}{\epsilon_g(2k_1+1-\epsilon_g)(2+\epsilon_g)}\right]G_\infty|p|^{k_1-\frac{1}{2}-\epsilon_g}\,, \qquad (21)$$

or

$$\frac{q_fQ_s}{4\pi} = -\pi\frac{\epsilon_g(2k_1+1-\epsilon_g)(2+\epsilon_g)}{2(1+\epsilon_g)}\,. \qquad (22)$$

Similarly at very large momenta (19b) simplifies: from (20) and (18a) it follows, respectively, that $|p|^{k_1-\frac{3}{2}-\epsilon_g} \ll p|^{k_1+\frac{1}{2}-\epsilon_f}$ and $|p|^{-k_1-\frac{1}{2}} \ll p|^{k_1+\frac{1}{2}-\epsilon_f}$. Thus

$$-F_\infty|p|^{k_1+\frac{1}{2}-\epsilon_f} = \frac{q_fQ_s}{4\pi}\frac{1}{\pi}\left[\frac{2(2k_1+2-\epsilon_f)}{\epsilon_f(2k_1+1-\epsilon_f)(2k_1+3-\epsilon_f)}\right]F_\infty|p|^{k_1+\frac{1}{2}-\epsilon_f} \qquad (23)$$

or

$$\frac{q_fQ_s}{4\pi} = -\pi\frac{\epsilon_f(2k_1+1-\epsilon_f)(2k_1+3-\epsilon_f)}{2(2k_1+2-\epsilon_f)}\,. \qquad (24)$$

Equating (22) and (24) yields the following relationship between ϵ_f and ϵ_g:

$$\frac{\epsilon_g(2k_1+1-\epsilon_g)(2+\epsilon_g)}{(1+\epsilon_g)} = \frac{\epsilon_f(2k_1+1-\epsilon_f)(2k_1+3-\epsilon_f)}{(2k_1+2-\epsilon_f)} \qquad (25)$$

If ϵ_f is specified, then (25) becomes a cubic equation for ϵ_g that is easily solved when written in the factored form

$$0 = (\epsilon_g+\epsilon_f-2k_1-1)[\epsilon_g^2(-\epsilon_f+2k_1+2)+\epsilon_g(\epsilon_f-2)(\epsilon_f-2k_1-2)+\epsilon_f(\epsilon_f-2k_1-3)] \quad (26)$$

The solutions are

$$\epsilon_g = 2k_1 + 1 - \epsilon_f \,, \tag{27a}$$

$$\epsilon_g = \frac{1}{2}\left[\epsilon_f - 2 \pm \sqrt{\frac{8(k_1+1) + 2(k_1+1)\epsilon_f^2 - \epsilon_f^3}{2(k_1+1) - \epsilon_f}}\right] . \tag{27b}$$

Expressing ϵ_g in terms of ϵ_f greatly simplifies the numerical search for solutions: instead of independently searching for solutions with values of ϵ_f and ϵ_g in their allowed ranges (18), it is only necessary to search for solutions with values of ϵ_f in its allowed region $0 < \epsilon_f < 2k_1 + 1$.

The constituents in a composite model of leptons and quarks would likely have charges with magnitudes on the order of e, or coupling constants with magnitudes on the order of the fine structure constant $\alpha = e^2/(4\pi) \simeq 1/137$. From (22) and (24), respectively, the magnitude of the coupling constant is small when either ϵ_g or $2k_1 + 1 - \epsilon_g$ is small and when either ϵ_f or $2k_1 + 1 - \epsilon_f$ is small. Solutions are normalizable only if they decrease faster than $|p|^{-3}$ at large momenta[5]. When ϵ_g is small, from (18b) solutions are normalizable only when $k_1 = 1/2$, implying that all possible solutions have spin-1/2. When ϵ_f is small, from (18a), the solution is not normalizable. Thus there are two cases where the boundary conditions allow normalizable solutions with small coupling constants: (a) When ϵ_g and $2k_1+1-\epsilon_f$ are small, (27a) is satisfied. Boundary conditions exist that are compatible with small coupling constants, and any solutions have spin-1/2. When $2k_1 + 1 - \epsilon_g$ and $2k_1 + 1 - \epsilon_f$ are small, (27b) is satisfied. Boundary conditions allow higher-spin solutions.

5. Numerical procedure for solving the bound-state Bethe-Salpeter equation

Although solving the two-body, bound-state Schrödinger equation numerically is significantly easier than solving the two-body, bound-state Bethe-Salpeter equation, the general method for solving both is the same. The two-body, bound-state Schrödinger equation (3) can be solved in two steps:

Step # 1: The solution $\psi(\mathbf{x})$ is expanded in terms of a finite set of basis functions $\{b_j(\mathbf{x})\}$,

$$\psi(\mathbf{x}) = \sum_{j=1}^{N} c_j b_j(\mathbf{x}) \,. \tag{28}$$

Using only a finite number of basis functions in the above expansion is a necessary approximation when carrying out a numerical calculation. By increasing the number of basis functions, the approximation is improved so that increasingly accurate solutions are obtained. Substituting the expansion for $\psi(\mathbf{x})$ into the Schrödinger equation (3),

$$\sum_{j=1}^{N} H(\mathbf{x}) b_j(\mathbf{x}) c_j = E \sum_{j=1}^{N} b_j(\mathbf{x}) c_j \,. \tag{29}$$

Step # 2: The eigenvalue equation is discretized and solved numerically by converting it into a generalized matrix eigenvalue equation using the Rayleigh-Ritz-Galerkin variational method[13, 14]: Multiplying both sides of the above equation by $b_i^\dagger(\mathbf{x})$ and integrating over $\mathbf{x}$,

$$\sum_{j=1}^{N} \int_{-\infty}^{\infty} \mathrm{d}^3x\, b_i^\dagger(\mathbf{x}) H(\mathbf{x}) b_j(\mathbf{x}) c_j = E \sum_{j=1}^{N} \int_{-\infty}^{\infty} \mathrm{d}^3x\, b_i^\dagger(\mathbf{x}) b_j(\mathbf{x}) c_j \,. \tag{30}$$

In (30) the Schrödinger equation has been rewritten as a $N \times N$ generalized matrix eigenvalue equation of the form

$$Ac = E\, Bc, \tag{31}$$

where the matrices A and B are given, respectively, by

$$A_{ij} = \int_{-\infty}^{\infty} \mathrm{d}^3x\, b_i^\dagger(\mathbf{x}) H(\mathbf{x}) b_j(\mathbf{x})\,, \qquad B_{ij} = \int_{-\infty}^{\infty} \mathrm{d}^3x\, b_i^\dagger(\mathbf{x}) b_j(\mathbf{x})\,. \tag{32}$$

All eigenvalues of a generalized matrix eigenvalue equation of the form (31) are real if A and B are Hermitian and if A, B or both are positive definite.

Although numerical solutions are obtained by discretizing the Schrödinger equation, the solutions of the discretized Schrödinger equation can be used to construct solutions to the Schrödinger equation itself. The eigenvalues E of the discretized equation are also the eigenvalues of the Schrödinger equation. Furthermore, the eigenvector $c = (c_1, c_2, \dots)$ of the discretized equation (31) is the set of expansion coefficients for the solution $\psi(\mathbf{x})$ in (28) of the Schrödinger equation (3).

The Bethe-Salpeter equation is solved similarly to the separated Schrödinger equation with three complicating factors:
(a) As can be seen from the Bethe-Salpeter equation (2), the kernel of the equation - and thus the solutions - are singular as a result of propagators in Minkowski space. The singularities are eliminated by rewriting the equation in Euclidean space[6].
(b) It is possible to discretize the Bethe-Salpeter equation and obtain a generalized matrix eigenvalue equation of the form

$$\text{Discretized Bethe-Salpeter Equation: } Ac = \frac{q_f Q_s}{4\pi}\, B\,. \tag{33}$$

Because a relativistic Hamiltonian does not exist, it is not usually possible to discretize the equation so that all eigenvalues are real. That is, it is not possible to create a generalized matrix eigenvalue equation of the form (33) where the matrices A and B are Hermitian and at least one is positive definite. However, solutions with real eigenvalues are obtained when the basis functions obey the boundary conditions.
(c) For the bound-state Bethe-Salpeter equation (2) that might describe leptons and quarks as bound states, the coupling constant determines the behavior of the solution at large momenta, which are boundary conditions. The boundary conditions at large momenta can be obtained analytically because the Bethe-Salpeter equation can be solved analytically in this limit. The discretized Bethe-Salpeter equation is non-linear in the coupling constant since the coupling constant appears both in the Bethe-Salpeter equation as the eigenvalue and in the basis functions.

Only spin-1/2 solutions are calculated. Since the angular momentum j satisfies $j \le k_1$, the index $k_1 = 1/2$. Spin-1/2 solutions are somewhat less difficult to calculate than higher-spin solutions and are the most interesting since, if leptons and quarks are composite, they would be spin-1/2 bound states.

The zero-energy solutions $F(|p|)$ and $G(|p|)$ in (12) are expanded in terms of basis functions as follows:

$$F(|p|) = \frac{|p|^{f_0}}{(|p|+1)^{f_0+2+\epsilon_f}} \sum_{j=1}^{N} f_j B_j\,, \tag{34a}$$

$$G(|p|) = \frac{|p|^{g_0}}{(|p|+1)^{g_0+3+\epsilon_g}} \sum_{j=1}^{N} g_j B_j\,. \tag{34b}$$

In (34) f_0 and g_0 are given in (17), and from (18) $f_\infty = 2 + \epsilon_f$ and $g_\infty = 3 + \epsilon_g$. Thus the factor preceding each summation symbol obeys the boundary conditions that $F(|p|)$ or $G(|p|)$, respectively, obey. The B_j are given by

$$B_j = \begin{cases} \text{cubic spline} & \text{for } j = 1, \ldots, N-4, \\ \text{boundary spline} & \text{for } j = N-3 \ldots N. \end{cases} \tag{35}$$

Cubic splines[15] are cubic polynomials that, along with their first two derivatives, are continuous. Cubic splines, one of which is depicted in Fig. 2(a), are nonzero on four contiguous intervals labeled #1, #2, #3 and #4.

At large momenta solutions are expanded in terms of boundary splines. Boundary splines and their first two derivatives are everywhere continuous. In addition, as the momentum approaches infinity, the boundary splines become constants. As a consequence all basis functions, and, therefore, all solutions obey the boundary conditions at large momenta. An example of each of the four types of boundary splines is depicted in Fig. 2(b).

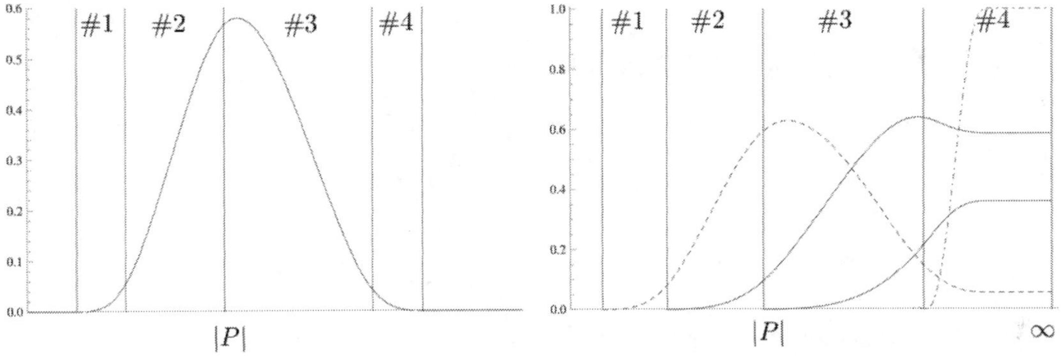

(a) Cubic Spline (b) Boundary Splines

Figure 2.

Because the discretized, Bethe-Salpeter equation is non-linear in the coupling constant, it is solved by systematically sweeping through the allowed values of ϵ_f in the range specified by (18a). The value of ϵ_g is then calculated from (27), and the input value of the coupling constant $q_f Q_s/(4\pi)_{\text{input}}$ is calculated from either (22) or (24). The coupling constant $q_f Q_s/(4\pi)_{\text{ouput}}$ is calculated as an eigenvalue of the discretized Bethe-Salpeter equation. When the quantity $[q_f Q_s/(4\pi)_{\text{input}} - q_f Q_s/(4\pi)_{\text{ouput}}]$ changes sign, a value of the coupling constant that is a solution to the Bethe-Salpeter equation lies between the two input values of the coupling constant that yield different signs for $[q_f Q_s/(4\pi)_{\text{input}} - q_f Q_s/(4\pi)_{\text{ouput}}]$.

6. Numerical solutions of the bound-state Bethe-Salpeter equation

The only solutions found satisfy (27a), namely, $\epsilon_f + \epsilon_g = 2.00$. In Fig. 3 the coupling constant $q_f Q_s/(4\pi)$ is plotted as a function of the dimensionless mass parameter Δ.

It is possible to understand why the coupling constant $q_f Q_s/(4\pi)$ plotted in Fig. 3 is weakly dependent on Δ. As discussed in §4, the behavior of the solutions at large $|p|$ determines the

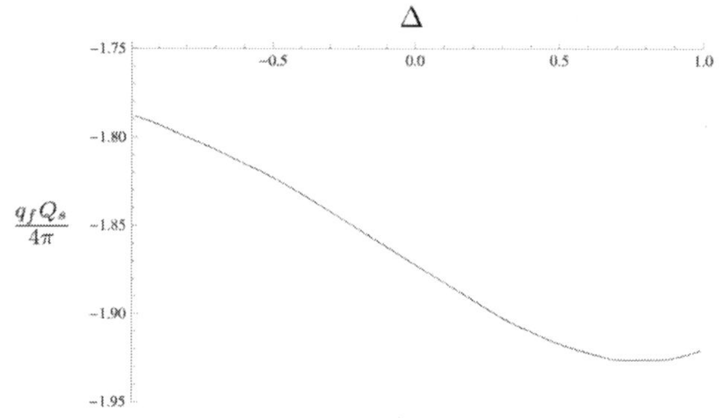

Figure 3. The coupling constant $q_f Q_s/(4\pi)$ as a function of the dimensionless mass parameter Δ when $j = 1/2$ and $E = 0$.

coupling constant. At large $|p|$ (12a) and (12b) become, respectively,

$$|p|^3\, G(|p|) \underset{|p|>>1}{=} -\frac{q_f Q_s}{(2\pi)^4} \int_0^\infty |q|^3 \left[|p| \Lambda^{(2)}_{k_1+1/2}(|p|,|q|) + |q| \Lambda^{(2)}_{k_1-1/2}(|p|,|q|) \right] G(|q|) \mathrm{d}|q| \,, \tag{36a}$$

$$|p|^3\, F(|p|) \underset{|p|>>1}{=} -\frac{q_f Q_s}{(2\pi)^4} \int_0^\infty |q|^3 \left[|p| \Lambda^{(2)}_{k_1-1/2}(|p|,|q|) + |q| \Lambda^{(2)}_{k_1+1/2}(|p|,|q|) \right] F(|q|) \mathrm{d}|q| \,. \tag{36b}$$

As can be seen from (36), at large $|p|$ the Bethe-Salpeter equation does not depend explicitly on Δ. But because Δ occurs in (11), when $|p|$ is small $F(|p|)$ and $G(|p|)$ depend significantly on Δ. For large $|p|$, $F(|p|)$ and $G(|p|)$ depend only implicitly on Δ through integrals over $F(|p|)$ and $G(|p|)$. That dependence is sufficiently small that the coupling constant depends weakly on Δ.

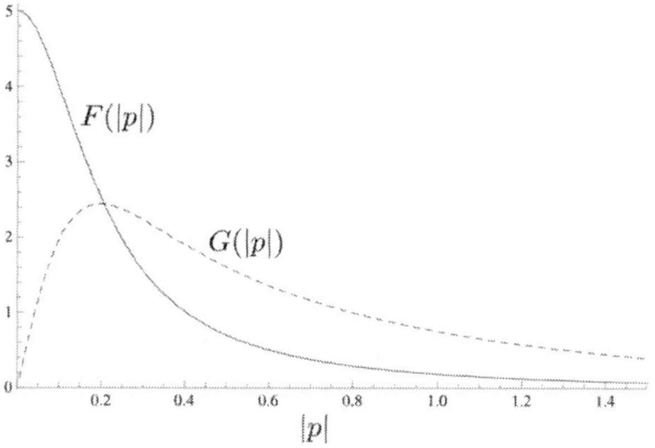

Figure 4. The components $F(|p|)$ and $G|p|)$ of a solution as a function of $|p|$ for $m_f/m_s = 1/9$, $\epsilon_f = 1.52$, $\epsilon_g = 0.483$ and $q_f Q_s/(4\pi) = -1.93$.

In Fig. 4 the components $F(|p|)$ and $G(|p|)$ of the solution are plotted as a function of $|p|$ when the mass parameter $\Delta = 0.8$, implying $m_f/m_s = 1/9$. A solution is found for $\epsilon_f = 1.52$, which implies that $\epsilon_g = 0.483$ and $q_f Q_s/(4\pi) = -1.93$. When m_f/m_s is substantially less than 1, as can be seen from Fig. 4, $F(|p|)$, which is the pure angular momentum $\ell = 0$ component, and $G(|p|)$, which is a mixed angular momentum $\ell = 0$ and $\ell = 1$ component, are roughly comparable in magnitude.

As the mass ratio m_f/m_s increases, however, the magnitude of the pure $\ell = 0$ component $F(|p|)$ becomes much larger than the magnitude of the mixed $\ell = 0$ and $\ell = 1$ component $G(|p|)$. In Fig. 5(a) the component $G(|p|)$ of a solution is plotted as a function of $|p|$ when the mass parameter $\Delta = -0.8$ or $m_f/m_s = 9$ and $\epsilon_f = 1.57$, implying that $\epsilon_g = 0.429$ and

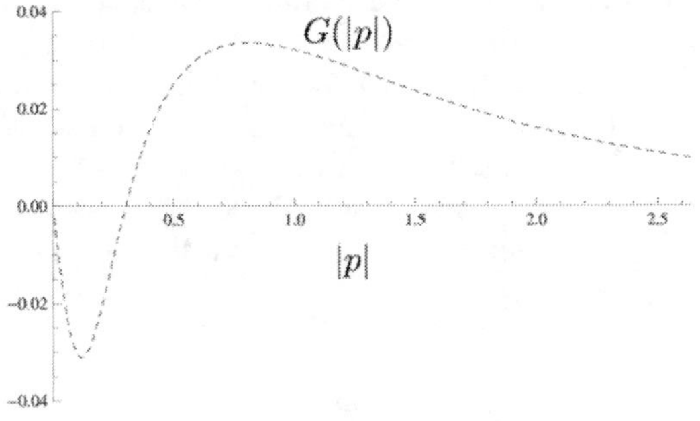

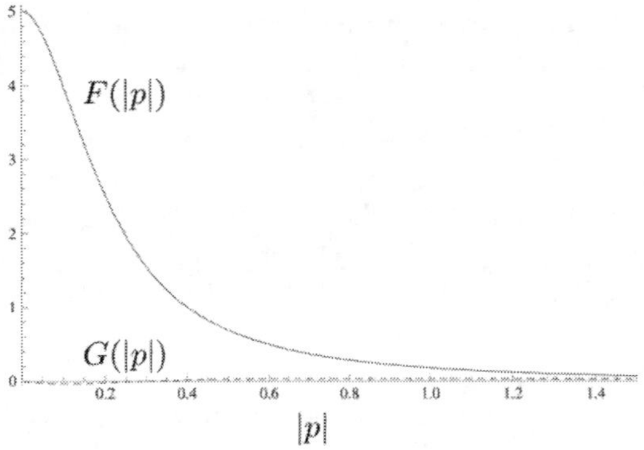

Figure 5. (a) The component $G|p|)$ and (b) the components $F(|p|)$ and $G|p|)$ of the same solution as a function of $|p|$ for $m_f/m_s = 9$, $\epsilon_f = 1.57$, $\epsilon_g = 0.429$ and $q_fQ_s/(4\pi) = -1.80$.

$q_fQ_s/(4\pi) = -1.80$. To provide a visual comparison of $F(|p|)$ and $G(|p|)$, both are plotted as a function of $|p|$ in Fig. 5(b). The magnitude of $G(|p|)$ is so much smaller than the magnitude of $F(|p|)$ that the dashed line depicting $G(|p|)$ is barely visible. If two spin-1/2, bound states are both pure $\ell = 0$ states, the first-order, electromagnetic decay of one into the other is forbidden by conservation of angular momentum. Thus if two states have large values of m_f/m_s, the first-order electromagnetic decay of one into the other is strongly suppressed.

7. Checking numerical solutions

Solutions calculated from the generalized matrix eigenvalue equation (33) are checked in the following ways: (a) As the number of basis functions is increased, the value of each eigenvalue must converge. (b) For each solution the left- and right-hand sides of the Bethe-Salpeter equation are compared in the physical region on a set of points spanning the space. The values of the left- and right-hand sides of the equation are written at the points where the absolute errors between the two sides of the equation are greatest. For solutions with $|\Delta| \leq 0.8$ the greatest percent difference between points with the largest absolute difference between the left- and right-hand sides of the equation is less than 0.08%. For $0.8 < |\Delta| < 0.95$, the percent difference increases substantially and is always only less than 4.6% although the left-and right-hand sides agree much more closely at almost all points. (c) The left-and right-hand sides of the equation must very nearly agree at the largest values of momenta. At the largest value of dimensionless momentum sampled, $|p| = 1.50 \times 10^8$, the maximum percent difference between the left- and right-hand sides of (12) is 3×10^{-3}% for $-0.99 \leq \Delta \leq 0.99$. (d) A reliability coefficient $r_{lhs-rhs}$[16], which is a statistical measure of how closely the left- and right-hand sides of (12) agree at the sampled points, is calculated. If the left- and right-hand sides agree exactly at every point, then $r_{lhs-rhs} = 1$. Except when $\Delta = -0.99$ calculations are refined until the reliability coefficient for each solution equals 1 to ten significant figures. (e) A second reliability coefficient $r^{\rm final}_{lhs-rhs}$ is calculated for the final $N_{\rm large}$ points that are sampled at large momenta. Calculations are always refined until the reliability coefficient for each solution equals 1 to ten significant figures.

8. Conclusions and future research

Strongly bound states with coupling constants with magnitudes on the order of the electromagnetic fine structure constant α are allowed by the boundary conditions and may exist. In the ladder approximation as the magnitude of coupling constant is decreased from 1.00

to the fine structure constant α, the input and output values of the coupling constant always have the same sign, implying that there are no solutions. However, when the magnitude of the input value of the coupling constant is α, the input and output values of the coupling constant differ by $4 \times 10^{-3}\%$, The additional attractive force created by the seagull and crossed fourth-order diagrams may be sufficient to create strongly bound states with a coupling constant on the order of $-\alpha$. The author is currently calculating these effects. For the class of solutions discussed here, all such strongly bound states would have spin-1/2 as do the leptons and quarks. Finally, a mechanism exists to suppress the decay $\mu \to e + \gamma$.

Acknowledgments

This work was supported by an allocation of computing time from the Ohio Supercomputer Center.

References

[1] Smolin L 2006 *The Trouble with Physics* (Boston: Houghton Mifflin Co.)
[2] Bennett G W *et al.* 2004 *Physical Review Letters* **92** 161802
[3] Bjorken J D and Drell S D 1965 *Relativistic Quantum Fields* (New York: McGraw Hill)
[4] Salpeter E E and Bethe H A 1951 *Physical Review* **84** 1232
[5] Mainland G B 2008 *Prog. Theor. Phys.* **119** 263
[6] Wick G C 1954 *Physical Review* **96** 1124
[7] Cutkosky R E 1954 *Physical Review* **96** 1135
[8] Mainland G B 2004 *Computat. Physics* **197** 610
[9] Sugano R and Munakata Y 1956 *Prog. Theor. Phys.* **16** 532
[10] Mainland G B 1986 *J. Math. Phys.* **27** 1344
[11] Hecke E 1918 *Math. Ann.* **78** 398
[12] Mainland G B 2003 *Few-Body Systems* **33** 71
[13] Delves L M and Walsh J 1974 *Numerical Solution of Integral Equations* (Oxford: Clarendon Press)
[14] Atkinson K E 1976 *A Survey of Numerical Methods for the Solution of Fredholm Integral Equations of the Second Kind* (Philadelphia: SIAM)
[15] de Boor C 1978 *A Practical Guide to Splines* (Berlin: Springer-Verlag)
[16] Winer B J 1962 *Statistical Principles in Experimental Design* (New York: McGraw Hill)

IARD 2014 IOP Publishing
Journal of Physics: Conference Series **615** (2015) 012007 doi:10.1088/1742-6596/615/1/012007

Pair production in classical Stueckelberg-Horwitz-Piron electrodynamics

Martin Land
Department of Computer Science, Hadassah College, 37 HaNeviim Street, Jerusalem
E-mail: martin@hadassah.ac.il

Abstract.
We calculate pair production from bremsstrahlung as a classical effect in Stueckelberg-Horwitz electrodynamics. In this framework, worldlines are traced out dynamically through the evolution of events $x^\mu(\tau)$ parameterized by a chronological time τ that is independent of the spacetime coordinates. These events, defined in an unconstrained 8D phase space, interact through five τ-dependent gauge fields induced by the event evolution. The resulting theory differs in its underlying mechanics from conventional electromagnetism, but coincides with Maxwell theory in an equilibrium limit. In particular, the total mass-energy-momentum of particles and fields is conserved, but the mass-shell constraint is lifted from individual interacting events, so that the Feynman-Stueckelberg interpretation of pair creation/annihilation is implemented in classical mechanics.

We consider a three-stage interaction which when parameterized by the laboratory clock x^0 appears as (1) particle-1 scatters on a heavy nucleus to produce bremsstrahlung, (2) the radiation field produces a particle/antiparticle pair, (3) the antiparticle is annihilated with particle-2 in the presence of a second heavy nucleus. When parameterized in chronological time τ, the underlying process develops as (1) particle-2 scatters on the second nucleus and begins evolving backward in time with negative energy, (2) particle-1 scatters on the first nucleus and releases bremsstrahlung, (3) particle-2 absorbs radiation which returns it to forward time evolution with positive energy.

1. Introduction

In the historical introduction to his book on quantum field theory [1], Weinberg devotes a paragraph to deprecation of Dirac's hole theory of antiparticles, observing that QFT had made the theory "unnecessary, even though it lingers on in many textbooks." It might also have been mentioned that the essential idea of hole theory also lingers on productively as the quasiparticle formalism in condensed matter physics and many-particle theory.[1] Still, as an interpretation of particle/antiparticle processes, the Feynman-Stueckelberg time reversal formalism, expressed in QFT through the Feynman propagator, has many conceptual advantages. One advantage not mentioned by Weinberg is that besides not requiring the Dirac sea, it does not actually require quantum field theory. The description of an antiparticle as a particle propagating

[1] More generally, Dirac's fundamental insight that the absence of a physical object can behave like the presence of an inverse object has been influential in many fields, especially psychology, and can be compared to the remark attributed to Miles Davis that, "Music is the space between the notes. It's not the notes you play; it's the notes you don't play."

backward in time was first proposed by Stueckelberg [2] in the context of classical relativity, without resort to quantum ideas or phenomena. In this model, a pair process is represented by a single worldline, generated dynamically by a classical event whose time coordinate advances or retreats with respect to the laboratory clock, as its instantaneous energy changes sign under interaction with a field.

In order to generate worldlines of this type, Stueckelberg constructed a covariant Hamiltonian theory of interacting spacetime events, in which the events evolve dynamically, as functions of a Poincaré invariant parameter τ. As shown in Figure 1, the particle worldline is traced out in terms of the values taken on by the four-vector $x^\mu(\tau)$ as the parameter proceeds monotonically from $\tau = -\infty$ to $\tau = \infty$.

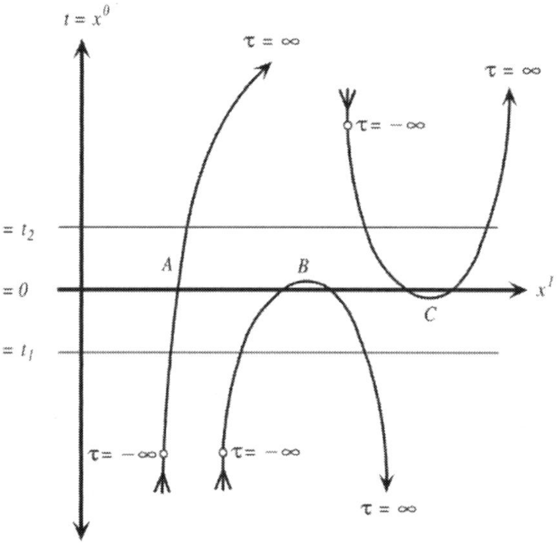

Figure 1: World Lines [2]
A: Usual type, with a unique solution to $t(\tau) = x^0$ for each x^0
B: Annihilation type, with two solutions to $t(\tau) = x^0$ for $x^0 \ll 0$ and no solution for $x^0 \gg 0$
C: Creation type, with two solutions to $t(\tau) = x^0$ for $x^0 \gg 0$ and no solution for $x^0 \ll 0$

By explicitly distinguishing the Einstein coordinate time $x^0 = t$ from the temporal order [3], the parameter time τ becomes formally similar to the Galilean invariant time in Newtonian theory, serving Stueckelberg's broader goal of generalizing the techniques of nonrelativistic classical and quantum mechanics to covariant form. Stueckelberg identified pair creation in worldlines of type C in Figure 1, because there are two solutions to $t(\tau) = t_2$, but no solution to $t(\tau) = t_1$. The observer will therefore first encounter no particle trajectories and then encounter two. It seems clear that the intrinsic electric charge should not change along the worldline, but to identify one part of the worldline as an antiparticle trajectory, requires that the measured charge reverse sign. Charge reflection may be grasped intuitively as in Dirac's hole model: carrying positive charge in one time direction should be equivalent to carrying negative charge in the opposite time direction. In standard QFT, charge conjugation is accomplished through the action of an operator with no classical analog. In Stueckelberg's classical formalism, the 0-component of the current includes the electric charge multiplied by $dx^0/d\tau$, which becomes negative when the event $x^\mu(\tau)$ evolves toward earlier values of $t = x^0$. Thus, particles and antiparticles do not appear as distinct classes of solutions to a defining equation, but as a single event whose qualitative behavior depends instantaneously on the dynamical value of its velocity.

A standard technique for pair creation in the laboratory is the two-step process by which Anderson [4] first observed positrons in 1932: high energy electrons are first scattered by heavy nuclei to produce bremsstrahlung radiation, and electron/positron pairs are then created from the radiation field. The Bethe-Heitler mechanism [5] describes this technique as the quantum process,

$$\begin{aligned} e^- + Z &\longrightarrow e^- + Z + \gamma \\ Z + \gamma &\longrightarrow Z + e^- + e^+ \end{aligned}$$

involving a quantized radiation field and the external Coulomb field of the nuclei. The Feynman diagrams describing the second step in QED are shown in Figure 2 [6].

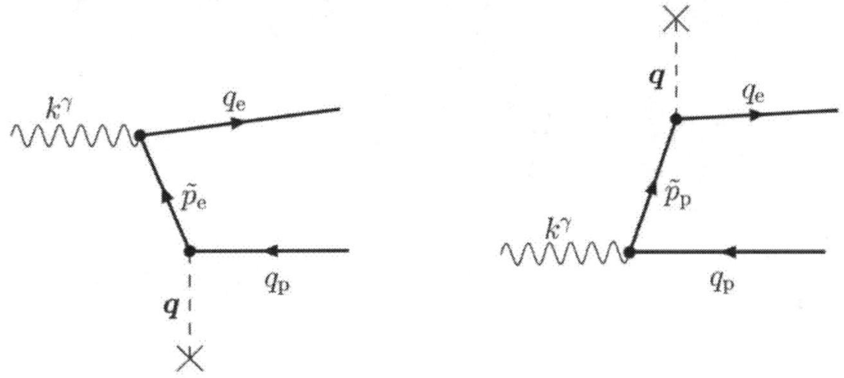

Figure 2: Bethe-Heitler mechanism in QED

The incoming real photon carries 4-momentum k^γ and the real outgoing electron and positron carry 4-momentum q_e and q_p. The intermediate virtual state carries 4-momentum $\tilde{p}_e$ or $\tilde{p}_p$, and the virtual photon exchanged with the nucleus is represented by the dashed line.

A modern experimental setup is shown in Figure 3 [7].

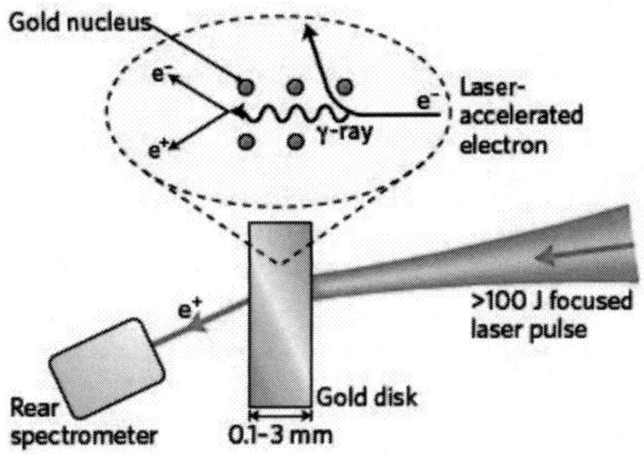

Figure 3: Bethe-Heitler mechanism in the laboratory [7]

Electrons are accelerated to high energy by focusing an intense laser pulse on a thin gold disk. The electrons are strongly deflected in the Coulomb field of the nuclei and emit bremsstrahlung represented in the illustration as γ-rays. In the Coulomb field of a nucleus, the photon can decay into an electron/positron pair.

Stueckelberg was not able to provide a classical account of pair processes, because the mass-shell constraint $p^2 = (M\dot{x})^2 = -M^2$ prevents continuous evolution of the event trajectory from the timelike region into the spacelike region on its way to time-reversed timelike motion. He considered adding a vector component to his Lorentz force that would overcome the constraint, but dropped the idea, finding no justification from first principles. Just such a vector field appears naturally in a gauge-invariant approach to Stueckelberg's theory [8, 9].

In this paper, we give a brief overview of Stueckelberg-Horwitz electrodynamics and use the formalism to provide a classical description of the pair creation process described by the Bethe-Heitler mechanism. Our goal is to calculate the classical trajectories that produce the two-step process shown in Figure 4.

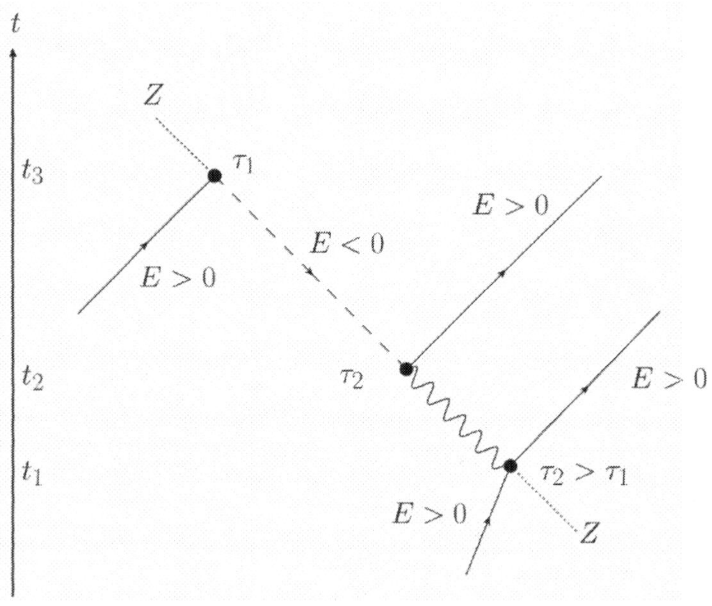

Figure 4: Bethe-Heitler mechanism in classical electrodynamics

At chronological time τ_1 positive energy particle-1 arrives at the laboratory time coordinate $t = t_3$ and scatters in the Coulomb field of nucleus Z. Particle-1 emerges with negative energy moving backward in t. At a subsequent time $\tau_2 > \tau_1$ positive energy particle-2 arrives at $t = t_1$ and scatters to positive energy in the field of another nucleus, emitting classical bremsstrahlung. This radiation impinges on particle-1 at $t = t_2$, providing sufficient energy to divert it back to positive energy evolution.

In the laboratory, where events are recorded in the order determined by clock t, the process appears as particle-2 scattering at $t = t_1$ and emitting bremsstrahlung, followed by the appearance at $t = t_2$ of a particle/antiparticle pair. Then at $t = t_3$, the antiparticle encounters another particle causing their mutual annihilation.

In section 2 we present those features of Stueckelberg-Horwitz electrodynamics required to describe this scattering process. In section 3 we provide a description of Coulomb scattering required for the events at $t = t_3$ and $t = t_1$. In section 4 we calculate the acceleration of a particle in the radiation field of a scattering event, providing a description of the pair creation event. In both the pair creation event at $t = t_2$ and the pair annihilation event at $t = t_3$, the basic requirement is that the classical interaction energy be greater than the masses of the created particles. Section 5 presents a discussion of the results and directions for further study.

2. Overview of Horwitz-Stueckelberg electrodynamics

The generalized Stueckelberg-Schrodinger equation

$$(i\partial_\tau + e_0\phi)\psi(x,\tau) = \frac{1}{2M}(p^\mu - e_0 a^\mu)(p_\mu - e_0 a_\mu)\psi(x,\tau) \tag{1}$$

describes the interaction of an event characterized by the wavefunction $\psi(x,\tau)$ with five gauge fields $a_\mu(x,\tau)$ and $\phi(x,\tau)$. Equation (1) is invariant under local gauge transformations

$$\begin{array}{lrcl} & \psi(x,\tau) & \rightarrow & e^{ie_0\Lambda(x,\tau)}\,\psi(x,\tau) \\ \text{Vector potential} & a_\mu(x,\tau) & \rightarrow & a_\mu(x,\tau) + \partial_\mu\Lambda(x,\tau) \\ \text{Scalar potential} & \phi(x,\tau) & \rightarrow & \phi(x,\tau) + \partial_\tau\Lambda(x,\tau) \end{array} \tag{2}$$

whose τ-dependence is the essential departure from Stueckelberg's work, and determines the structure of the resulting theory [8, 9]. The corresponding global gauge invariance leads to the conserved Noether current

$$\partial_\mu j^\mu + \partial_\tau \rho = 0 \tag{3}$$

where

$$j^\mu = -\tfrac{i}{2M}\left\{\psi^*(\partial^\mu - ie_0 a^\mu)\psi - \psi(\partial^\mu + ie_0 a^\mu)\psi^*\right\} \qquad \rho = \left|\psi(x,\tau)\right|^2 . \tag{4}$$

Adopting the formal designations

$$x^5 = \tau \qquad \partial_5 = \partial_\tau \qquad j^5 = \rho \qquad a_5 = \phi \tag{5}$$

and the index convention

$$\lambda, \mu, \nu = 0,1,2,3 \qquad \alpha, \beta, \gamma = 0,1,2,3,5 \tag{6}$$

the gauge and current conditions can be written

$$a_\alpha \rightarrow a_\alpha + \partial_\alpha \Lambda \qquad \partial_\alpha j^\alpha = 0 \ . \tag{7}$$

The classical mechanics of a relativistic event is found by rewriting the Stueckelberg-Schrodinger equation in the form

$$i\partial_\tau\psi(x,\tau) = \left[\frac{1}{2M}(p - e_0 a)^2 - e_0 a_5\right]\psi(x,\tau) = K\psi(x,\tau) \tag{8}$$

and transforming the classical Hamiltonian to Lagrangian as

$$L = \dot{x}^\mu p_\mu - K = \frac{1}{2}M\dot{x}^\mu\dot{x}_\mu + e_0\dot{x}^\alpha a_\alpha \tag{9}$$

Journal of Physics: Conference Series **615** (2015) 012007 doi:10.1088/1742-6596/615/1/012007

from which the Euler-Lagrange equations

$$\frac{d}{d\tau}\frac{\partial L}{\partial \dot{x}_\mu} - \frac{\partial L}{\partial x_\mu} = 0 \tag{10}$$

are

$$\frac{d}{d\tau}\left[M\dot{x}^\mu + e_0 a^\mu(x,\tau)\right] = e_0 \dot{x}^\alpha \partial^\mu a_\alpha(x,\tau) \tag{11}$$

leading to the Lorentz force

$$M\ddot{x}^\mu = e_0\left[\dot{x}^\alpha \partial^\mu a_\alpha - (\dot{x}^\nu \partial_\nu + \partial_\tau) a^\mu\right] = e_0 f^\mu{}_\alpha(x,\tau)\dot{x}^\alpha \tag{12}$$

where

$$f^\mu{}_\alpha = \partial^\mu a_\alpha - \partial_\alpha a^\mu \qquad\qquad \dot{x}^5 = \dot{\tau} = 1\ . \tag{13}$$

As required for time reversal, particles may exchange mass with fields

$$\frac{d}{d\tau}(-\tfrac{1}{2}M\dot{x}^2) = -M\dot{x}^\mu \ddot{x}_\mu = -e_0\, \dot{x}^\mu (f_{\mu 5} + f_{\mu\nu}\dot{x}^\nu) = -e_0\, \dot{x}^\mu f_{\mu 5} \tag{14}$$

and in this formalism, the mass shell is demoted from the status of constraint to that of conservation law for interactions in which $f_{\mu 5} = 0$. Analysis of the mass-energy-momentum tensor shows that the total mass, energy, and momentum of the particles and fields are conserved. To write an electromagnetic action requires the choice of a kinetic term for the gauge field, which must be both gauge and O(3,1) invariant. We write

$$S_{\text{em}} = \int d^4x d\tau \left\{ e_0 j^\alpha(x,\tau) a_\alpha(x,\tau) - \int ds\, \frac{\lambda}{4}\left[f^{\alpha\beta}(x,\tau)\Phi(\tau - s) f_{\alpha\beta}(x,s)\right]\right\} \tag{15}$$

where the local event current

$$j^\alpha(x,\tau) = \dot{X}^\alpha(\tau)\delta^4(x - X(\tau)) \tag{16}$$

has support at the instantaneous location $X(\tau)$ of the event. The τ-integral of (16) along the worldline concatenates the event current into the Maxwell particle current in the usual form. Taking the field interaction kernel to be [8]

$$\Phi(\tau) = \delta(\tau) - \lambda^2 \delta''(\tau) = \int \frac{d\kappa}{2\pi}\left[1 + (\lambda\kappa)^2\right] e^{-i\kappa\tau}\ , \tag{17}$$

the inverse function becomes

$$\varphi(\tau) = \Phi^{-1} = \int \frac{d\kappa}{2\pi} \frac{e^{-i\kappa\tau}}{1 + (\lambda\kappa)^2} = \frac{1}{2\lambda} e^{-|\tau|/\lambda} \tag{18}$$

which will be seen to spread the current along the worldline. The classical action can be written using (17) as

$$S = \int d\tau\, \frac{1}{2} M\dot{x}^\mu \dot{x}_\mu + \int d^4x d\tau \left\{ e_0\, a_\alpha j^\alpha - \frac{\lambda}{4} f_{\alpha\beta} f^{\alpha\beta} - \frac{\lambda^3}{4}\left(\partial_\tau f^{\alpha\beta}\right)\left(\partial_\tau f_{\alpha\beta}\right)\right\} \tag{19}$$

in which the gauge and O(3,1) invariance are manifest. The τ derivatives in the last term explicitly break any formal higher symmetry in the electromagnetic terms. Varying the action in the form (15) with respect to the fields and applying (18) leads to the field equations

$$\partial_\beta f^{\alpha\beta}(x,\tau) = \frac{e_0}{\lambda}\int ds\, \varphi(\tau - s)\, j^\alpha(x,s) = e\, j^\alpha_\varphi(x,\tau) \tag{20}$$

$$\partial_\alpha f_{\beta\gamma} + \partial_\gamma f_{\alpha\beta} + \partial_\beta f_{\gamma\alpha} = 0 \tag{21}$$

which are formally similar to 5D Maxwell equations with $e = e_0/\lambda$. The source $j^\alpha_\varphi(x,\tau)$ of the field in (21) is the instantaneous current $j^\alpha(x,\tau)$ defined in (16) with its support along the worldline smoothed by the kernel function $\varphi(\tau)$. For λ very small, φ becomes a delta function which narrows the source to a small neighborhood around the event inducing the current. The parameter λ plays the role of a correlation length, characterizing the range of the electromagnetic interaction.

Rewriting the field equations in vector and scalar components, they take the form

$$\begin{array}{ll} \partial_\nu f^{\mu\nu} - \partial_\tau f^{5\mu} = e\, j^\mu_\varphi & \partial_\mu f^{5\mu} = e\, j^5_\varphi = e\rho_\varphi \\ \partial_\mu f_{\nu\rho} + \partial_\nu f_{\rho\mu} + \partial_\rho f_{\mu\nu} = 0 & \partial_\nu f_{5\mu} - \partial_\mu f_{5\nu} + \partial_\tau f_{\mu\nu} = 0 \end{array} \tag{22}$$

which may be compared with the 3-vector form of Maxwell equations

$$\begin{array}{ll} \nabla \times \mathbf{B} - \partial_0 \mathbf{E} = e\mathbf{J} & \nabla \cdot \mathbf{E} = eJ^0 \\ \nabla \cdot \mathbf{B} = 0 & \nabla \times \mathbf{E} + \partial_0 \mathbf{B} = 0 \end{array} \tag{23}$$

with $f_{5\mu}$ playing the role of the vector electric field and $f^{\mu\nu}$ playing the role of the magnetic field. It is sometimes notationally convenient to further expand the field into 3-vector components as

$$(\mathbf{e})^i = f^{0i} \qquad (\mathbf{h})_{in} = \epsilon_{ijk} f^{jk} \qquad (\mathbf{f}^5)^i = f^{5i}\ . \tag{24}$$

The connection with Maxwell theory is found, as seen for the instantaneous event current, by concatenation — integration over τ along the worldline,

$$\left.\begin{array}{c} \partial_\beta f^{\alpha\beta}(x,\tau) = e j^\alpha_\varphi(x,\tau) \\ \partial_{[\alpha} f_{\beta\gamma]} = 0 \\ \partial_\alpha j^\alpha = 0 \end{array}\right\} \xrightarrow[\int d\tau]{} \left\{\begin{array}{c} \partial_\nu F^{\mu\nu}(x) = eJ^\mu(x) \\ \partial_{[\mu} F_{\nu\rho]} = 0 \\ \partial_\mu J^\mu(x) = 0 \end{array}\right. \tag{25}$$

where

$$A^\mu(x) = \int d\tau\, a^\mu(x,\tau) \qquad F^{\mu\nu}(x) = \int d\tau\, f^{\mu\nu}(x,\tau) \qquad J^\mu(x) = \int d\tau\, j^\mu(x,\tau)\ . \tag{26}$$

The field equations (22) are called pre-Maxwell equations, and together with the Lorentz force (12) describe a microscopic event dynamics for which Maxwell theory can be understood as an equilibrium limit. Since $e_0 a^\mu$ must have the dimensions of eA^μ, it follows that e_0 and λ have the dimension of time and $e = e_0/\lambda$ is dimensionless. The pre-Maxwell equations lead to the wave equation

$$\partial_\alpha \partial^\alpha a^\beta(x,\tau) = \left(\partial_\mu \partial^\mu - \partial_\tau^2\right) a^\beta(x,\tau) = -e j^\beta_\varphi(x,\tau) \tag{27}$$

whose solutions may respect 5D symmetries broken by the O(3,1) symmetry of the event dynamics. The principal part Green's function is

$$\begin{aligned} G(x,\tau) &= -\frac{1}{2\pi}\delta(x^2)\delta(\tau) - \frac{1}{2\pi^2}\frac{\partial}{\partial x^2}\frac{\theta(x^2-\tau^2)}{\sqrt{x^2-\tau^2}}, \qquad x^2 = x^\mu x_\mu \\ &= -D(x)\,\delta(\tau) - G_{correlation}(x,\tau) \end{aligned} \tag{28}$$

where $D(x)$ is the 4D Maxwell Green's function and $G_{correlation}$ vanishes under concatenation. In this paper we neglect the correlation term. The 'static' Coulomb potential in this framework is induced by an isolated event moving uniformly along the t axis. Writing the event as

$$x(\tau) = (\tau, 0, 0, 0) \tag{29}$$

produces the currents

$$j^0(x,\tau) = j^5(x,\tau) = \delta(t-\tau)\,\delta^4(x) \qquad \mathbf{j}(x,\tau) = 0 \tag{30}$$

$$j^0_\varphi(x,\tau) = j^5_\varphi(x,\tau) = \varphi(t-\tau)\,\delta^4(x) \qquad \mathbf{j}_\varphi(x,\tau) = 0 \tag{31}$$

inducing the Yukawa-type potential

$$a^0(x,\tau) = a^5(x,\tau) = \frac{e}{4\pi|\mathbf{x}|}\varphi(\tau - (t - |\mathbf{x}|)) \tag{32}$$

and recovering the standard Coulomb potential

$$A^0(x) = \int d\tau\, a^0(x,\tau) = \frac{e}{4\pi|\mathbf{x}|} \tag{33}$$

under concatenation. A test event on the lightcone of this event will experience the force

$$M\ddot{\mathbf{x}} = e^2\,\nabla\left[\frac{e^{-|\mathbf{x}|/\lambda}}{4\pi\,|\mathbf{x}|}\right] \tag{34}$$

in which $1/\lambda$ represents the mass spectrum of the pre-Maxwell field. If λ is small (so that φ approaches a delta function and the current narrows to around the event) the mass spectrum becomes wide. If λ is large, the support of the current spreads along the worldline and the potential becomes Coulomb-like.

An arbitrary event $X^\mu(\tau)$ induces the current

$$j^\mu_\varphi(x,\tau) = \int ds\,\varphi(\tau - s)\,\dot{X}^\alpha(s)\,\delta^4[x - X(s)] \tag{35}$$

leading to the Liénard-Wiechert potential

$$a^\beta(x,\tau) = -e\int d^4x' d\tau' D(x-x')\,\delta(\tau-\tau')\,j^\mu_\varphi(x',\tau') \tag{36}$$

$$= \frac{e}{2\pi}\int ds\,\varphi(\tau-s)\,\dot{r}^\alpha(s)\,\delta\left((x - X(s))^2\right)\theta^{ret} \tag{37}$$

Using identity

$$\int d\tau f(\tau)\,\delta[g(\tau)] = \frac{f(\tau_R)}{|g'(\tau_R)|}, \tag{38}$$

where τ_R is the retarded time found from

$$g(\tau) = (x - X(\tau_R))^2 = 0 \qquad \theta^{ret} = \theta\left(x^0 - X^0(\tau_R)\right) , \tag{39}$$

provides

$$a^\beta(x,\tau) = \frac{e}{4\pi}\varphi(\tau - \tau_R)\frac{\dot{X}^\beta(\tau_R)}{(x^\mu - X^\mu(\tau_R))\dot{X}_\mu(\tau_R)} . \tag{40}$$

Notice that the τ-dependence is limited to the smoothing kernel $\varphi(\tau - \tau_R)$ and again λ plays the role of a correlation length that localizes the interaction to the neighborhood $\tau_R \pm \lambda$. Using this potential and writing

$$u^\mu = \dot{X}^\mu(\tau) \qquad z^\mu = x^\mu - X^\mu(\tau) \tag{41}$$

we find the field strengths, separated into the retarded and radiation parts, as

$$f^{\mu\nu}_{ret}(x,\tau) = -e\varphi(\tau - \tau_R)\frac{(z^\mu u^\nu - z^\nu u^\mu)u^2}{4\pi(u \cdot z)^3} \sim \frac{1}{\mathbf{z}^2} \tag{42}$$

$$f^{5\mu}_{ret}(x,\tau) = e\varphi(\tau - \tau_R)\frac{z^\mu u^2 - u^\mu(u \cdot z)}{4\pi(u \cdot z)^3} \sim \frac{1}{\mathbf{z}^2} \tag{43}$$

$$f^{\mu\nu}_{rad}(x,\tau) = -e\varphi(\tau - \tau_R)\frac{(z^\mu \dot{u}^\nu - z^\nu \dot{u}^\mu)(u \cdot z) - (z^\mu u^\nu - z^\nu u^\mu)(\dot{u} \cdot z)}{4\pi(u \cdot z)^3}$$
$$-\frac{e}{4\pi}\left[\frac{z^\mu u^\nu - z^\nu u^\mu}{(u \cdot z)^2}\right]\frac{d}{d\tau_R}\varphi(\tau - \tau_R) \sim \frac{1}{|\mathbf{z}|} \tag{44}$$

$$f^{5\mu}_{rad}(x,\tau) = e\varphi(\tau - \tau_R)\frac{(\dot{u} \cdot z)z^\mu}{4\pi(u \cdot z)^3} - e\frac{z^\mu - u^\mu(u \cdot z)}{4\pi(u \cdot z)^2}\frac{d}{d\tau_R}\varphi(\tau - \tau_R) \sim \frac{1}{|\mathbf{z}|} \tag{45}$$

3. Coulomb scattering

We begin by analyzing the scattering at τ_1 in Figure 4. Initially (at time $\tau \to -\infty$) the target nucleus Z and incoming particle are widely separated. The nucleus is at rest in the laboratory frame,

$$X_Z(\tau) = (t_Z, \mathbf{x}_Z) = (1, \mathbf{0})\,\tau \tag{46}$$

and the incoming particle approaches on the trajectory

$$X_{in}(\tau) = (t,x,y,z) = u\tau + s = \dot{t}_{in}(1,v,0,0)\,\tau + (s_t, 0, s_y, 0) \tag{47}$$

where

$$u = \frac{d}{d\tau}(t,x,y,z) \qquad \frac{dx}{d\tau} = \frac{dx}{dt}\dot{t}_{in} = v\dot{t}_{in} \qquad \dot{t}_{in} = \frac{dt}{d\tau} = \frac{1}{\sqrt{1-v^2}} . \tag{48}$$

The scattering takes place in the plane $z = 0$ so that the spatial distance between the incoming particle and the target is

$$R(\tau) = |\mathbf{x}| = \sqrt{x^2 + y^2} = \sqrt{(v\dot{t}_{in}\tau)^2 + s_y^2} . \tag{49}$$

It is convenient take the correlation length λ to be small so that the support of the fields is narrowly centered around the retarded time τ_1. Taking $\lambda \approx R(\tau_1)$ allows us to approximate $\varphi(\tau - \tau_1) \approx \delta(\tau - \tau_1)$, so that τ_1 is determined from the causality conditions for the initial trajectories,

$$[X_{in}(\tau_1) - X_Z(\tau_1)]^2 = 0 \qquad X^0_{in}(\tau_1) - X^0_Z(\tau_1) > 0 \tag{50}$$

These equations have the solution

$$\tau_1 = \frac{1}{v\dot{t}_{in}(1-\eta_v^2)}\left(\eta_v s_t + \sqrt{s_t^2 - s_y^2(1-\eta_v^2)}\right) \xrightarrow[v\ll 1]{} \frac{\sqrt{s_t^2 - s_y^2}}{v} \tag{51}$$

where it is convenient to introduce the smooth parameter

$$\eta_v = \frac{1}{v}\left(1 - \frac{1}{\dot{t}_{in}}\right) \longrightarrow \begin{cases} 0, & v = 0 \\ 1, & v = 1 \end{cases} . \tag{52}$$

Notice that the 0-component s_t of the impact parameter must be positive in order for the interaction to take place. The location of the incoming particle at the time of interaction is found to be

$$\mathbf{x}(\tau_1) = R\,\hat{\mathbf{R}} \qquad t(\tau_1) = \dot{t}_{in}\tau_1 + s_t \tag{53}$$

where

$$R = \frac{1}{1-\eta_v^2}\left(\eta_v\sqrt{s_t^2 - s_y^2(1-\eta_v^2)} + s_t\right) \xrightarrow[v\ll 1]{} s_t \tag{54}$$

$$\hat{\mathbf{R}} = \frac{\left(\eta_v s_t + \sqrt{s_t^2 - s_y^2(1-\eta_v^2)},\, (1-\eta_v^2)\, s_y, 0\right)}{s_t + \eta_v\sqrt{s_t^2 - s_y^2(1-\eta_v^2)}} \xrightarrow[v\ll 1]{} \left(\sqrt{1 - \frac{s_y^2}{s_t^2}}, \frac{s_y}{s_t}, 0\right) . \tag{55}$$

From equation (32) the potential induced by the target nucleus is

$$a^0(x,\tau) = a^5(x,\tau) = \frac{Ze}{4\pi R}\delta(\tau - \tau_1) \qquad a^i = 0 \tag{56}$$

so that the nonzero field strengths can be written

$$e^i = \partial^0 a^i - \partial^i a^0 \qquad f^{5i} = \partial^5 a^i - \partial^i a^5 \qquad f^{50} = \partial^5 a^0 - \partial^0 a^5 \tag{57}$$

$$\mathbf{e} = -\nabla a^0 \qquad \mathbf{f}^5 = -\nabla a^5 = \mathbf{e} \qquad f^{50} = -\left(1 + \frac{1}{\dot{t}_{in}}\right)\partial_\tau a^0 . \tag{58}$$

Using these expressions in (12) provides the Lorentz force on the incoming particle in the form

$$\ddot{t} = -\frac{e_0}{M}\mathbf{e}\cdot\dot{\mathbf{x}} - \frac{e_0}{M}f^{50} = \frac{\lambda e}{M}\left(\dot{\mathbf{x}}\cdot\nabla + \left(1 + \frac{1}{\dot{t}_{in}}\right)\partial_\tau\right) a^0(x,\tau) \tag{59}$$

$$\ddot{\mathbf{x}} = -\frac{e_0}{M}\mathbf{e}\dot{t} + \frac{e_0}{M}\mathbf{f}^5 = \frac{\lambda e}{M}(\dot{t}+1)\nabla a^0(x,\tau) . \tag{60}$$

The delta function in (56) enables immediate integration of the force equations as

$$\begin{aligned}
\dot{t}_f - \dot{t}_{in} &= \frac{\lambda}{M} \int_{\tau_1-\lambda/2}^{\tau_1+\lambda/2} d\tau \left(\dot{\mathbf{x}} \cdot \nabla + \left(1 + \frac{1}{\dot{t}_{in}}\right) \partial_\tau \right) \frac{Ze^2}{4\pi R} \delta\left(\tau - \tau_1\right) \\
&= \frac{\lambda}{M} \dot{\mathbf{x}}\left(\tau_1\right) \cdot \nabla \frac{Ze^2}{4\pi R} \\
&= -\frac{\lambda}{M} \frac{Ze^2}{4\pi R^2} \dot{\mathbf{x}}\left(\tau_1\right) \cdot \hat{\mathbf{R}}
\end{aligned} \tag{61}$$

$$\begin{aligned}
\dot{\mathbf{x}}_f - \dot{\mathbf{x}}_{in} &= \frac{\lambda}{M} \int_{\tau_1-\lambda/2}^{\tau_1+\lambda/2} d\tau \left(\dot{t} + 1\right) \nabla \frac{Ze^2}{4\pi R} \delta\left(\tau - \tau_1\right) \\
&= -\frac{\lambda}{M} \frac{Ze^2}{4\pi R^2} \left(\dot{t}\left(\tau_1\right) + 1\right) \hat{\mathbf{R}}
\end{aligned} \tag{62}$$

where the velocities are evaluated at the interaction point as

$$\left(\dot{t}, \dot{\mathbf{x}}\right)\left(\tau_1\right) = \frac{1}{2}\left[\left(\dot{t}, \dot{\mathbf{x}}\right)_f + \left(\dot{t}, \dot{\mathbf{x}}\right)_{in}\right] \ . \tag{63}$$

We introduce the dimensionless parameter

$$g_e = \frac{\lambda}{M} \frac{\dot{Z}e^2}{4\pi R^2} = \frac{\lambda}{R} \times \frac{Ze^2}{4\pi R} \frac{1}{M} = \frac{\text{correlation length}}{\text{impact parameter}} \times \frac{\text{interaction energy}}{\text{mass energy}} \tag{64}$$

which appears in (61) and (62) as the factor controlling the strength of the interaction. Writing

$$\alpha_x = \frac{1}{2} g_e \hat{R}_x \qquad\qquad \alpha_y = \frac{1}{2} g_e \hat{R}_y \tag{65}$$

we can expand the Lorentz force as components in the form

$$\begin{bmatrix} 1 & \alpha_x & \alpha_y \\ \alpha_x & 1 & 0 \\ \alpha_y & 0 & 1 \end{bmatrix} \begin{bmatrix} \dot{t}_f \\ \dot{x}_f \\ \dot{y}_f \end{bmatrix} = \begin{bmatrix} 1 & -\alpha_x & 0 \\ -\alpha_x & 1 & 0 \\ -\alpha_y & 0 & 0 \end{bmatrix} \begin{bmatrix} \dot{t}_{in} \\ v\dot{t}_{in} \\ 0 \end{bmatrix} - 2 \begin{bmatrix} 0 \\ \alpha_x \\ \alpha_y \end{bmatrix} \tag{66}$$

and solve for the final velocity

$$\begin{bmatrix} \dot{t}_f \\ \dot{x}_f \\ \dot{y}_f \end{bmatrix} = \frac{1}{1 - \frac{1}{4} g_e^2} \left\{ \begin{bmatrix} \dot{t}_{in} \\ v\dot{t}_{in} \\ 0 \end{bmatrix} - g_e \begin{bmatrix} \dot{t}_{in} v \hat{R}_x \\ \left(\dot{t}_{in} + 1\right) \hat{R}_x \\ \left(\dot{t}_{in} + 1\right) \hat{R}_y \end{bmatrix} + \frac{1}{4} g_e^2 \begin{bmatrix} \dot{t}_{in} + 2 \\ \left(\hat{R}_x^2 - \hat{R}_y^2\right) v\dot{t}_{in} \\ 2\hat{R}_x \hat{R}_y v\dot{t}_{in} \end{bmatrix} \right\} \ . \tag{67}$$

Before considering pair annhilation, we examine the low velocity and low interaction energy limit of this result. Taking

$$|\dot{\mathbf{x}}| = v \ll 1 \qquad \dot{t}_{in} \to 1 \qquad \eta_v \to 0 \qquad g_e \ll 1 \tag{68}$$

the initial velocity reduces to

$$\dot{X}_{in}\left(\tau\right) \to \left(1, v, 0, 0\right) \ , \tag{69}$$

the final velocity becomes

$$\dot{t}_f \approx \dot{t}_{in} \qquad \dot{\mathbf{x}}_f \approx \dot{\mathbf{x}} - 2g_e\hat{\mathbf{R}} \qquad \hat{\mathbf{R}} = \left(\sqrt{1-\frac{s_y^2}{s_t^2}}, \frac{s_y}{s_t}, 0\right) \qquad R = s_t \tag{70}$$

and the scattering angle can be found as

$$\cos\theta = \frac{\dot{\mathbf{x}}_f \cdot \dot{\mathbf{x}}}{|\dot{\mathbf{x}}_f|\,|\dot{\mathbf{x}}|} = \frac{\dot{\mathbf{x}}^2 - 2g_e\hat{\mathbf{R}}\cdot\dot{\mathbf{x}}}{|\dot{\mathbf{x}}_f|\,|\dot{\mathbf{x}}|} = \frac{v - 2g_e\hat{R}_x}{|\dot{\mathbf{x}}_f|} \ . \tag{71}$$

If we impose the nonrelativistic condition for conservation of energy, we obtain a new constraint in the form

$$\dot{\mathbf{x}}^2 = v^2 = \dot{\mathbf{x}}_f^2 = \left[\dot{\mathbf{x}} - 2g_e\hat{\mathbf{R}}\right]^2 \quad \Rightarrow \quad v\hat{R}_x = g_e \tag{72}$$

in which case

$$\cos\theta = \frac{1}{|\dot{\mathbf{x}}_f|}\left[v - g_e\hat{R}_x\right] = 1 - 2\hat{R}_x^2 \tag{73}$$

Now, using (64) we find

$$\cot\frac{\theta}{2} = \sqrt{\frac{1+\cos\theta}{1-\cos\theta}} = \frac{\hat{R}_y}{\hat{R}_x} = \frac{s_y}{s_t}\frac{v}{g_e} = \frac{s_t}{\lambda v} \times \frac{4\pi M v^2 s_y}{Ze^2} \tag{74}$$

which recovers the Rutherford scattering formula if

$$\frac{s_t}{\lambda v} = 1 \ . \tag{75}$$

But from (68) we have $s_t = R(\tau_1)$ which we assumed to be comparable to λ. Since $\lambda v \ll \lambda$ in this low velocity case, (75) cannot be maintained. This result is unsurprising because the short-range potential cannot provide an adequate model of nonrelativistic Rutherford scattering.

Removing these restrictions and returning to the relativistic case, the condition for pair annihilation is that particle-1 scatters to negative energy, that is $\dot{t}_f < 0$ for some value of g_e. From (61)

$$\dot{t}_f = \frac{\dot{t}_{in}\left(1 - g_e v\hat{R}_x\right) + \frac{1}{4}g_e^2\left(\dot{t}_{in} + 2\right)}{1 - \frac{1}{4}g_e^2} \tag{76}$$

and we see that for small values of g_e,

$$\dot{t}_f \longrightarrow \dot{t}_{in} \geq 1 \ . \tag{77}$$

Since $v < 1$ and $R_x < 1$, the discriminant of the numerator satisfies

$$(vR_x)^2 - \left[1 + \frac{2}{\dot{t}_{in}}\right] < 0 \tag{78}$$

so that the numerator is positive definite. The denominator becomes negative when

$$1 - \frac{1}{4}g_e^2 < 0 \quad \Rightarrow \quad g_e = \frac{\text{correlation length}}{\text{impact parameter}} \times \frac{\text{interaction energy}}{\text{mass energy}} > 2 \tag{79}$$

and since we take the correlation length λ approximately equal to the impact parameter R, the requirement for pair annihilation is

$$\frac{Ze^2}{4\pi R} > 2\,M \tag{80}$$

meaning that the interaction energy is greater than the mass energy of the annihilated particles. As g_e approaches 2 from below $\dot{t}_f$ becomes very large. After g_e passes this critical value, $\dot{t}_f$ decreases from large negative values, taking the limiting value

$$\dot{t}_f \xrightarrow[g_e\to\infty]{} -(\dot{t}_{in}+2) \qquad \Longrightarrow \qquad E_f = -(E_{in}+2M) \tag{81}$$

so that the outgoing trajectory is timelike for all values of g_e.

4. Bremsstrahlung

Having found the condition for pair annihilation at time τ_1 we now apply the general expression (67) to describe the scattering at time τ_2. Particle-2 approaches a second nucleus along some trajectory $x^\mu_{in}(\tau)$ and emerges from the interaction along trajectory $x^\mu_f(\tau)$ with positive energy. To find the radiation field emitted by the scattering and acceleration of particle-2, at a point y^μ along a line of observation

$$z = y - x(\tau_2)\ , \tag{82}$$

we write the initial and final 4-velocities as

$$u_{in} = \dot{x}_{in} \qquad\qquad u_f = \dot{x}_f \tag{83}$$

so that

$$\Delta u = u_f - u_{in} \tag{84}$$

$$u\,(\tau) = u_{in} + \Delta u\;\theta\,(\tau-\tau_2) \tag{85}$$

$$\dot{u}\,(\tau) = \Delta u\;\delta\,(\tau-\tau_2) \tag{86}$$

$$u\,(\tau_2) = \bar{u} = \frac{1}{2}\left[u_f + u_{in}\right]\ . \tag{87}$$

From (44) and (45) we rewrite the radiation fields produced by an arbitrary trajectory in the form

$$f^{\mu\nu}_{rad} = -e\varphi(\tau-\tau_2)\mathcal{F}^{\mu\nu}\,(z,u,\dot{u}) - e\varphi'(\tau-\tau_2)\mathcal{G}^{\mu\nu} \tag{88}$$

$$f^{5\mu}_{rad} = e\varphi(\tau-\tau_2)\mathcal{F}^{5\mu}\,(z,u,\dot{u}) - e\varphi'(\tau-\tau_2)\mathcal{G}^{5\mu} \tag{89}$$

where

$$\mathcal{F}^{\mu\nu} = \left[\frac{(z\wedge\dot{u})\,(u\cdot z) - (z\wedge u)\,(\dot{u}\cdot z)}{4\pi\,(u\cdot z)^3}\right]^{\mu\nu} \qquad \mathcal{G}^{\mu\nu} = \frac{1}{4\pi}\left[\frac{z\wedge u}{(u\cdot z)^2}\right]^{\mu\nu} \tag{90}$$

$$\mathcal{F}^{5\mu} = \left[\frac{(\dot{u}\cdot z)\,z}{4\pi\,(u\cdot z)^3}\right]^{\mu} \qquad \mathcal{G}^{5\mu} = \left[\frac{z-u\,(u\cdot z)}{4\pi\,(u\cdot z)^2}\right]^{\mu}\ . \tag{91}$$

As pictured in Figure 4, the radiation emitted by the scattering of particle-2 is absorbed by the negative energy particle-1 arriving at y^μ. Using the Lorentz force equations (12) we calculate the change in velocity to particle-1 caused by the incoming radiation. Since we take λ to be small, we may approximate the smoothing kernel as

$$\varphi(\tau-\tau_2)=\frac{1}{2\lambda}\left[\theta\left(\tau-(\tau_2-\lambda)\right)-\theta\left(\tau-(\tau_2+\lambda)\right)\right] \ . \tag{92}$$

The τ integrations over $\varphi'(\tau-\tau_2)$ vanish, leaving the change in velocity $\dot{y}^\mu(\tau_2)$ of particle-1 in the form

$$\begin{aligned}\Delta\dot{y}^\mu &= \frac{\lambda e}{M}\int_{-\infty}^{\infty}d\tau\ \left[f^{\mu\nu}_{rad}\dot{y}_\nu+f^{\mu5}_{rad}\dot{y}_5\right]=-\frac{e^2}{2M}\int_{\tau_2-\lambda}^{\tau_2+\lambda}d\tau\ \left[\mathcal{F}^{\mu\nu}\dot{y}_\nu+\mathcal{F}^{\mu5}\dot{y}_5\right]\\ &=-\frac{e^2}{2M}\left[\mathcal{F}^{\mu\nu}\left(z,\bar{u},\Delta u\right)\dot{y}_\nu+\mathcal{F}^{5\mu}\left(z,\bar{u},\Delta u\right)\right]\end{aligned} \tag{93}$$

expressed in terms of the velocity change Δu and average velocity $\bar{u}$, which are found from (67) to be

$$\Delta u=-\frac{g_e}{1-\frac{1}{4}g_e^2}\begin{bmatrix} v\dot{t}_{in}\hat{R}_x\\ \left(\dot{t}_{in}+1\right)\hat{R}_x\\ \left(\dot{t}_{in}+1\right)\hat{R}_y\end{bmatrix}+\frac{\frac{1}{2}g_e^2}{1-\frac{1}{4}g_e^2}\begin{bmatrix}\left(\dot{t}_{in}+1\right)\\ v\dot{t}_{in}\hat{R}_x^2\\ v\dot{t}_{in}\hat{R}_x\hat{R}_y\end{bmatrix} \tag{94}$$

$$\bar{u}=\frac{1}{1-\frac{1}{4}g_e^2}\begin{bmatrix}\dot{t}_{in}\\ v\dot{t}_{in}\\ 0\end{bmatrix}-\frac{\frac{1}{2}g_e}{1-\frac{1}{4}g_e^2}\begin{bmatrix} v\dot{t}_{in}\hat{R}_x\\ \left(\dot{t}_{in}+1\right)\hat{R}_x\\ \left(\dot{t}_{in}+1\right)\hat{R}_y\end{bmatrix}+\frac{\frac{1}{4}g_e^2}{1-\frac{1}{4}g_e^2}\begin{bmatrix}1\\ -v\dot{t}_{in}\hat{R}_y^2\\ v\dot{t}_{in}\hat{R}_x\hat{R}_y\end{bmatrix} \ . \tag{95}$$

Since the support of $\varphi(\tau-\tau_2)$ is narrowly centered on τ_2, the line of observation z^μ must be a lightlike vector, which we write as

$$z^\mu=y^\mu-x^\mu\left(\tau_2\right)=\rho\hat{n}^\mu \qquad \hat{n}=\left(1,\hat{\mathbf{n}}\right)\ ,\ \hat{\mathbf{n}}^2=1\ . \tag{96}$$

From (93) the Lorentz force acting on particle-1 at τ_2 can be written

$$\begin{aligned}\dot{y}^\mu_f &+\frac{e^2}{4M}\frac{\left(z^\mu\Delta u_\nu-z_\nu\Delta u^\mu\right)\left(\bar{u}\cdot z\right)-\left(z^\mu\bar{u}_\nu-z_\nu\bar{u}^\mu\right)\left(\Delta u\cdot z\right)}{4\pi\left(\bar{u}\cdot z\right)^3}\dot{y}^\nu_f\\ &=\dot{y}^\mu_i-\frac{1}{4}\frac{e^2}{M}\frac{\left(z^\mu\Delta u_\nu-z_\nu\Delta u^\mu\right)\left(\bar{u}\cdot z\right)-\left(z^\mu\bar{u}_\nu-z_\nu\bar{u}^\mu\right)\left(\Delta u\cdot z\right)}{4\pi\left(\bar{u}\cdot z\right)^3}\dot{y}^\nu_i\\ &\quad+\frac{e^2}{2M}\frac{\left(\Delta u\cdot z\right)z^\mu}{4\pi\left(\bar{u}\cdot z\right)^3}\end{aligned} \tag{97}$$

Making the simplifying choice $\hat{\mathbf{R}}\cdot\hat{\mathbf{n}}=0$, we find

$$\bar{u}\cdot z=-\rho\gamma\left[1-v\left(\hat{n}_x+\frac{1}{2}g_e\hat{R}_x\right)\right] \qquad \Delta u\cdot z=g_e\gamma v\rho\hat{R}_x \tag{98}$$

so that taking $v\ll 1$ and neglecting g_e^2, the Lorentz force equations become

$$\dot{y}^0_f-g_eg_R\hat{\mathbf{R}}\cdot\mathbf{y}_f=\dot{y}^0_{in}+g_eg_R\hat{\mathbf{R}}\cdot\dot{\mathbf{y}}_{in} \tag{99}$$

$$\dot{\mathbf{y}}_f + g_e g_R \left[\left(\hat{\mathbf{n}} \cdot \dot{\mathbf{y}}_f - \dot{y}_f^0 \right) \hat{\mathbf{R}} - \hat{\mathbf{n}} \left(\hat{\mathbf{R}} \cdot \dot{\mathbf{y}}_f \right) \right] = \dot{\mathbf{y}}_{in} - g_e g_R \left[\left(\hat{\mathbf{n}} \cdot \dot{\mathbf{y}}_{in} - \dot{y}_{in}^0 \right) \hat{\mathbf{R}} - \hat{\mathbf{n}} \left(\hat{\mathbf{R}} \cdot \dot{\mathbf{y}}_{in} \right) \right] \quad (100)$$

where

$$g_R = \frac{1}{2M} \frac{e^2}{4\pi\rho} \quad . \quad (101)$$

We write the velocity of incoming negative energy particle-1 as

$$\dot{y}_{in}^0 < -1 \qquad \dot{\mathbf{y}}_{in} \cdot \hat{\mathbf{n}} = 0 \;\Rightarrow\; \dot{\mathbf{y}}_{in} = |\dot{\mathbf{y}}_{in}| \, \hat{\mathbf{R}} \quad (102)$$

and write the Lorentz force in components, with $g = g_e g_R$, as

$$\begin{bmatrix} 1 & -g\hat{R}_x & -g\hat{R}_y \\ -g\hat{R}_x & 1 - g\hat{n}_x\hat{R}_x & -g\hat{n}_x\hat{R}_y \\ -g\hat{R}_y & -g\hat{n}_y\hat{R}_x & 1 - g\hat{n}_y\hat{R}_y \end{bmatrix} \begin{bmatrix} \dot{y}_f^0 \\ \dot{y}_{xf} \\ \dot{y}_{yf} \end{bmatrix} = \begin{bmatrix} 1 & 0 & 0 \\ g\hat{R}_x & 1 & 0 \\ g\hat{R}_y & 0 & 1 \end{bmatrix} \begin{bmatrix} \dot{y}_i^0 \\ \dot{y}_{xi} \\ \dot{y}_{yi} \end{bmatrix} + g \, |\dot{\mathbf{y}}_i| \begin{bmatrix} 1 \\ \hat{n}_x \\ \hat{n}_y \end{bmatrix} \quad (103)$$

so that the final velocity of particle-1 after absorbing the radiation is

$$\begin{bmatrix} \dot{y}_f^0 \\ \dot{\mathbf{y}}_f \end{bmatrix} = \frac{1}{1-g^2} \begin{bmatrix} \dot{y}_{in}^0 \\ |\dot{\mathbf{y}}_{in}| \, \hat{\mathbf{R}} \end{bmatrix} + \frac{2g}{1-g^2} \begin{bmatrix} |\dot{\mathbf{y}}_{in}| \\ \dot{y}_{in}^0 \hat{\mathbf{R}} + |\dot{\mathbf{y}}_{in}| \, \hat{\mathbf{n}} \end{bmatrix}$$

$$+ \frac{g^2}{1-g^2} \begin{bmatrix} \dot{y}_{in}^0 + |\dot{\mathbf{y}}_{in}| \\ 2\dot{y}_{in}^0 \hat{\mathbf{n}} + 2 \, |\dot{\mathbf{y}}_{in}| \, \hat{\mathbf{R}} \end{bmatrix} + \frac{g^3}{1-g^2} \begin{bmatrix} |\dot{\mathbf{y}}_{in}| \\ |\dot{\mathbf{y}}_{in}| \, \hat{\mathbf{n}} \end{bmatrix} \quad . \quad (104)$$

The 0-component is

$$\dot{y}_f^0 = \frac{1+g^2}{1-g^2} \dot{y}_{in}^0 + g \frac{2+g+g^2}{1-g^2} \, |\dot{\mathbf{y}}_{in}| \quad (105)$$

approximated at low velocity as

$$\dot{y}_f^0 \approx \frac{1+g^2}{1-g^2} \, \dot{y}_{in}^0 = -\alpha \dot{y}_{in}^0 \quad (106)$$

where

$$g^2 = \frac{\alpha+1}{\alpha-1} \quad \Rightarrow \quad \alpha = -\frac{1+g^2}{1-g^2} \quad (107)$$

is written so that $\alpha > 1$ for a positive energy timelike particle. The exact final velocity of the scattered particle is

$$\begin{bmatrix} \dot{y}_f^0 \\ \dot{\mathbf{y}}_f \end{bmatrix} = -\frac{\alpha-1}{2} \begin{bmatrix} \dot{y}_{in}^0 \\ |\dot{\mathbf{y}}_{in}| \, \hat{\mathbf{R}} \end{bmatrix} + \sqrt{\alpha^2-1} \begin{bmatrix} |\dot{\mathbf{y}}_{in}| \\ \dot{y}_{in}^0 \hat{\mathbf{R}} + |\dot{\mathbf{y}}_{in}| \, \hat{\mathbf{n}} \end{bmatrix}$$

$$-\frac{\alpha+1}{2} \begin{bmatrix} \dot{y}_{in}^0 + |\dot{\mathbf{y}}_{in}| \\ 2\dot{y}_{in}^0 \hat{\mathbf{n}} + 2 \, |\dot{\mathbf{y}}_{in}| \, \hat{\mathbf{R}} \end{bmatrix} - \frac{\alpha+1}{2} \sqrt{\frac{\alpha+1}{\alpha-1}} \begin{bmatrix} |\dot{\mathbf{y}}_{in}| \\ |\dot{\mathbf{y}}_{in}| \, \hat{\mathbf{n}} \end{bmatrix} \quad (108)$$

with 0-component

$$\dot{y}_f^0 = -\alpha \dot{y}_{in}^0 - \frac{\alpha+1}{2} \left[1 + \frac{3-\alpha}{\sqrt{\alpha^2-1}} \right] |\dot{\mathbf{y}}_{in}| \quad . \quad (109)$$

A pair creation event is observed at τ_2 for $\alpha > 1$ which requires that $g^2 = g_e^2 g_R^2 > 1$. In this case, g_e is the interaction strength for the scattering of particle-2 to positive energy, so $g_e < 2$. From (101) the requirement for pair creation is then

$$\frac{e^2}{4\pi\rho} > \frac{2M}{g_e} \quad . \quad (110)$$

5. Conclusions

In this paper we have shown that a classical equivalent of the Bethe-Heitler mechanism is permitted in Stueckelberg-Horwitz electrodynamics. Although Stueckelberg proposed his model with the goal of providing such a description, the calculation has not been previously carried out in detail. The process begins at τ_1 with a pair annihilation event produced by the scattering of an incoming particle in the Coulomb field of a target nucleus. The general solution for the change in velocity leads to a requirement for this pair process

$$g_e = \frac{\lambda}{R} \times \frac{Ze^2}{4\pi R}\frac{1}{M} > 2 \quad \Rightarrow \quad \frac{\lambda}{R} \times \frac{Ze^2}{4\pi R} > 2M \tag{111}$$

which is reasonable on relativistic grounds and can be seen as consistent with the QED requirement of interaction energy greater than the total mass energy of the particle pair. The next stage in the mechanism is the scattering to positive energy ($g_e < 2$ for this interaction) of a second particle in the Coulomb field of another nucleus. The outgoing 4-velocity is found from the general solution for Coulomb scattering, and this allows us to calculate the radiation emitted by the accelerating particle. Using the Lorentz force on the first particle produced by absorption of bremsstrahlung we find the requirement on pair creation to be

$$g_e g_R > 1 \quad \Rightarrow \quad \frac{e^2}{4\pi\rho} > \frac{2M}{g_e} \tag{112}$$

which is similarly reasonable on relativistic grounds.

Further work is required to obtain a realistic description of the classical Bethe-Heitler mechanism. To describe the interaction for the long range Coulomb force, it is necessary to solve the nonlinear differential equations that arise from the smoothing function

$$\varphi(\tau) = \frac{1}{2\lambda} e^{-|\tau|/\lambda} \tag{113}$$

with a large enough λ to accurately model Rutherford scattering. It is also necessary to examine the mass transfer in the pair processes and check mass conservation among particles and fields.

References

[1] Weinberg S 2005 *The Quantum Theory of Fields, Volume 1: Foundations* (Cambridge: Cambridge University Press) p. 14.
[2] Stueckelberg E C G 1941 *Helv. Phys. Acta* **14** 322
Stueckelberg E C G 1941 *Helv. Phys. Acta* **14** 588
[3] Horwitz L P, Arshansky R I and Elitzur A C 1988 *Found. of Phys.* **18** 1159.
[4] Anderson C D 1932 *Phys. Rev.* **41** 405.
[5] Bethe H A and Heitler W 1934 *Proc. R. Soc. London* **A 146** 83.
[6] Lötstedt E et. al. 2009 *New J. Phys.* **11** 013054.
[7] Müller C 2010 *High-energy collision processes involving intense laser fields*, EMMI Workshop.
[8] Horwitz L P and Piron C 1973 *Helv. Phys. Acta* **48** 316
Arshansky R L, Horwitz P and Lavie Y 1983 *Found. of Phys.* **13** 1167
Land M C and Horwitz L P 1991 *Found. of Phys. Lett.* **4** 61
Land M C and Horwitz L P 1991 *Found. of Phys.* **21** 299
Land M C, Shnerb N and Horwitz L P 1995 *J. Math. Phys.* **36** 3263
Land M C and Horwitz L P 1998 Land M C **A239** 135
Land M C 1996 *Found. of Phys.* **27** 19
Land M C 2001 *Found. of Phys.* **31** 967

Land M C 2003 *Found. of Phys.* **33** 1157
Land M C 2011 *J. Phys.: Conf. Ser.* **330** 012015
Land M 2013 *J. Phys.: Conf. Ser.* **437** 012012
[9] Saad D, Horwitz L P and Arshansky R I 1989 *Found. of Phys.* **19** 1126

IARD 2014 IOP Publishing
Journal of Physics: Conference Series **615** (2015) 012008 doi:10.1088/1742-6596/615/1/012008

Field signature for apparently superluminal particle motion

Martin Land
Department of Computer Science, Hadassah College, 37 HaNeviim Street, Jerusalem
E-mail: martin@hadassah.ac.il

Abstract. In the context of Stueckelberg's covariant symplectic mechanics, Horwitz and Aharonovich [1] have proposed a simple mechanism by which a particle traveling below light speed almost everywhere may exhibit a transit time that suggests superluminal motion. This mechanism, which requires precise measurement of the particle velocity, involves a subtle perturbation affecting the particle's recorded time coordinate caused by virtual pair processes. The Stueckelberg framework is particularly well suited to such problems, because it permits pair creation/annihilation at the classical level. In this paper, we study a trajectory of the type proposed by Horwitz and Aharonovich, and derive the Maxwell 4-vector potential associated with the motion. We show that the resulting fields carry a signature associated with the apparent superluminal motion, providing an independent test for the mechanism that does not require direct observation of the trajectory, except at the detector.

1. Introduction

The interpretation of antiparticles as negative energy particles propagating backward in time was proposed by Stueckelberg [2] in the context of his covariant Hamiltonian theory of interacting spacetime events $x^\mu(\tau)$ evolving as functions of a Poincaré-invariant parameter τ. His goal was to represent a particle/antiparticle process by a single worldline whose time coordinate advances and retreats with respect to the laboratory clock as its instantaneous energy changes sign under interaction with gauge fields. In order to obtain a well-posed electrodynamic theory, Sa'ad, Horwitz, and Arshansky [3] generalized Stueckelberg's formalism by introducing five τ-dependent gauge fields. The resulting pre-Maxwell theory differs from conventional electrodynamics, but reduces to Maxwell theory at τ equilibrium. An overview of pre-Maxwell electrodynamics can be found in [4].

In the context of Stueckelberg electrodynamics, Horwitz and Aharonovich [1] have proposed a simple mechanism by which a particle traveling below light speed almost everywhere may exhibit a transit time that suggests superluminal motion. This mechanism is analogous to a mountain trail with a switchback, on which a hiker may walk for 2 hours covering 12 km, but arrive at a point only 8 km from the starting point, with an apparent average speed of 4 km/hr instead of 6 km/hr. Exchanging time and space, an event may briefly reverse its time direction without affecting its space motion. In this case, the particle will continue its progress in space coordinates from the source to the detector, but the laboratory clock will report a difference between time-of-start and time-of-arrival that is less than expected. In this paper, we study the Coulomb field expected from motion of this type and indicate a small deviation from the field expected for continuous linear motion.

2. Pre-Maxwell electrodynamics

In pre-Maxwell theory, interactions take place between events in spacetime rather than between worldlines. Each event occurring at τ induces a current density in spacetime that disperses for large τ for free particles (and hence asymptotically). Conversely, the current density induces an electromagnetic field that acts on the events. The *five* gauge fields are τ-dependent and invariant under local gauge transformations that also depend on τ. The combined theory can be derived from a unique scalar action and is integrable. It is convenient to write $x^5 = \tau$ and adopt the index convention

$$\lambda, \mu, \nu = 0,1,2,3 \qquad \alpha, \beta, \gamma = 0,1,2,3,5\ . \tag{1}$$

The fields act on the events through the Lorentz force given by

$$M\,\ddot{x}^\mu = e_0\, f^\mu{}_\alpha(x,\tau)\,\dot{x}^\alpha \qquad \frac{d}{d\tau}(-\frac{1}{2}M\dot{x}^2) = e_0\, f_{5\alpha}\dot{x}^\alpha \tag{2}$$

where the gauge invariant fields are formed from the potentials

$$f_{\alpha\beta}(x,\tau) = \partial_\alpha a_\beta(x,\tau) - \partial_\beta a_\alpha(x,\tau)\ . \tag{3}$$

The fields are induced by event currents through the pre-Maxwell equations

$$\partial_\beta f^{\alpha\beta}(x,\tau) = \frac{e_0}{\lambda} j^\alpha(x,\tau) = e j^\alpha(x,\tau) \qquad \epsilon^{\alpha\beta\gamma\delta\epsilon}\partial_\alpha f_{\beta\gamma}(x,\tau) = 0 \tag{4}$$

in which the parameter λ acts as a coherence length. The classical current for an event $\xi^\mu(\tau)$ is

$$j^\alpha\,(x,\tau) = e\,\dot{\xi}^\alpha(\tau)\,\delta^4\,(x - \xi(\tau)) \tag{5}$$

where $\xi^5 = 1$. Under the boundary conditions

$$a^5\,(x,\tau) \xrightarrow[\tau\to\pm\infty]{} 0 \qquad j^5\,(x,\tau) \xrightarrow[\tau\to\pm\infty]{} 0 \tag{6}$$

the standard Maxwell theory is extracted as the equilibrium limit of (4) by integration over the worldline

$$\left.\begin{array}{c} \partial_\beta f^{\alpha\beta}\,(x,\tau) = e j^\alpha\,(x,\tau) \\ \partial_{[\alpha} f_{\beta\gamma]} = 0 \end{array}\right\} \xrightarrow[\int d\tau]{} \left\{\begin{array}{c} \partial_\nu F^{\mu\nu}\,(x) = e J^\mu\,(x) \\ \partial_{[\mu} F_{\nu\rho]} = 0 \end{array}\right. \tag{7}$$

where

$$F^{\mu\nu}(x) = d\tau\; f^{\mu\nu}\,(x,\tau) \qquad A^\mu(x) = \int d\tau\; a^\mu\,(x,\tau) \qquad J^\mu(x) = d\tau\; j^\mu\,(x,\tau)\,. \tag{8}$$

This integration has been called concatenation [5] and links the event current $j^\mu\,(x,\tau)$ with the divergenceless Maxwell particle current $J^\mu(x)$ defined on the entire worldline. The field equations lead to a wave equation

$$\partial_\alpha\partial^\alpha a^\beta\,(x,\tau) = \left(\partial_\mu\partial^\mu - \partial_\tau^2\right) a^\beta\,(x,\tau) = -e j^\beta\,(x,\tau) \tag{9}$$

and the principal part Green's function [6] is

$$G(x,\tau) = -\frac{1}{2\pi}\delta(x^2)\delta(\tau) - \frac{1}{2\pi^2}\frac{\partial}{\partial x^2}\frac{\theta(x^2-\tau^2)}{\sqrt{x^2-\tau^2}} = D(x)\,\delta(\tau) - G_{correlation}(x,\tau). \tag{10}$$

The first term has support on the lightcone at instantaneous τ, and recovers the standard Maxwell Green's function under concatenation. The second term has spacelike support ($x^2 > \tau^2 \geq 0$) and vanishes under concatenation, so it may contribute to correlations but not to Maxwell potentials. Terms of this type have been studied in [7].

3. Piecewise linear trajectories

We define a particle trajectory as a set of piecewise linear event segments

$$\xi(\tau) = \sum_{i=1}^{n} \xi_i(\tau)\,\Theta_i(\tau) = \sum_{i=1}^{n}(u_i\tau + q_i)\,\Theta_i(\tau) \tag{11}$$

where u_i and q_i are constant 4-vectors and $\Theta_i(\tau)$ is some combination of θ functions that defines the support of segment i. The velocities satisfy

$$u_i = \dot{\xi}_i(\tau) = \frac{d\xi_i}{d\tau} = \frac{1}{\sqrt{1-\mathbf{v}^2}}(1,\mathbf{v}) \qquad u_i^2 = -1 \ . \tag{12}$$

The trajectory is associated with an instantaneous event current

$$j(x,\tau) = e\sum_{i=1}^{n} u_i\Theta_i(\tau)\,\delta^4(x - q_i - u_i\tau) \tag{13}$$

with delta function support. To handle the delta functions the instantaneous current is spread along the worldline through

$$\begin{aligned}
j_\varphi(x,\tau) &= e\int ds\,\varphi(\tau-s)\,j(x,s)\\
&= e\sum_{i=1}^{n}\int ds\,\varphi(\tau-s)\,u_i\Theta_i(s)\,\delta^3(\mathbf{x}-\mathbf{q}_i-\mathbf{u}_i s)\,\delta\left(t-q_i^0-u_i^0 s\right)\\
&= e\sum_{i=1}^{n}\int ds\,\varphi(\tau-s)\,\frac{u_i\Theta_i(s)}{|u_i^0|}\delta^3(\mathbf{x}-\mathbf{q}_i-\mathbf{u}_i s)\,\delta\left(\frac{t-q_i^0}{u_i^0}-s\right)\\
&= e\sum_{i=1}^{n}\varphi\left(\tau-\frac{t-q_i^0}{u_i^0}\right)\frac{u_i}{|u_i^0|}\Theta_i\left(\frac{t-q_i^0}{u_i^0}\right)\delta^3\left(\mathbf{x}-\mathbf{q}_i-\mathbf{u}_i\frac{t-q_i^0}{u_i^0}\right)
\end{aligned} \tag{14}$$

where

$$\varphi(\tau) = \frac{1}{2\lambda}e^{-|\tau|/2\lambda} \qquad \int_{\infty}^{\infty} d\tau\,\varphi(\tau) = 1 \ . \tag{15}$$

The Maxwell current can be found by integrating this event current over τ, summing the event contributions along worldline and extracting the equilibrium current

$$\begin{aligned}
J(x) &= e\int d\tau\sum_{i=1}^{n}\varphi\left(\tau-\frac{t-q_i^0}{u_i^0}\right)\frac{u_i}{|u_i^0|}\Theta_i\left(\frac{t-q_i^0}{u_i^0}\right)\delta^3\left(\mathbf{x}-\mathbf{q}_i-\mathbf{v}_i\left(t-q_i^0\right)\right)\\
&= e\sum_{i=1}^{n}\frac{u_i}{|u_i^0|}\Theta_i\left(\frac{t-q_i^0}{u_i^0}\right)\delta^3\left(\mathbf{x}-\mathbf{q}_i-\mathbf{v}_i\left(t-q_i^0\right)\right) \ .
\end{aligned} \tag{16}$$

The Liénard-Wiechert potential is found from the principal part Green's function

$$G(x-x',\tau-\tau') = -\delta\left[(x-x')^2\right]\delta(\tau-\tau') \tag{17}$$

and the event current, so that

$$\begin{aligned} a_\varphi(x,\tau) &= -\frac{e}{2\pi}\int d^4x'd\tau'\,\delta\left[(x-x')^2\right]\delta(\tau-\tau')\; j_\varphi(x',\tau') \\ &= -\frac{e}{2\pi}\int d^4x'\,\delta\left[(x-x')^2\right]\; j_\varphi(x',\tau)\ . \end{aligned} \tag{18}$$

Inserting (14) and performing the x' integration leads to

$$a_\varphi(x,\tau) = -\frac{e}{2\pi}\sum_{i=1}^{n}\int d\tau'\;\varphi(\tau-\tau')\;u_i\Theta_i(\tau')\,\delta\left[(x-q_i-u_i\tau')^2\right]\ . \tag{19}$$

Using the identity

$$\int d\tau f(\tau)\,\delta\left[g(\tau)\right] = \frac{f(s)}{|g'(s)|} \tag{20}$$

we may perform the integral over τ' to find

$$a_\varphi(x,\tau) = -\frac{e}{4\pi}\sum_{i=1}^{n}\frac{u_i\Theta_i(s_i)\,\varphi(\tau-s_i)}{|u_i\cdot(x-q_i-u_is_i)|} \tag{21}$$

where s_i is determined by the fixed observation point x and the *a priori* trajectory $\xi_i(\tau)$ through

$$(x-\xi_i(s_i))^2 = (x-q_i-u_is_i)^2 = 0 \qquad\qquad x^0-q_i^0-u_i^0s_i^0>0\ . \tag{22}$$

Using $u_i^2=-1$ we find the retarded times as

$$s_i = -u_i\cdot(x-q_i)\mp\sqrt{(u_i\cdot(x-q_i))^2+(x-q_i)^2}\ . \tag{23}$$

The choice of sign is associated with the retarded and advanced times of the event evolution, and depends upon the direction of the time evolution. Consider the "static" event for which $\mathbf{u}=0$ and $q=0$. For the forward time trajectory $u=(1,\mathbf{0})$

$$s_\pm = t\mp\sqrt{(-t)^2+\mathbf{x}^2-t^2} = t\mp R\ . \tag{24}$$

At the observation point $(t,\mathbf{x})$ the event observed must have been produced by the event current at a time earlier than t, and so we choose the upper sign and retarded time s_-. The event produced at time s_- is located at $x=(t-R,\mathbf{0})$ and as expected is at future lightlike separation from the observation point. But for the "static" event evolving backward in time with $u=(-1,\mathbf{0})$

$$s_\pm = -t\mp\sqrt{(-t)^2+\mathbf{x}^2-t^2} = -t\mp R \tag{25}$$

and we must choose the lower sign. Of the two times $s_\pm$ it is clear that the event evolving toward negative times will reach the point s_+ before reaching the point s_- and so this is the retarded time for that trajectory. At time s_+ the event is located at

$$x = us_+ = (-1,0)\,(-t+R) = (t-R,0)\ . \tag{26}$$

Adding the lightlike vector $(R, \mathbf{x})$ takes this to the observation point $(t, \mathbf{x})$ as for the forward evolving event. The choice of the upper sign can also be understood as choosing the T-reversed picture of the forward time evolution, respecting the discrete Lorentz symmetry. We summarize these cases as

$$s_i = -u_i \cdot (x - q_i) - \varepsilon\left(u_i^0\right)\sqrt{\left(u_i \cdot (x - q_i)\right)^2 + (x - q_i)^2} \tag{27}$$

and it follows that

$$\left|u_i \cdot (x - q_i - u_i s_i)\right| = \sqrt{\left(u_i \cdot (x - q_i)\right)^2 + (x - q_i)^2} \tag{28}$$

Finally, the Liénard-Wiechert potential takes the form

$$a_\varphi(x, \tau) = -\frac{e}{4\pi}\sum_{i=1}^{n} \frac{u_i \Theta_i\left(s_i\right)\varphi(\tau - s_i)}{\sqrt{\left(u_i \cdot (x - q_i)\right)^2 + (x - q_i)^2}} \tag{29}$$

which under concatenation provides the Maxwell potential in the form

$$A(x) = -\frac{e}{4\pi}\sum_{i=1}^{n} \frac{u_i \Theta_i\left(s_i\right)}{\sqrt{\left(u_i \cdot (x - q_i)\right)^2 + (x - q_i)^2}} \quad . \tag{30}$$

4. Field from pair annihilation

We first consider a trajectory on which an event moves uniformly forward in time until $\tau = 0$ and then moves uniformly backward in time. In the laboratory, this will we seen as pair annihilation. The event trajectory is described by

$$\xi\left(\tau\right) = \begin{cases} \left(u^0, \mathbf{u}\right)\tau & , \quad \tau < 0 \\ \\ \left(-u^0, \mathbf{u}\right)\tau & , \quad \tau > 0 \end{cases} \tag{31}$$

from which

$$u_1 = \left(u^0, \mathbf{u}\right) \qquad u_2 = \left(-u^0, \mathbf{u}\right) \qquad q_1 = q_2 = 0 \tag{32}$$

$$\Theta_1 = \left[1 - \theta\left(\tau\right)\right] = \theta\left(-\tau\right) \qquad \Theta_2 = \theta\left(\tau\right) \tag{33}$$

This trajectory is shown in Figure 1.

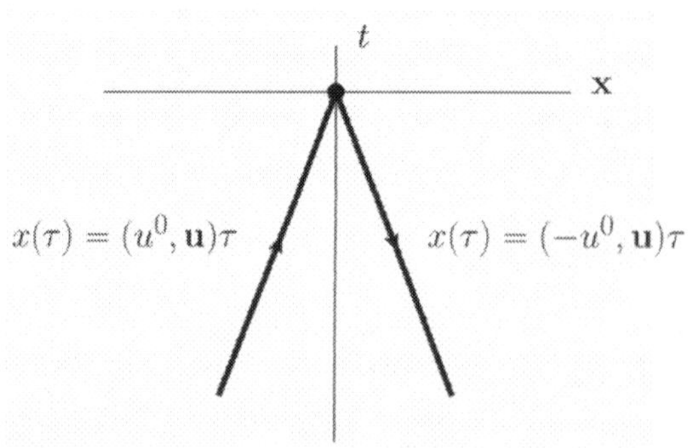

Figure 1: Pair annihilation event

The event current takes the form

$$j(x,\tau) = e\left(u^0,\mathbf{u}\right)\left[1-\theta(\tau)\right]\delta^4\left(x-\left(u^0,\mathbf{u}\right)\tau\right) + \left(-u^0,\mathbf{u}\right)\theta(\tau)\,\delta^4\left(x-\left(-u^0,\mathbf{u}\right)\tau\right) \quad (34)$$

and the spread current is

$$j_\varphi(x,\tau) = e\sum_{i=1}^{n}\varphi\left(\tau-\frac{t}{u_i^0}\right)\frac{u_i}{|u_i^0|}\Theta_i\left(\frac{t}{u_i^0}\right)\delta^3\left(\mathbf{x}-\mathbf{u}_i\frac{t}{u_i^0}\right) \quad (35)$$

$$= e\Big[\varphi(\tau-\frac{t}{u^0})\,(1,\mathbf{v})\,\delta^3(\mathbf{x}-\mathbf{v}t) + \varphi(\tau+\frac{t}{u^0})\,(-1,\mathbf{v})\,\delta^3(\mathbf{x}+\mathbf{v}t)\Big]\,\theta(-t)\;. \quad (36)$$

In this form, the first term describes a particle located at $\mathbf{x} = \mathbf{v}t$ and the second term describes an antiparticle located at $\mathbf{x} = -\mathbf{v}t$, until they mutually annihilate at $t = 0$. Similarly, the concatenated current is

$$J(x,\tau) = e\left[(1,\mathbf{v})\;\delta^3(\mathbf{x}-\mathbf{v}t) + (-1,\mathbf{v})\;\delta^3(\mathbf{x}+\mathbf{v}t)\right]\theta(-t) \quad (37)$$

The retarded times are

$$\begin{aligned} s_1 &= -\left(u^0,\mathbf{u}\right)\cdot x - \sqrt{\left(\left(u^0,\mathbf{u}\right)\cdot x\right)^2 + x^2} \\ &= u^0 t - \mathbf{u}\cdot\mathbf{x} - R\sqrt{1+\left(\frac{t}{R}\right)^2\left[\left(\frac{\mathbf{u}\cdot\mathbf{x}}{t}-u^0\right)^2-1\right]} \end{aligned} \quad (38)$$

$$\begin{aligned} s_2 &= -\left(-u^0,\mathbf{u}\right)\cdot x + \sqrt{\left(\left(-u^0,\mathbf{u}\right)\cdot x\right)^2 + x^2} \\ &= -u^0 t - \mathbf{u}\cdot\mathbf{x} + R\sqrt{1+\left(\frac{t}{R}\right)^2\left[\left(\frac{\mathbf{u}\cdot\mathbf{x}}{t}+u^0\right)^2-1\right]} \end{aligned} \quad (39)$$

and the potential takes the form

$$a^\beta(x,\tau) = -\frac{e}{4\pi R}\left[\frac{\left(u^0,\mathbf{u}\right)\varphi(\tau-s_1)\left[1-\theta(s_1)\right]}{\sqrt{1+\left(\frac{t}{R}\right)^2\left[\left(\frac{\mathbf{u}\cdot\mathbf{x}}{t}-u^0\right)^2-1\right]}} + \frac{\left(-u^0,\mathbf{u}\right)\varphi(\tau-s_2)\theta(s_2)}{\sqrt{1+\left(\frac{t}{R}\right)^2\left[\left(\frac{\mathbf{u}\cdot\mathbf{x}}{t}+u^0\right)^2-1\right]}}\right]\;. \quad (40)$$

Under concatenation, the Maxwell potential is found to be

$$A(x) = -\frac{e}{4\pi R}\left[\frac{(u^0,\mathbf{u})\,\theta(-s_1)}{\sqrt{1+\left(\frac{t}{R}\right)^2\left[\left(\frac{\mathbf{u}\cdot\mathbf{x}}{t}-u^0\right)^2-1\right]}} + \frac{(-u^0,\mathbf{u})\,\theta(s_2)}{\sqrt{1+\left(\frac{t}{R}\right)^2\left[\left(\frac{\mathbf{u}\cdot\mathbf{x}}{t}+u^0\right)^2-1\right]}}\right] \tag{41}$$

We notice that the first term is nonzero for $s_1 < 0$ which holds in a region of spacetime for which

$$-(u^0,\mathbf{u})\cdot x - \sqrt{((u^0,\mathbf{u})\cdot x)^2 + x^2} < 0 \tag{42}$$

$$\left[(u^0,\mathbf{u})\cdot x\right]^2 < ((u^0,\mathbf{u})\cdot x)^2 + x^2 \tag{43}$$

$$0 < x^2 \tag{44}$$

and the second term is nonzero for $s_2 > 0$ which requires

$$-(-u^0,\mathbf{u})\cdot x + \sqrt{((-u^0,\mathbf{u})\cdot x)^2 + x^2} > 0 \tag{45}$$

$$((u^0,\mathbf{u})\cdot x)^2 + x^2 > \left[(u^0,\mathbf{u})\cdot x\right]^2 \tag{46}$$

$$x^2 > 0 \tag{47}$$

so that

$$\theta(-s_1) = \theta(s_2) = \theta(R-t)\ . \tag{48}$$

The field (41) thus describes the combined field of two opposite charges that when observed from a fixed distance R are seen to mutually annihilate at $t = R$. Put another way, the field is observable only at spacelike separation from the annihilation event $\xi = 0$. At any point in the future of the annihilation event, there is no observed field. We notice that if $\mathbf{u}\cdot\mathbf{x} = 0$ the component A^0 vanishes identically, describing a combined charge distribution built from a pair of opposite charges viewed symmetrically from the observation point x.

5. Switchback in time

We now consider an event evolving according to

$$\xi(\tau) = \begin{cases} (u^0\tau,\mathbf{u}\tau) & ,\quad \tau < 0 \\ (-u^0\tau,\mathbf{u}\tau) & ,\quad 0 \le \tau \le T \\ (u^0(\tau-2T),\mathbf{u}\tau) = (u^0\tau,\mathbf{u}\tau) + q_3 & ,\quad \tau > T \end{cases} \tag{49}$$

where $q_3 = (-2u^0T,\mathbf{0})$. This trajectory is shown in Figure 2 — its laboratory phenomenology is discussed in section 6.

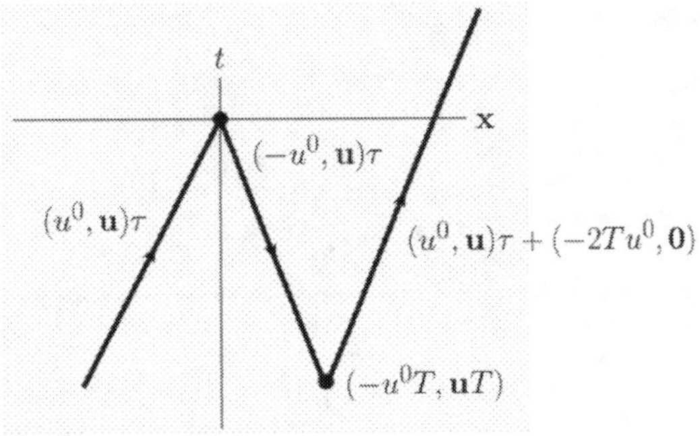

Figure 2: Switchback in time

As described in the Horwitz-Aharonovich model, this event proceeds moves linearly and continuously in space, but experiences a brief switchback in its time coordinate. For example, if the particle is observed at $\tau = -\tau_0 < 0$ and observed again at $\tau = \tau_0 > T$, the particle is seen to travel the distance $2\,|\mathbf{u}|\,\tau_0$ during an elapsed time $2u^0\tau_0 - 2T$ with speed

$$v = \frac{|\mathbf{u}|}{u^0\left(1 - \dfrac{T}{\tau_0}\right)} < \frac{|\mathbf{u}|}{u^0} \; . \tag{50}$$

The four-velocity for this trajectory is

$$\dot{\xi}(\tau) = \begin{cases} (u^0, \mathbf{u}) & , \quad \tau < 0 \\ (-u^0, \mathbf{u}) & , \quad 0 \le \tau \le T \\ (u^0, \mathbf{u}) & , \quad \tau > T \end{cases} \tag{51}$$

and current is

$$j(x,\tau) = e \begin{cases} (u^0, \mathbf{u})\, \delta^3(\mathbf{x} - \mathbf{u}\tau)\, \delta(t - u^0\tau) & , \quad \tau < 0 \\ (-u^0, \mathbf{u})\, \delta^3(\mathbf{x} - \mathbf{u}\tau)\, \delta(t + u^0\tau) & , \quad 0 \le \tau \le T \\ (u^0, \mathbf{u})\, \delta^3(\mathbf{x} - \mathbf{u}\tau)\, \delta(t - u^0(\tau - 2T)) & , \quad \tau > T \end{cases} \; . \tag{52}$$

which can be put into the form (11) as

$$\begin{aligned} j(x,\tau) \;=\; & e\,(u^0, \mathbf{u})\, \delta^3(\mathbf{x} - \mathbf{u}\tau)\, \delta(t - u^0\tau)\, \theta(-\tau) \\ & + e\,(-u^0, \mathbf{u})\, \delta^3(\mathbf{x} - \mathbf{u}\tau)\, \delta(t + u^0\tau)\, [\theta(\tau) - \theta(\tau - T)] \\ & + e\,(u^0, \mathbf{u})\, \delta^3(\mathbf{x} - \mathbf{u}\tau)\, \delta(t - u^0(\tau - 2T))\, \theta(\tau - T) \; . \end{aligned} \tag{53}$$

Using (27) we find the retarded times

$$s_1 = u^0 t - \mathbf{u}\cdot\mathbf{x} - \sqrt{[\mathbf{u}\cdot\mathbf{x} - u^0 t]^2 + R^2 - t^2} \tag{54}$$

$$s_2 = -u^0 t - \mathbf{u}\cdot\mathbf{x} + \sqrt{[\mathbf{u}\cdot\mathbf{x} + u^0 t]^2 + R^2 - t^2} \tag{55}$$

$$s_3 = u^0(t + 2u^0 T) - \mathbf{u}\cdot\mathbf{x} - \sqrt{[\mathbf{u}\cdot\mathbf{x} - u^0(t + 2u^0 T)]^2 + R^2 - (t + 2u^0 T)^2} \tag{56}$$

so that the potential is

$$a_\varphi(x,\tau) = -\frac{e}{4\pi R}\left\{\frac{(u^0,\mathbf{u})\,\varphi(\tau-s_1)\theta(-s_1)}{\sqrt{1+\left(\frac{t}{R}\right)^2\left[\left(\frac{\mathbf{u}\cdot\mathbf{x}}{t}-u^0\right)^2-1\right]}}\right.$$

$$+\frac{(-u^0,\mathbf{u})\,\varphi(\tau-s_2)\,[\theta(s_2)-\theta(s_2-T)]}{\sqrt{1+\left(\frac{t}{R}\right)^2\left[\left(\frac{\mathbf{u}\cdot\mathbf{x}}{t}+u^0\right)^2-1\right]}}$$

$$\left.+\frac{(u^0,\mathbf{u})\,\varphi(\tau-s_3)\theta(s_3-T)}{\sqrt{1+\left(\frac{t+2u^0T}{R}\right)^2\left[\left(\frac{\mathbf{u}\cdot\mathbf{x}}{(t+2u^0T)}-u^0\right)^2-1\right]}}\right\}. \tag{57}$$

Under concatenation, the Maxwell potential becomes

$$A(x) = -\frac{e}{4\pi R}\left\{\frac{(u^0,\mathbf{u})\,\theta(-s_1)}{\sqrt{1+\left(\frac{t}{R}\right)^2\left[\left(\frac{\mathbf{u}\cdot\mathbf{x}}{t}-u^0\right)^2-1\right]}}\right.$$

$$+\frac{(-u^0,\mathbf{u})\,[\theta(s_2)-\theta(s_2-T)]}{\sqrt{1+\left(\frac{t}{R}\right)^2\left[\left(\frac{\mathbf{u}\cdot\mathbf{x}}{t}+u^0\right)^2-1\right]}}$$

$$\left.+\frac{(u^0,\mathbf{u})\,\theta(s_3-T)}{\sqrt{1+\left(\frac{t+2u^0T}{R}\right)^2\left[\left(\frac{\mathbf{u}\cdot\mathbf{x}}{(t+2u^0T)}-u^0\right)^2-1\right]}}\right\}. \tag{58}$$

As seen in the pair annihilation problem, the functions $\theta(-s_1)$ and $\theta(s_2)$ become $\theta(R-t)$ and we must similarly convert $\theta(s_2-T)$ and $\theta(s_3-T)$ to functions on spacetime. Using (55) the condition $s_2 > T$ becomes

$$-u^0t-\mathbf{u}\cdot\mathbf{x}+\sqrt{[\mathbf{u}\cdot\mathbf{x}+u^0t]^2+R^2-t^2} > T \tag{59}$$

which can be put into the form

$$t^2+2Tu^0t+T^2+2T\mathbf{u}\cdot\mathbf{x}-R^2 < 0\ . \tag{60}$$

Since $u_i^2 = -1$ this can be rewritten as

$$-(u_2T - x)^2 < 0 \ , \tag{61}$$

which is equivalent to the requirement that $u_2T - x$ be spacelike. For this term to be nonzero the observation point x must not be in the future of the event u_2T, the second turn-around point at which the trajectory returns to forward time evolution. Using (56) the condition $s_3 > T$ is expressed as

$$u^0\left(t + 2u^0T\right) - \mathbf{u}\cdot\mathbf{x} - \sqrt{\left[\mathbf{u}\cdot\mathbf{x} - u^0\left(t + 2u^0T\right)\right]^2 + R^2 - \left(t + 2u^0T\right)^2} > T \tag{62}$$

equivalent to the requirement

$$-\left[(u_3T - q_3) - x\right]^2 > 0 \tag{63}$$

where again $q_3 = \left(-2u^0T, \mathbf{0}\right)$. Since

$$u_3T - q_3 = u_2T \tag{64}$$

this requires that the observation point be in the future of the second turn-around point. It is convenient to write

$$\theta\left(s_3 - T\right) = \theta\left(-\left(x - u_2T\right)^2\right) = 1 - \theta\left(\left(x - u_2T\right)^2\right) \tag{65}$$

which splits the potential into 3 pieces

$$A(x) = A_1(x) + A_2(x) + A_3(x) \tag{66}$$

where

$$A_1 = -\frac{e}{4\pi R}\frac{\left(u^0, \mathbf{u}\right)\theta(t + u^0T)}{\sqrt{1 + \left(\frac{t + 2u^0T}{R}\right)^2\left[\left(\frac{\mathbf{u}\cdot\mathbf{x}}{(t + 2u^0T)} - u^0\right)^2 - 1\right]}} \tag{67}$$

appears most like the usual Liénard-Wiechert potential with support on the future of the second turn-around point, and

$$A_2 = -\frac{e}{4\pi R}\left[\frac{(u^0,\mathbf{u})}{\sqrt{1+\left(\frac{t}{R}\right)^2\left[\left(\frac{\mathbf{u}\cdot\mathbf{x}}{t}-u^0\right)^2-1\right]}}\right.$$
$$\left.+\frac{(-u^0,\mathbf{u})}{\sqrt{1+\left(\frac{t}{R}\right)^2\left[\left(\frac{\mathbf{u}\cdot\mathbf{x}}{t}+u^0\right)^2-1\right]}}\right]\theta(R-t) \qquad (68)$$

$$A_3 = \frac{e}{4\pi R}\left[\frac{(u^0,\mathbf{u})}{\sqrt{1+\left(\frac{t+2u^0T}{R}\right)^2\left[\left(\frac{\mathbf{u}\cdot\mathbf{x}}{(t+2u^0T)}-u^0\right)^2-1\right]}}\right.$$
$$\left.+\frac{(-u^0,\mathbf{u})}{\sqrt{1+\left(\frac{t}{R}\right)^2\left[\left(\frac{\mathbf{u}\cdot\mathbf{x}}{t}+u^0\right)^2-1\right]}}\right]\theta(R-(t+u_2T)) \qquad (69)$$

have support in the present (spacelike separation) of the "pull-back" in time.

6. Discussion

In the previous section we derived the Maxwell potential associated with a particle trajectory that includes a "pull-back" in time. Under the Stueckelberg-Feynman interpretation, the resulting zig-zag in the time coordinate will be observed in the laboratory as the following sequence:

(i) Particle-1 moves on a linear trajectory with energy $E = mu^0$ and velocity $\mathbf{u}/u^0$,

(ii) At clock time $t = -u^0T$, particle-1 reaches $\mathbf{x} = -\mathbf{u}T$, while a pair creation process at $\mathbf{x} = \mathbf{u}T$ produces particle-2 and an antiparticle,

(iii) At clock time $t = 0$, particle-1 and the antiparticle mutually annihilate,

(iv) Particle-2 continues on a linear trajectory congruent with the original path of particle-1.

The observation of such a process and the measurement of the time "pull-back" would require many detectors along the particle trajectory. Given the good fortune to have placed detectors in (spacelike) proximity to the pair processes, one may also observe the unusual contributions (68) and (69) to the Maxwell potential — and hence the electromagnetic fields.

However, in addition to these signatures, the duration T of the "pull-back" in time will be embedded, through equation (67), in the form of the electromagnetic fields at any point along the trajectory, specifically when the particle arrives at the detector. If the line of observation is transverse to the particle velocity, so that $\mathbf{u} \cdot \mathbf{x} = 0$, then at high energy, with $(u^0)^2 \gg 1$, one may have $t \simeq 2u^0 T$ causing the potential

$$A_1 \simeq -\frac{e}{4\pi R} \frac{(1, \mathbf{v})}{\sqrt{1 + \left(\frac{t + 2u^0 T}{R}\right)^2}} \tag{70}$$

to deviate measurably from the expected value (found from equation (70) with $T = 0$).

References

[1] L P Horwitz and I Aharonovich 2013 *J. Phys.: Conf. Ser.* **437** 012021
[2] Stueckelberg E C G 1941 *Helv. Phys. Acta* **14** 322
Stueckelberg E C G 1941 *Helv. Phys. Acta* **14** 588
[3] Saad D, Horwitz L P and Arshansky R I 1989 *Found. of Phys.* **19** 1126
[4] Land M 2013 *J. Phys.: Conf. Ser.* **437** 012012
[5] Arshansky R L, Horwitz P and Lavie Y 1983 *Found. of Phys.* **13** 1167
[6] Land M C and Horwitz L P 1991 *Found. of Phys.* **21** 299
[7] Aharonovich I and Horwitz L P 2012 *J. Math. Phys.* **53** 032902

Journal of Physics: Conference Series **615** (2015) 012009 doi:10.1088/1742-6596/615/1/012009

Implications of the Lorentz signature of space-time: from bound states to clock synchronization, non-inertial frames, particle localization, relativistic entanglement, dark side of the universe

Luca Lusanna

Sezione INFN di Firenze, Polo Scientifico, Via Sansone 1, 50019 Sesto Fiorentino (FI), Italy

E-mail: lusanna@fi.infn.it

Abstract.

In my contribution to IARD 2012 [1] I made a detailed review of my formulation of relativistic classical and quantum mechanics (RCM and RQM) of massive [2] and massless [3] particles in SR. See Refs.[4] for extensive reviews of the obtained results in special and general relativity (SR and GR). Here I will sketch the main conceptual points of it and I will speak about new developments in this framework.

The Lorentz signature of the space-times of SR and GR is a source of problems absent in non-relativistic theories. The classical mechanics (CM) of Newton gives a deterministic description of objects (particles, bodies) supposed to have a reality in an inertial frame of the Galilei space-time centered on an inertial mathematical observer playing no dynamical role beyond defining Cartesian coordinates. This space-time is assumed to be a given background container of the real objects, whose world-lines are described in terms of an absolute notion of time. At each instant there is an absolute Euclidean 3-space where the objects are localized. The inertial frames are connected by the transformations of the Galilei group. This description can be extended to non-inertial frames centered on mathematical accelerated observers.

This realistic description of the world-lines of particles is preserved in special relativity (SR). However, now they are described in the inertial frames of Minkowski space-time centered on inertial mathematical relativistic observers and the Poincaré group describes the transformations connecting the inertial frames. However, in SR there is no notion of absolute time and of absolute 3-space: only the whole Minkowski space-time is absolute and only the conformal structure (i.e. the light-cone describing the locus of incoming and outgoing radiation in every point) has an intrinsic meaning. As a consequence, we must introduce a *convention of clock synchronization* to define an instantaneous 3-space.

Usually RCM is formulated in inertial frames, whose Euclidean 3-spaces are defined by Einstein's convention [1]. Only with this convention does the 1-way velocity of light between two observers

[1] The inertial observer A sends a ray of light at x_i^o towards the (in general accelerated) observer B; the ray is

Journal of Physics: Conference Series **615** (2015) 012009 doi:10.1088/1742-6596/615/1/012009

(it depends on how their clocks are synchronized) coincide with the 2-way velocity of light c of an inertial observer (it replaces the unit of length in relativistic metrology [5]).

However this description of RCM is still incomplete for interacting systems due to the following problems:

(i) There is not a unique notion of relativistic center of mass of a system of particles like in Newtonian mechanics;
(ii) There is the problem of the elimination of relative times in relativistic bound states (time-like excitations are not seen in spectroscopy);
(iii) It is highly non trivial to find the explicit form of the Poincaré generators (especially the Lorentz boosts) for interacting particles in the instant form of dynamics;
(iv) There is no accepted global formulation of non-inertial frames without the pathologies of the rotating disk and of Fermi coordinates.

The absence of time-like excitations in relative times in spectroscopy and the need of a sound formulation of the Cauchy problem in field theory require a metrological convention about the instantaneous 3-spaces, namely a clock synchronization convention. This can be done in a Wigner covariant way with a 3+1 splitting of space-time, the use of radar 4-coordinates centered on a time-like observer and the separation of the non-local (non-measurable) relativistic center of mass of the 3-universe, both in special relativistic inertial frames and in (either special or general) relativistic non-inertial ones of globally hyperbolic space-times asymptotically Minkowskian without super-translations.

A solution to all these problems has been given in Refs.[6, 7, 8] (see also the reviews in Ref.[4]). In these papers there is the first general theory of global non-inertial frames in Minkowski space-time. In the recent paper [9] there is the most updated construction of explicit non-inertial frames and a discussion on accelerated time-like observers avoiding the uniformly accelerated Rindler ones (they disappear in the non-relativistic limit).

The main differences from non-relativistic CM are the *non-local nature* of the relativistic collective variables proposed for the relativistic center of mass (implying their non-measurability with local measurements) and a *spatial non-separability* of the particles, which must be described by means of suitable Wigner-covariant relative 3-variables. Only relative variables are measurable due to this basic spatial non-separability already present at the classical level.

This formulation of RCM allows one to get a consistent definition of RQM of particles with an associated notion of relativistic entanglement as an extension of non-relativistic quantum mechanics (NRQM) [2] avoiding all the known relativistic pathologies. This was done in Ref.[2]. However in RQM the spatial non-separability is an obstruction to the identification of subsystems [2, 10] and leads to a theory of relativistic entanglement strongly different from the non-relativistic one, in which Alice and Bob cannot be decoupled observers.

reflected towards A at a point P of B world-line and then reabsorbed by A at x^o_f; by convention P is synchronous with the mid-point between emission and absorption on A's world-line, i.e. $x^o_P = x^o_i + \frac{1}{2}\,(x^o_f - x^o_i) = \frac{1}{2}\,(x^o_i + x^o_f)$. This convention selects the Euclidean instantaneous 3-spaces $x^o = ct = const.$ of the inertial frames centered on A. However, if the observer A is accelerated, the convention can breaks down due the possible appearance of coordinate singularities.

[2] Unlikely from the transition from Galilei space-time to the Minkowski one, the transition from CM to NRQM can be done only in an operational way due to the big unsolved foundational problems of NRQM (see for instance Ref.[11, 12])

Journal of Physics: Conference Series **615** (2015) 012009 doi:10.1088/1742-6596/615/1/012009

This framework for RCM has been extended to classical field theory (CFT) in Refs.[7] [4] both for classical fields and fluids, but the extension of the approach to quantum field theory (QFT) has still to be done. In field theory the separation of the relativistic center of mass may help [7] in avoiding Haag theorem, but leads to extra problems for the definition of particles besides the known ones in non-inertial frames and in curved space-times.

As shown in Ref.[10] the non-separability and non-locality due to the relativistic center of mass (i.e. the Lorentz signature of the space-time) both at the classical and quantum levels reduce the relevance of the still debated quantum non-locality of NRQM. As shown in Ref.[3] these properties are present also in CFT and their extension to QFT is a difficult open problem, which adds to the existing problems with the notion of particle in QFT [13, 14] (instability under either unitary or non-unitary Bogoliubov transformations). The relativistic non-separability and non-locality point towards non-local QFT and rise the problem of the validity of *micro-causality* (quantum operators at space-like distances commute) at the relativistic level. This issue has already been raised by Busch [15] with the notion of *unsharp observables* (a local operator not measurable with local actions in a given 3-region). According to him sharp spatial localization is an *operationally meaningless idealization* (it requires an infinite amount of energy with unavoidable pair production; the quantum nature of the constituents of the detectors should be taken into account; and so on).

We have shown [10] the status of the particle localization problem both at the relativistic and non-relativistic levels showing the existing theoretical problems and some of the experimental limitations. One would need the identification of some relevant (well mathematically defined) position basis of wave functions with compact support (or like the over-complete coherent state basis for the harmonic oscillator) for the position operator to be used for every type of localization problem.

By taking into account the quantum nature of both the preparing apparatus and of the detectors we have shown that also at the non-relativistic level the real quasi-isolated system with a decoupled center of mass is the full set of "apparatus + detectors + observed system" and that the prepared and detected particles, moving along mean classical Newton trajectories, are described by relative variables [10].

Since, in any case, electromagnetism is always present, this implies that the relativistic picture is valid at every level in experiments.

We have shown the *non-factual* nature of the mathematical time-like observers and of their mathematical synchronization conventions for building reference frames (see also Ref.[16] in the framework of quantum information). It seems quite difficult to develop a theory in which Alice and Bob are dynamical observers [3] and exchange information by using a dynamical electromagnetic field! Instead Alice and Bob are present in every protocol for relativistic quantum information (see Ref.[18] for an old review emphasizing the problems with special relativity), a quickly developing theory, in which the study of the spatio-temporal trajectories of the investigated quantum systems is nearly always lacking.

The main open problem is that if one describes the free massive Klein-Gordon field in non-

[3] First steps in this direction are done with *quantum metrology* (see Ref.[17], ch.7), in which a quantum system is described by means of the relative variables with respect to another quantum system (the observer) without using variables defined with respect to an external classical reference frame (this relational approach is consistent with relativistic non-separability). It is under investigation what is the change in the information and in the entanglement if one goes from the description with respect to a quantum observer to the one with respect to another quantum observer.

inertial frames (like it is done in the Tomonaga-Schwinger formulation on arbitrary space-like hyper-surfaces of Minkowski space-time) then generically the time evolution is not unitarily implementable (the implied Bogoliubov transformation is not of the Hilbert-Schmidt type). How to avoid this no-go theorem, already present before going to GR where only non-inertial frames are allowed by the equivalence principle?

The described state of affairs [10] is in accord with Bohr's point of view according to which we need a classical description of the experimental apparatus. It seems that all the realizable experiments must admit a quasi-classical description not only of the apparatus but also of the quantum particles: they are present in the experimental area as classical effective particles with a mean trajectory and a mean value of 4-momentum (measured with time-of-flight methods).

As shown in Ref.[21], if one assumes that the wave function describes the given quantum system (no ensemble interpretation), the statement of Bohr can be justified by noting that the wave functions used in the preparation of particle beams (semi-classical objects with a mean classical trajectory and a classical mean momentum determined with time-of-flight methods) are a special subset of the wave functions solutions of the Schroedinger equation for the given particles. Their associated density matrix, pervading the whole 3-space, admits a *multi-polar expansion around a classical trajectory* having *zero dipole*. This implies that in this case the equations of the Ehrenfest theorem give rise to the Newton equations for the Newton trajectory (the monopole; it is not a Bohmian trajectory) with a classical force augmented by forces of a quantum nature coming from the quadrupole and the higher multipoles (they are proportional to powers of the Planck constant). As shown in Ref.[21] the mean trajectories of the prepared beams of particles and of the particles revealed by the detectors are just these classically emerging Newton trajectories impled by the Ehrenfest theorem for wave functions with zero dipole. Also all the intuitive descriptions of experiments in atomic physics are compatible with this emergence of classicality. In these descriptions an atom is represented as a classical particle delocalized in a small sphere, whose origin can be traced to the effect of the higher-multipole forces in the emerging Newton equations for the atom trajectory. The wave functions without zero dipole do not seem to be implementable in feasible experiments. No explanation is given of the probabilistic Born rule, but it is suggested that the random unique outcomes have a quasi-classical localization given by these Newton trajectories.

An extremely important (till now unnoticed) point is that the fixation of the gauge freedom of GR (and of every generally covariant theory of gravity), i.e. the choice of the non-inertial frame and of its 4-coordinates, is nothing else that *the establishment of conventions for relativistic metrology* [5], an operation done from atomic physicists, NASA engineers and astronomers. In Ref.[22, 23] there is the Hamiltonian formulation of ADM tetrad gravity in asymptotically Minkowskian space-times admitting global non-inertial frames. It turns out that the trace of the extrinsic curvature (the York time) of the non-Euclidean 3-spaces is a gauge variable to be fixed with a metrological convention (it is the remnant of the gauge freedom in clock synchronization in GR). As shown in the third paper of Refs.[20], the inertial gauge freedom in the York time is connected with the existence of dark matter at the cosmological level. It is possible that at least part of dark matter is a relativistic inertial effect eliminable with a suitable convention for the ICRS celestial reference system. Moreover the York time is also connected with dark energy.

References

[1] L. Lusanna, *On Relativistic Quantum Mechanics in the Rest-Frame Instant Form of Dynamics*, in Proceedings of the *8th Biennial Conference on Classical and Quantum Relativistic Dynamics of Particles and Fields*

(IARD 2012), Florence, Italy, May 29-June 1, 2012, eds. L. P. Horwitz, L. Lusanna, T. Gill, M. Land, P. Salucci, J. Phys. Conf. Ser. **437** (2013) 012015.

[2] D. Alba, H. W. Crater and L. Lusanna, *Relativistic Quantum Mechanics and Relativistic Entanglement in the Rest-Frame Instant Form of Dynamics*, J. Math. Phys. **52**, 062301 (2011) (arXiv 0907.1816).

[3] D. Alba,H. W. Crater and L. Lusanna, *Massless Particles plus Matter in the Rest-Frame Instant Form of Dynamics*, *J. Phys.* A **43** 405203 (2011) (arXiv 1005.5521).

[4] L. Lusanna, *From Clock Synchronization to Dark Matter as a Relativistic Inertial Effect*, Lecture at the *Black Objects in Supergravity* School BOSS2011, Frascati, 9-13 May 2011 (arXiv 1205.2481), Springer Proc. Phys. **144**, pp.267-343 (Spinger, Berlin, 2013).

L. Lusanna, *Non-Inertial Frames in Special and General Relativity*, to appear in the Proceedings of the School *Gravity: where do we stand?*, 11-15 May 2009, Villa Olmo, Como, Italy, ed. R. Peron(arXiv:1310.4465).

[5] L. Lusanna, *Relativistic Metrology: from Earth to Astrophysics*, In-Tech E-Book *Modern Metrology Concerns*, 2012 (ISBN 978-953-51-0584-8).

[6] D. Alba and L. Lusanna, *Charged Particles and the Electro-Magnetic Field in Non-Inertial Frames: I. Admissible 3+1 Splittings of Minkowski Spacetime and the Non-Inertial Rest Frames*, Int. J. Geom. Methods in Physics **7**, 33 (2010) (arXiv 0908.0213), *II. Applications: Rotating Frames, Sagnac Effect, Faraday Rotation, Wrap-up Effect*, Int. J. Geom. Methods in Physics, **7**, 185 (2010) (arXiv 0908.0215)

Generalized Radar 4-Coordinates and Equal-Time Cauchy Surfaces for Arbitrary Accelerated Observers (2005), Int. J. Mod. Phys. **D16**, 1149 (2007) (arXiv gr-qc/0501090).

[7] D. Alba, H. W. Crater and L. Lusanna, *Towards Relativistic Atom Physics. I. The Rest-Frame Instant Form of Dynamics and a Canonical Transformation for a system of Charged Particles plus the Electro-Magnetic Field*, Canad. J. Phys. **88**, 379 (2010) (arXiv 0806.2383), *II. Collective and Relative Relativistic Variables for a System of Charged Particles plus the Electro-Magnetic Field*, Canad. J. Phys. **88**, 425 (2010) (arXiv 0811.0715).

[8] D. Alba, H. W. Crater and L. Lusanna, *Hamiltonian Relativistic Two-Body Problem: Center of Mass and Orbit Reconstruction*, J. Phys. **A40**, 9585 (2007) (gr-qc/0610200).

[9] H. W. Crater and L. Lusanna, *Non-Inertial Frames in Minkowski Space-Time, Accelerated either Mathematical or Dynamical Observers and Comments on Non-Inertial Relativistic Quantum Mechanics*, Int. J. Geom. Meth. Mod. Phys. **11**, 1450086 (2014) (arXiv 1405.3257).

[10] H. W. Crater and L. Lusanna, *On Relativistic Entanglement and Localization of Particles and on their Comparison with the Non-Relativistic Theory*, Int. J. Mod. Phys. **A29**, 1450163 (2014) (arXiv 1306.6524).

[11] M. Dickson, *Non-Relativistic Quantum Mechanics*, in *Philosophy of Physics*, Part A, p.275, eds. J. Butterfield and J. Earman (Elsevier, Amsterdam, 2007) (Philsci-archive.pitt.edu/3321/).

[12] N. P. Landsman, *Between Classical and Quantum*, in *Handbook of the Philosophy of Science, Vol.2: Philosophy of Physics*, eds. J. Butterfield and J. Earman, p. 417-553 (Elsevier, Amsterdam, 2006) (arXiv quant-ph/0506082).

[13] H. Halvorson and R. Clifton, *No Place for Particles in Relativistic Quantum Theories?*, Philosophy of Science **69**, 1 (2002) (quant-ph/0103041).

[14] D. Fraser, *The Fate of 'Particles' in Quantum Field Theories with Interactions*, Studies in the History and Philosophy of Modern Physics **39**, 841 (2008).

[15] P. Busch, *Unsharp Localization and Causality in Relativistic Quantum Theory*, J. Phys. **A32**, 6535 (1999).

[16] S. D. Bartlett, T. Rudolph and R. W. Spekkens, *Reference Frames, Superselection Rules and Quantum Information*, Rev. Mod. Phys. **79**, 555 (2007) (arXiv quant-ph/0610030).

[17] M. C. Palmer, *Relativistic Quantum Information Theory and Quantum References Frames*, Thesis for the degree of Doctor of Philosophy, School of Physics, Sydney University, 2013 (arXiv 1312.4780).

[18] A. Peres and D. R. Terno, *Quantum Information and Relativity Theory*, Rev. Mod. Phys. **76**, 93 (2004).

[19] C. G. Torre and M. Varadarajan, *Quantum Fields at Any Time* , Phys. Rev. **D58**, 064007 (1998).

[20] A. Arageorgis, J. Earman and L. Ruetsche, *Weyling the Time Away: the Non-Unitary Implementability of Quantum Field Dynamics on Curved Spacetime*, Studies in History and Philosophy of Modern Physics **33**, 151 (2002).

A. D. Helfer, *The Hamiltonians of Linear Quantum Fields: I. Existence Theory for Scalar Fields*, (hep-th/9908011); *II. Classically Positive Hamiltonians* (hep-th/9908012).

[21] L. Lusanna and M. Pauri, *On the Transition from the Quantum to the Classical Regime for Massive Scalar Particles: A Spatiotemporal Approach*, Eur. Phys. J. Plus **129** (2014) 178 (arXiv 1207.1248).

[22] D. Alba and L. Lusanna, *The Einstein-Maxwell-Particle System in the York Canonical Basis of ADM Tetrad Gravity: I) The Equations of Motion in Arbitrary Schwinger Time Gauges.*, Canad. J. Phys. **90**, 1017 (2012) (arXiv 0907.4087); *II) The Weak Field Approximation in the 3-Orthogonal Gauges and Hamiltonian Post-Minkowskian Gravity: the N-Body Problem and Gravitational Waves with Asymptotic Background.*, Canad. J. Phys. **90**, 1077 (2012) (arXiv 1003.5143); *III) The Post-Minkowskian N-Body Problem, its Post-*

Newtonian Limit in Non-Harmonic 3-Orthogonal Gauges and Dark Matter as an Inertial Effect. , Canad. J. Phys. **90**, 1131 (2012) (arXiv 1009.1794).

[23] L. Lusanna and M. Villani, *Hamiltonian Expression of Curvature Tensors in the York Canonical Basis: I) Riemann Tensor and Ricci Scalars*, Int. J. Geom. Meth. Mod. Phys. **11**, 1450052 (2014) (arXiv 1401.1370); *II) Weyl Tensor, Weyl Scalars, Weyl Eigenvalues and the Problem of the Observables of the Gravitational Field*, Int. J. Geom. Meth. Mod. Phys. **11**, 1450053 (2014) (arXiv 1401.1375).

IARD 2014 IOP Publishing
Journal of Physics: Conference Series **615** (2015) 012010 doi:10.1088/1742-6596/615/1/012010

Quantum group perturbative formalism for affine Toda models

Alexander Zuevsky
Institute of Mathematics, Academy of Sciences of the Czech Republic, Prague
E-mail: zuevsky@yahoo.com

Abstract. In the case of non-compact space we consider the quantum affine Toda systems in the operator approach. We introduce quantum analogues of the systems of equations as well as elements of quantum connections. For the solution of the quantum affine Toda systems we use the light-cone as well as Yang-Feldman formalisms of quantization. We then propose solutions to the quantum affine Toda equations based on the quantum-group approach.

1. Introduction
Two-dimensional integrable field theories always attract much attention [1–8]. Such systems appear to be a very important tool in understanding of main non-perturbative aspects of mathematical physics. The importance of two-dimensional integrable models consists also in their beautiful algebraic structure. Consideration of algebraic foundations of classical and quantum two-dimensional exactly solvable models leads also to interesting results either in the theory of classical (infinite dimensional) Lie algebras and groups and in the theory of q-deformed structures. For conformal and affine Toda models in classical and quantum regions [1–4,9,12–19] we consider the most important properties of two-dimensional exactly solvable models by applying methods based on deformed universal enveloping algebras of Lie algebras. Toda models have applications in many fields of applied mathematics and mathematical physics [2,20].

2. Affine Toda systems in the classical region
Let $\mathcal{M}$ be two-dimensional manifold $\mathbb{R}^2$ with standard coordinates $z^{\pm} = t \pm x$ and derivatives $\partial_{\pm}$. Let $\mathfrak{g}$ be a rank r simple Lie algebra with simple roots α_i, $i = 1, \ldots, r$. Let $\widehat{\mathfrak{g}}$ be an affine Lie algebra corresponding to $\mathfrak{g}$ endowed with $\mathbb{Z}$-grading, and $\widehat{G}$ be corresponding infinite-dimensional Lie group. In the principle grading, the subspace $\widehat{\mathfrak{g}}_0$ in the decomposition $\widehat{\mathfrak{g}} = \oplus_{m \in Z} \widehat{\mathfrak{g}}_m$ is abelian. The affine Toda system fields $\phi = \sum_{i=1}^{r} h_i \phi_i$ (where h_i denotes Cartan elements of the Lie algebra $\widehat{\mathfrak{g}}$) satisfying equations

$$\partial_+ \partial_- \phi + \frac{4\eta^2}{\beta} \left(\sum_{i=1}^{r} m_i \frac{\alpha_i}{\alpha_i^2} e^{\beta \alpha_i \cdot \phi} - \frac{\psi}{\psi^2} e^{-\beta \psi \cdot \phi} \right) = 0, \tag{1}$$

where α_i, $i = 1, ..., r$, are simple roots of $\mathfrak{g}$, and $\psi = -\alpha_0$, is the highest root

$$\frac{\psi}{\psi^2} = \sum_{i=1}^{r} m_i \frac{\alpha_i}{\alpha_i^2}, \tag{2}$$

which determines m_i. Here η denotes a real valued inverse length constant and β is a complex interaction constant. The equations (1) appear from the zero curvature condition on a Lax pair, related to the Lie algebra $\widehat{\mathfrak{g}}$ in the principal grading. Note that the equations of the affine Toda systems with pure imaginary β ($\beta \in \mathbb{C} \setminus \mathbb{R}$) possess solitonic solutions while when $\beta \in \mathbb{C} \cap \mathbb{R}$ such solutions are missing. The coefficients in the equations (1) have been chosen in such a way that $\phi = 0$ gives a solution.

The formal general solution to the equations (1) was obtain in [9]

$$e^{-\beta\lambda_j\cdot\phi} = \frac{\langle\lambda_j|\gamma_+^{-1}\mu_+^{-1}\mu_-\gamma_-|\lambda_j\rangle}{\langle\lambda_0|\gamma_+^{-1}\mu_+^{-1}\mu_-\gamma_-|\lambda_0\rangle^{m_j}}, \tag{3}$$

where where λ_j, $0 \le j \le r$ are $r+1$ fundamental weights of $\widehat{\mathfrak{g}}$, and $|\lambda_j>$, the corresponding highest–weight states for the fundamental representations. These have as levels the integers m, – labels on Dynkin diagram defined by (2). Here we use the irreducible highest-weight representations, which are generated by the action of arbitrary products of the negative step operators on a state called the highest-weight state. This state is annihilated by the step operators corresponding to positive roots. The representation is characterised by the action of the Cartan subalgebra on this weight, which must be an eigenvector under this action (for details see Appendix A in ref. [9]).

Holomorphic and anti-holomorphic $\gamma_\pm(z^\pm) : \mathcal{M} \longrightarrow \widehat{G}_0$ and $\mu_\pm(z^\pm) : \mathcal{M} \longrightarrow \widehat{G}_\pm$ map points of the manifold $\mathcal{M}$ to the subgroups $\widehat{G}_0$ and $\widehat{G}_\pm$ of the infinite-dimensional group $\widehat{G}$. The general theory of infinite-dimensional Lie algebras can be found in [10, 11].

The sine–Gordon equation

$$\partial_+\partial_-\phi + \frac{4\eta^2}{\beta}\left(\frac{\alpha_1}{\alpha_1^2}e^{\beta\alpha_1\cdot\phi} - \frac{\psi}{\psi^2}e^{-\beta\psi\cdot\phi}\right) = 0, \tag{4}$$

is a particular case of affine Toda systems connected with the Lie algebra $\widehat{\mathfrak{g}} = \widehat{\mathfrak{sl}_2}$ in the principal and homogeneous gradings. In both gradings the general solution to the equation (4) has the same form (3) but it is assumed that the group elements are certain exponentials of the Lie algebra elements associated to the corresponding grading. The mappings $\mu_\pm(z^\pm)$ satisfy the conditions

$$\partial_\pm\mu_\pm = \mu_\pm\kappa_\pm, \tag{5}$$

where

$$\kappa_\pm(z^\pm) = \pm\eta\sum_{i=0}^{r}\Psi^0_{\pm i}\, x_{\pm i}, \tag{6}$$

($x_{\pm i}$, $i = 0, ..., r$, are Chevalley generators of $\widehat{\mathfrak{g}}$),

$$\Psi^0_{\pm i} = \sqrt{m_i}\, e^{\beta\sum\limits_{i=0}^{r}k_{ij}\phi^0_{\pm i}}. \tag{7}$$

Here k_{ij} are elements of the Cartan matrix, and $\phi^0_{\pm i}$ are free fields satisfying $\partial_+\partial_-\phi^0_\pm = 0$, for $\phi^0_\pm = \sum\limits_{i=1}^{r} h_i\phi^0_{\pm i}$. The elements (6) can be represented in the form

$$\kappa_\pm(z^\pm) = \pm\eta\gamma_\pm^{\mp 1}\widehat{E}_{\pm 1}\gamma_\pm^{\pm 1}, \tag{8}$$

$$\widehat{E}_{\pm 1} = \sum_{i=0}^{r}\sqrt{m_i}x_{\pm i}. \tag{9}$$

The mappings $\gamma_\pm(z^\pm) : \mathcal{M} \longrightarrow \widehat{G}_0$ in (3) have the form

$$\gamma_\pm = e^{-\beta\sum\limits_{i=0}^{r}\phi^0_{\pm i}h_i}. \tag{10}$$

Journal of Physics: Conference Series **615** (2015) 012010 doi:10.1088/1742-6596/615/1/012010

3. Quantization in the light cone formalism: affine Toda systems

In this paper we quantize the conformal and affine Toda systems in the case of non-compact space (plane) is the formalism of the light-cone quantization. It was applied in [24, 25] to the conformal Toda systems. In this section we apply the formalism of the light-cone quantization of affine Toda models in the particular case of the sine–Gordon equation. Let $\phi(z^+, z^-)$ be the Heisenberg fields satisfying the canonical the commutation relations. We construct the quantum sine–Gordon equation and quantum Lax pair on the basis of the affine Lie algebra $\widehat{\mathfrak{sl}_2}$. Let's use the loop affinization [10] of the Lie algebra $\mathfrak{sl}_2$ with a parameter λ. Let $\Phi = \sqrt{2}\phi$ and $\psi = \frac{\sqrt{2}}{2}\beta\phi$. Denote

$$\rho = \frac{1}{2}\beta\phi. \tag{11}$$

Then, following [23], one can introduce the quantum Lax operators. The zero curvature condition applied to the Lax pair results in the quantum sine–Gordon equation.

Introduce the operators

$$\widehat{\omega}_+ = h\partial_+\rho + \alpha(x_+ + \lambda x_-) + \varsigma\Omega, \qquad \widehat{\omega}_- = \alpha\sigma\left(: e^{\beta\phi} : x_- + \lambda^{-1} : e^{-\beta\phi} : x_+\right), \tag{12}$$

where h, x_+, x_- are generators of the Lie algebra $\mathfrak{sl}_2$, and α, ς, σ are constants while $\Omega = \frac{1}{2}h^2$. In (12) the normal ordering is defined in the light-cone $I_\tau^\pm = \{z^\pm = t/2, z^\mp \geq t/2\}$, with respect to the associated Fock space oscillator operators

$$a_j(p) = \int\limits_{I_\tau^+} dz^- e^{i\vec{p}\cdot\vec{x}} i\hat{\partial}_-\phi_i + \int\limits_{I_\tau^-} dz^+ e^{i\vec{p}\cdot\vec{x}} i\hat{\partial}_+\phi_i,$$

where $a\hat{\partial}_\pm b = a(\partial_\pm b) - (\partial_\pm a)b$.

If we apply the zero curvature condition for the operators $\widehat{\omega}_\pm$

$$[\partial_+ + \widehat{\omega}_+, \, \partial_- + \widehat{\omega}_-] = 0, \tag{13}$$

we obtain

$$\begin{aligned} &\alpha\sigma\left(\partial_+\left(: e^{\beta\phi} :\right)x_- + \lambda^{-1}\partial_+\left(: e^{-\beta\phi} :\right)x_+\right) - \partial_+\partial_-\rho h \\ &+\alpha\sigma\left[\partial_+\rho h, : e^{\beta\phi} : x_-\right] + \alpha\sigma\lambda^{-1}\left[\partial_+\rho h, : e^{-\beta\phi} : x_+\right] \\ &+\alpha^2\sigma\left(: e^{\beta\phi} : - : e^{-\beta\phi} :\right)h - \alpha\sigma\varsigma : e^{\beta\phi} : \{h, x_-\} \\ &+\alpha\sigma\lambda^{-1}\varsigma : e^{-\beta\phi} : \{h, x_+\} = 0. \end{aligned} \tag{14}$$

The last equality is equivalent to the following two:

$$\begin{aligned} &\alpha\sigma\partial_+\left(: e^{\beta\phi} :\right)x_- - \partial_+\partial_-\rho h + \alpha\sigma\left[\partial_+\rho h, : e^{\beta\phi} : x_-\right] \\ &+\alpha^2\sigma\left(: e^{\beta\phi} : - : e^{-\beta\phi} :\right)h - \alpha\sigma\varsigma : e^{\beta\phi} : \{h, x_-\} = 0, \\ &\partial_+\left(: e^{-\beta\phi} :\right)x_+ + \left[\partial_+\rho h, : e^{-\beta\phi} : x_+\right] + \varsigma : e^{-\beta\phi} : \{h, x_+\} = 0. \end{aligned} \tag{15}$$

Here we use the property of Ω:

$$[\Omega, x_\pm] = \pm\{h, x_\pm\}, \tag{16}$$

and the identity

$$[a\,A, b\,B] = \frac{1}{2}\left(\,[a,b]\,\{A,B\} + \{a,b\}\,[A,B]\,\right), \tag{17}$$

in which the operators A, B commute with a, b (in our case a, b are operators, depending on the fields and A, B are Lie algebra elements of $\mathfrak{sl}_2$). Using the commutation relations it follows that the commutator $\partial_+\Phi_i$ and Φ_i commute at coinciding points with Φ_i, and thus

$$\partial_+ e^{\beta\Phi_i} = \frac{1}{2}\left\{\partial_+\beta\Phi_i, e^{\beta\Phi_i}\right\}. \tag{18}$$

The same is true for the normal ordered exponent

$$\partial_+ : e^{\beta\Phi_i} := \frac{1}{2}\left\{\partial_+\beta\Phi_i, : e^{\beta\Phi_i} :\right\}. \tag{19}$$

From (19) it follows that

$$\partial_+ : e^{\pm\beta\phi} := \pm\left\{\partial_+\beta\rho, : e^{\pm\beta\phi} :\right\}. \tag{20}$$

Besides this, using the commutation relations we conclude that

$$\left[\partial_+\psi_i(z^+,z^-), : e^{\pm\beta\Phi_j(\tilde{z}^+,z^-)} :\right] = \mp\frac{1}{2}i\hbar\beta^2\delta_{ij}\delta(z^+-\tilde{z}^+) : e^{\pm\beta\Phi_j(\tilde{z}^+,z^+)} :, \tag{21}$$

from which it follows that

$$\left[\partial_+\rho(z^+,z^-), : e^{\pm\beta\phi(\tilde{z}^+,z^-)} :\right] = \mp\frac{1}{2}i\hbar\beta^2\delta(z^+-\tilde{z}^+) : e^{\pm\beta\phi(\tilde{z}^+,z^+)} : . \tag{22}$$

The comparison of summands in (15), (19) and the formulae (22) determines the choice of the constant ς in (12)

$$\varsigma = -\frac{i\hbar\beta^2}{4}\delta(0), \tag{23}$$

(the infinite constant ς vanishes in the final expressions). From (15) we obtain the quantum sine–Gordon equation.

$$\partial_+\partial_-\phi = \frac{2\alpha^2\sigma}{\beta}\left(: e^{\beta\phi} : - : e^{-\beta\phi} :\right). \tag{24}$$

Since α is an arbitrary constant then it is possible to let $\alpha^2 = \frac{2\eta}{\sigma}$ so that coefficients in the quantum equation (24) coincide with the coefficients of the sine–Gordon equation in the classical region (4).

4. Group-theoretical solution

In this section we construct solutions to the equation (24). Let g be $\widehat{G}$ group element. The zero curvature condition (13) implies the gradient form

$$\partial_\pm g^{-1} = g^{-1}\widehat{\omega}_\pm, \quad \partial_\pm g = -\widehat{\omega}_\pm g. \tag{25}$$

Therefore

$$\partial_-\langle\lambda_i|\, g = -\alpha\sigma\langle\lambda_i|\left(: e^{\beta\phi} : f_1 + : e^{-\beta\phi} : f_0\right) g = 0, \tag{26}$$

where $f_1 = x_-$, $f_0 = \lambda^{-1}x_+$, are the lowering operators and $\langle\lambda_i|$, $i = 0, 1$, are the highest weight vectors of the i–th fundamental weight of the Lie algebra $\widehat{\mathfrak{sl}_2}$ representation. Note that

$$\begin{aligned}\partial_+\left(g^{-1} : e^{-\rho h} :\right) &= \left(\partial_+ g^{-1}\right) : e^{-\rho h} : + g^{-1}\partial_+\left(: e^{-\rho h} :\right)\\ &= \left(g^{-1}\widehat{\omega}_+\right) : e^{-\rho h} : -\tfrac{1}{2}g^{-1}\left\{\partial_+\rho h, : e^{-\rho h} :\right\}\\ &= g^{-1}\left(\widehat{\omega}_+ - \partial_+\rho h\right) : e^{-\rho h} : + \frac{1}{2}g^{-1}\left[\partial_+\rho h, : e^{-\rho h} :\right]\\ &= g^{-1}\left(\widehat{\omega}_+ - \partial_+\rho h - \varsigma\Omega\right) : e^{-\rho h} :\\ &= g^{-1}\left(\alpha\left(e_0 + e_1\right)\right) : e^{-\rho h} :,\end{aligned} \tag{27}$$

where $e_1 = x_+$, $e_0 = \lambda x_-$ are the raising operators of the Lie algebra $\widehat{\mathfrak{sl}_2}$. Therefore

$$\partial_+\left(g^{-1} : e^{-\rho h} : |\lambda_i\rangle\right) = g^{-1}\alpha\,(e_0 + e_1) : e^{-\rho h} : |\lambda_i\rangle = 0, \tag{28}$$

$i = 0, 1$. We conclude that the right hand side of the expression

$$: e^{-\rho(z^+,z^-)} :=\frac{\langle\lambda_1|\; g(z^+,\widetilde{z}^-)\; g^{-1}(\widetilde{z}^+,z^-)\; : e^{-\rho(\widetilde{z}^+,z^-)h} : |\lambda_1\rangle}{\langle\lambda_0|\; g(z^+,\widetilde{z}^-)\; g^{-1}(\widetilde{z}^+,z^-)\; |\lambda_0\rangle}, \tag{29}$$

does not depend on $\widetilde{z}^\pm$ and equals to $: e^{-\rho(z^+,z^-)} :$ when $\widetilde{z}^\pm = z^\pm$. In the classical limit, the quantum expression (29) returns to the general solution to the affine Toda system of equations (3). Thus, at the point $P = (z^+, z^-)$ inside the light-cone and two points on the branches $I^\pm_\tau$ we have an expression for the Heisenberg operator.

We compute then the group elements in (29). Since the right hand side of the expression does not depend on $\widetilde{z}^\pm$, with no loss of generality, one can set $\widetilde{z}^+ = \widetilde{z}^- = 0$. From (25) it follows that

$$g^{-1}(0,z^-) = g^{-1}(0,0)\;\bar{T}exp\int_{(0,0)}^{(0,z^-)} \widehat{\omega}_-\; dz_1^-. \tag{30}$$

Thus, on the basis of (27) we obtain

$$\partial_+\left(: e^{\rho h} : g\right) = -\alpha\left(: e^{\rho h} : g\right)(x_+ + \lambda x_-)g. \tag{31}$$

From the commutation relations of the Lie algebra $\mathfrak{sl}_2$ we obtain the identity:

$$e^{\rho h} x_\pm = x_\pm\; e^{\pm 2\rho}\; e^{\rho h}. \tag{32}$$

Nevertheless, exponents of the operator ρ in (32) are not normal ordered and do not have a finite action on the Fock space$\mathcal{F}(I_\tau)$. Let us replace the operators of the fields in (32) by the regularized expressions and finally let us take $p_i \longrightarrow \infty$, $i = 1, 2$. Using the commutation relation we obtain the following identities

$$e^{\pm\rho h} =: e^{\pm\rho h} :\; e^{\frac{1}{2}\hbar\beta^2\Delta^{reg}(0)\Omega}, \tag{33}$$

$$e^{\beta\phi} =: e^{\beta\phi} :\; e^{\hbar\beta^2\Delta^{reg}(0)}. \tag{34}$$

Then we have

$$\begin{aligned} : e^{\rho h} : x_+ &= u^{-1} x_+ u\,(: e^{2\rho} :)\left(: e^{\rho h} :\right) e^{\hbar\beta^2\Delta^{reg}(0)},\\ \lambda : e^{\rho h} : \lambda x_- &= u^{-1}\lambda x_- u\,(: e^{-2\rho} :)\left(: e^{\rho h} :\right) e^{\hbar\beta^2\Delta^{reg}(0)}, \end{aligned} \tag{35}$$

where

$$u = e^{\frac{1}{2}\hbar\beta^2\Delta^{reg}(0)\Omega}. \tag{36}$$

Therefore

$$\partial_+\left(u : e^{\rho h} : g\right) = \widehat{B}_+ \cdot \left(u : e^{\rho h} : g\right), \tag{37}$$

where

$$\widehat{B}_+ = -\alpha\left(x_+ : e^{2\rho} : + \lambda x_- : e^{-2\rho} :\right) e^{\hbar\beta^2\Delta^{reg}(0)}. \tag{38}$$

The equation(37) can be integrated to obtain

$$u : e^{\rho h} : g(z^+,0) = \left(Texp \int\limits_{(0,0)}^{(z^+,0)} \widehat{B}_+ \, dz_1^+ \right) \cdot u : e^{\rho(0,0)h} : g(0,0). \tag{39}$$

As a result, we are able to rewrite (29) as

$$: e^{-\rho(z^+,z^-)} : = \frac{\langle \lambda_1 | \, G : e^{-\rho(0,z^-)h} : \, | \lambda_1 \rangle}{\langle \lambda_0 | \, G \, | \lambda_0 \rangle}, \tag{40}$$

where

$$G =: e^{-\rho(z^+,0)h} : u^{-1} \left(Texp \int\limits_{(0,0)}^{(z^+,0)} \widehat{B}_+ \, dz_1^+ \right) u \cdot : e^{\rho(0,0)h} : \left(\bar{T}exp \int\limits_{(0,0)}^{(0,z^-)} \widehat{\omega}_- \, dz_1^- \right). \tag{41}$$

Note that this expression formally coincides (when the normal ordering is not taken into consideration) with the expression for the general solution to the classical affine Toda system obtained according to the Leznov–Saveliev method [1,9].

5. Yang–Feldman approach

The Yang–Feldman perturbative method is a way to compute field operators in a quantum theory in the expansion over free fields [1,30]. In this section we recall, following [21,22] results of computation of orders in the expansion of the Heisenberg operators satisfying the quantum conformal and affine Toda equations over field operators in absence of interactions.

Let us start with the quantum conformal Toda system of equations. Let φ_α, $\alpha = 1, ..., r$, be field operators satisfying the quantum analogues of the conformal Toda system. The asymptotic fields $\varphi^0_{\pm\alpha}$, $\alpha = 1, ..., r$, satisfy the canonical commutation relations

$$[\varphi^0_{\pm\alpha}(z^\pm), \varphi^0_{\pm\beta}(\tilde{z}^\pm)] = -\frac{i\hbar}{4}(k_{\alpha\beta})^{-1} w_\beta^{-1} \epsilon(z^\pm - \tilde{z}^\pm), \quad [\varphi^0_{\pm\alpha}(z^\pm), \varphi^0_{\mp\beta}(\tilde{z}^\mp)] = 0, \tag{42}$$

(here ϵ is the standard sign function). Then let

$$\psi^0_{\pm\alpha} = \sum_{\beta=1}^{r} k_{\alpha\beta} \varphi^0_{\pm\beta}, \quad \varphi^0_{\pm\alpha} = \sum_{\beta=1}^{r} (k_{\alpha\beta})^{-1} \psi^0_{\pm\beta}, \tag{43}$$

where $k_{\alpha\beta}$ is the Cartan matrix element corresponding to the Lie algebra $\mathfrak{g}$, and w_α is a symmetrizer (a diagonal matrix) satisfying the relation

$$w_\alpha k_{\alpha\beta} = w_\beta k_{\beta\alpha}, \tag{44}$$

α, $\beta = 1, ..., r$. Then, correspondingly

$$\begin{aligned} [\psi^0_{\pm\alpha}(z^\pm), \psi^0_{\pm\beta}(\tilde{z}^\pm)] &= -\frac{i\hbar}{4}(\widehat{k}_{\alpha\beta})^{-1} w_\beta^{-1} \epsilon(z^\pm - \tilde{z}^\pm), \\ [\psi^0_{\pm\alpha}(z^\pm), \varphi^0_{\mp\beta}(\tilde{z}^\mp)] &= -\frac{i\hbar}{4 w_\beta} \delta_{\alpha\beta} \epsilon(z^\pm - \tilde{z}^\pm), \quad \widehat{k}_{\alpha\beta} = k_{\alpha\beta} w_\beta^{-1}. \end{aligned} \tag{45}$$

In the Yang–Feldman formalism [1,21], the n-th order of the exponential Heisenberg operator is given by the expression

$$\left(e^{-\varphi_\alpha}\right)_{(n)} = \frac{1}{(i\hbar)^n} \int\limits_{-\infty}^{+\infty} dz_1 ... dz_n \theta(t - t_1)..\theta(t_{n-1} - t_n) \left[e^{-\varphi^0_\alpha}, V_{11}, ..., V_{nn}\right], \tag{46}$$

where $\theta(z - \tilde{z}) = \theta(z^+ - \tilde{z}^+)\,\theta(z^- - \tilde{z}^-)$ is the ordinary step-function and

$$V_{ij} = V(z_i^+, z_j^-) = \sum_{\alpha=1}^{r} 2w_\alpha e^{\psi^0_{+\alpha}(z_i^+)} \cdot e^{\psi^0_{-\alpha}(z_j^-)}. \tag{47}$$

In (46) we use the notation

$$[A, B, ..., C] = [[[A, B], ...], C]. \tag{48}$$

The expression (46) can be rewritten in the form

$$\left(e^{-\varphi_\alpha}\right)_{(n)} = \frac{1}{(i\hbar)^n} \int\limits_{-\infty}^{+\infty} dz_1 ... dz_n \theta(z - z_1) ... \theta(z_{n-1} - z_n) \sum_{P(k_1,...,k_n)} \left[e^{-\varphi^0_\alpha}, V_{1k_1}, ..., V_{nk_n}\right], \tag{49}$$

where $P(k_1, ..., k_n)$ denotes a permutation of indexes $(k_1, ..., k_n)$. The operator of the field exponent in the first thee orders of the Yang–Feldman procedure (49) has the form

$$\begin{aligned}
\left(e^{-\varphi_\alpha}\right)_{(0)} &= e^{-\varphi^0_\alpha}, \\
\left(e^{-\varphi_\alpha}\right)_{(1)} &= e^{-\varphi^0_\alpha} \frac{2w_\alpha}{i\hbar} \left[1 - e^{-\frac{i\hbar}{2w_\alpha}}\right] \Phi^+_\alpha \Phi^-_\alpha, \\
\left(e^{-\varphi_\alpha}\right)_{(2)} &= e^{-\varphi^0_\alpha} \frac{2w_\alpha}{i\hbar} \left[1 - e^{-\frac{i\hbar}{2w_\alpha}}\right] \sum_{\substack{\gamma=1 \\ \gamma\neq\alpha}}^{r} \frac{2w_\alpha}{i\hbar} \left[1 - e^{\frac{i\hbar}{2} k_{\alpha\gamma}}\right] \Phi^+_{\alpha\gamma} \Phi^-_{\alpha\gamma},
\end{aligned} \tag{50}$$

where $\alpha = 1, ..., r$, and

$$\Phi^\pm_{\alpha_1,...,\alpha_m}(z^\pm) = \int\limits_{-\infty}^{z^\pm} dz_1^\pm e^{\psi^0_{\alpha_1}(z_1^\pm)} \int\limits_{-\infty}^{z_1^\pm} dz_2^\pm e^{\psi^0_{\alpha_2}(z_2^\pm)} ... \int\limits_{-\infty}^{z^\pm_{m-1}} dz_m^\pm e^{\psi^0_{\alpha_m}(z_m^\pm)}. \tag{51}$$

Thus,

$$\Phi^\pm_\alpha(z^\pm) = \int\limits_{-\infty}^{z^\pm} dz_1^\pm e^{\psi^0_{\alpha_1}(z_1^\pm)}, \tag{52}$$

$$\Phi^\pm_{\alpha_1,\beta}(z^\pm) = \int\limits_{-\infty}^{z^\pm} dz_1^\pm e^{\psi^0_{\alpha_1}(z_1^\pm)} \int\limits_{-\infty}^{z_1^\pm} dz_2^\pm e^{\psi^0_\beta(z_2^\pm)}. \tag{53}$$

Note that in the quantum conformal Toda system case the series (49) is finite, i.e., terminates after certain order. The expression for n-th order of expression for the solution (49) can be found in [21].

The Yang–Feldman procedure is also applicable to the affine Toda system of equations. In that case it is convenient to use the loop realization of the affine Lie algebra $\mathfrak{g}$. As in the case of conformal Toda system, the Heisenberg operators of the field φ_α satisfy the canonical commutation relations (42) as well as to the quantum equations, and coincide in the form with (1). The classical fields in (1) are formally substituted by the Heisenberg operators. Nevertheless, instead of the perturbation operator (47), one has to use another operator. For example, in a partial case the solution of the system of the affine Toda equations, the sine–Gordon model, we have

$$\widehat{V}_{ij} = V(z_i^+, z_j^-) = 2w\left(e^{\psi_+^0(z_i^+)}e^{\psi_-^0(z_j^-)} - e^{-\psi_+^0(z_i^+)}e^{-\psi_-^0(z_j^-)}\right), \tag{54}$$

(here w is a constant) which results in

$$\left(e^{-\varphi}\right)_{(0)} = e^{-\varphi^0}, \tag{55}$$

$$\left(e^{-\varphi}\right)_{(1)} = e^{-\varphi^0}\frac{2w}{i\hbar}\left[\left(1 - e^{-\frac{i\hbar}{2w}}\right)\Phi_+^+\Phi_+^- - \left(1 - e^{\frac{i\hbar}{2w}}\right)\Phi_-^+\Phi_-^-\right], \tag{56}$$

where we introduce

$$\Phi_\pm^+(z^+) = \int_{-\infty}^{z^+} dz_1^+ e^{\pm\psi^0(z_1^+)}, \quad \Phi_\pm^-(z^-) = \int_{-\infty}^{z^-} dz_1^- e^{\pm\psi^0(z_1^-)}, \tag{57}$$

and ψ^0 is defined in the expression (43). Then the n-the order of the exponent of the field operator is given by the expression (46) with the interaction operator (54). Note that in the case of the affine Toda system we obtain an infinite series of the expansion of the Heisenberg operators. Nevertheless, as in the case of conformal Toda systems, one can find the general formula for the exponent of the Heisenberg operator at any order.

6. Quantum group solutions for Toda systems

Originally, a relation between the quantization of the conformal Toda system of equations and quantum groups was found in [22]. Indeed, consider the Toda fields φ_i, $i = 1, ..., r$, as Heisenberg operators satisfying a certain quantum analogue of the conformal Toda system of equations in which the classical fields are replaced by field operators. Let us apply, as in [21] (see also [1]), the perturbative Yang–Feldman procedure (see the previous section). We arrive at the exact expressions for the exponents of the Heisenberg operators φ_i in the form of finite series over free field operators φ_i^0, satisfying the canonical commutation relations. In this section we show that the formal expressions constructed on the basis of classical solutions in the Leznov–Saveliev approach, though containing quantum group generators, coincide with the perturbative solutions for Toda systems of equations obtained in frames of the Yang–Feldman procedure.

Let us take the general solution to the conformal Toda system of equations and formally replace the group elements in the right and side by elements of the quantum universal enveloping algebra $U_q(\mathfrak{g})$ of the Lie algebra $\mathfrak{g}$, on the basis of which the system of equations was constructed. Let us replace also the highest weight vectors $|\lambda_i\rangle$ of the fundamental highest weight representation by the highest weight vectors $|\lambda_i^q\rangle_q$ of the fundamental representation of the quantum universal enveloping algebra $U_q(\mathfrak{g})$ of $\mathfrak{g}$, and the mappings $\kappa_\pm$ by the quantum mappings ${}_q\kappa_\pm : \mathcal{M} \longrightarrow U_q(\mathfrak{g})$

$${}_q\kappa_\pm(z^\pm) = \pm\eta\sum_{i=1}^{r}\Psi_{\pm i}^0 \, {}_qx_{\pm i}, \tag{58}$$

$$\Psi_{\pm i}^0 = e^{\sum\limits_{i=1}^{r} k_{ij}\,\varphi_{\pm i}^0}, \tag{59}$$

where $U_q(\mathfrak{g})$.

The generators of the quantum universal enveloping algebra satisfy the deformed commutation relations in the Jimbo–Drinfeld form [26, 27]:

$$[{}_qh_i, {}_qh_j] = 0, \tag{60}$$

$$[{}_qh_i, {}_qx_{\pm j}] = \pm k_{ji\,q}x_{\pm j}, \tag{61}$$

$$[{}_qx_i, {}_qx_j] = \delta_{ij}\frac{q_i^{{}_qh_i} - q_i^{-{}_qh_i}}{q_i - q_i^{-1}}, \tag{62}$$

where $q_i = e^{d_i\hbar}$, $\hbar$ is the Planck constant, and d_i – adjoint prime integers such that dk is symmetric matrix. Then one obtains [22] the following formal expression in the Leznov–Saveliev form

$$e^{-\varphi_i} = {}_q\langle\lambda_i^q|\, {}_qM_+^{-1} \cdot {}_qM_-\, |\lambda_i^q\rangle_q, \tag{63}$$

where ${}_qM_\pm = e^{-\sum\limits_{i=1}^{r} {}_qh_i\varphi^0_{\pm i}} \cdot {}_q\mu_\pm$ are quantum group $U_q(\mathfrak{g})$ elements, and h_i, $i = 1, ..., r$, Cartan elements of $U_q(\mathfrak{g})$. Here $|\lambda_i^q\rangle$, $1 \leq j \leq r$ are elements of the finite dimensional highest vector representation [22] for $U_q(\mathfrak{g})$ with the conditions ${}_qh_i|\lambda_j^q\rangle = \delta_{ij}l_j|\lambda_j^q\rangle$, ${}_qx_i|\lambda_j^q\rangle = 0$, for some l_j. The remaining basis elements are determined by the action of the lowering operators x_{-i} on the highest vectors $|\lambda_j^q\rangle$ [22].

The group elements ${}_q\mu_\pm$ satisfy the condition

$$\partial_\pm\, {}_q\mu_\pm = {}_q\mu_\pm\, {}_q\kappa_\pm. \tag{64}$$

At the same time, in the classical case, due to the properties of the fundamental highest weight representations of $U_q(\mathfrak{g})$, the number of summands in the series of the expansion of (63) is finite and exactly equals to the dimension of the i-th fundamental representation of $U_q(\mathfrak{g})$. It is easy to find, for example, the first three orders in the expansion(63):

$$\left(e^{-\varphi_i}\right)_{(0)} = {}_q\langle\lambda_i^q|e^{-\sum\limits_{j=1}^{r} {}_qh_j(\varphi^0_{+j}+\varphi^0_{-j})}|\lambda_i^q\rangle_q = e^{-\varphi_i^0}, \tag{65}$$

$$\begin{aligned}\left(e^{-\varphi_i}\right)_{(1)} &= -e^{-\varphi^0_{+i}} \sum_{\theta,p=1}^{r} \Phi_\theta^+\Phi_p^- \cdot {}_q\langle\lambda_i^q|{}_qx_{+\theta}\ {}_qx_{-p}|\lambda_i^q\rangle_q\ e^{-\varphi^0_{-i}} \\ &= -e^{-\varphi_i^0}\frac{2w_i}{i\hbar}\left[1 - e^{-\frac{i\hbar}{2w_i}}\right]\Phi_i^+\Phi_i^-,\end{aligned} \tag{66}$$

$$\begin{aligned}\left(e^{-\varphi_i}\right)_{(2)} &= -e^{-\varphi^0_{+i}} \sum_{\theta,\tau,p,q=1}^{r} {}_q\langle\lambda_i^q|{}_qx_{+\theta}\ {}_qx_{+\tau}\ {}_qx_{-p}\ {}_qx_{-q}\ |\lambda_i^q\rangle_q\Phi^+_{\theta\tau}\widetilde{\Phi}^-_{pq}\ e^{-\varphi^0_{-i}} \\ &= e^{-\varphi^0_{+i}}\sum_{\theta=1}^{r} e^{\frac{i\hbar\widehat{k}_{\theta i}}{4}}\frac{2w_i}{i\hbar}\sinh\left(\frac{i\hbar}{4w_i}\right)\frac{2w_\theta}{i\hbar}\sinh\left(\frac{i\hbar}{4}\left(2\delta_{i\theta} - k_{i\theta}\right)\right)\Phi^+_{\theta i}\Phi^-_{\theta i}e^{-\varphi^0_{-i}}.\end{aligned} \tag{67}$$

Here

$$\widetilde{\Phi}^-_{pq} = \int^{z^-} dz_1^- \int^{z^-} dz_2^-\,\theta(z_2^- - z_1^-)e^{q\psi_p^-}e^{q\psi_p^-}... = e^{-\frac{\hbar}{4i}\widehat{k}_{pq}}\Phi^-_{pq}, \tag{68}$$

$i = 1, ..., r$, and $\widehat{k}_{pq}$, w_i were determined in (44) and (45). The operators $\Phi_i^\pm$, $\Phi_{ij}^\pm$ are the same as in (52), (53). The expression for the n-th order in the expansion (63) can be found in [22]. One can see that the first three orders of (65) – (67) coincide with (50). The same is true for any order. Thus as it was shown in [22], the perturbative expressions found in the Yang–Feldman approach for the quantum conformal Toda systems of equations coincides with the quantum group expressions containing generators of the quantum universal enveloping algebra $U_q(\mathfrak{g})$ instead of ordinary group elements.

By analogy with the conformal Toda system case, we construct in this paper quantum group solutions in the approach similar to [22] for affine Toda systems. In the general solution formula (3) for the classical system of equations for the affine Toda one can replace the group elements and vectors in the $\widehat{\mathfrak{g}}$ Lie algebra representations by the elements and vectors of the quantum group $U_q(\widehat{\mathfrak{g}})$ representations. Then we obtain

$$e^{-\varphi_i} = \frac{{}_q\langle\lambda_i^q|{}_qM_+^{-1}\cdot{}_qM_-\,|\lambda_i^q\rangle_q}{{}_q\langle\lambda_0^q|\,{}_qM_+^{-1}\cdot{}_qM_-\,|\lambda_0^q\rangle_q^{m_i}}, \tag{69}$$

where ${}_qM_\pm = e^{-\sum\limits_{i=0}^{r} {}_qh_i\varphi^0_{\pm i}}\cdot{}_q\mu_\pm$, $i = 1,...,r$. When in the finite dimensional case the series representing the solutions (63) terminates and it is easy to compute it order by order, in the case of the affine Toda system of equations the nominator and denominator as well as the solution itself (69) are infinite series. The summands of series in the nominator and denominator are corresponding orders of the action of generators of the quantum universal enveloping algebra of corresponding Lie algebra on the highest vectors of i-th and 0-th fundamental representations. Nevertheless, one can consider the ratio of the numerator and denominator series and use a formula for the n-th order for such a series [31]. Then one can find the general formula for $(e^{-\varphi_i})_{(n)}$. It turns out that in the case of the affine Toda system of equations the expressions, obtained on the basis of the Yang–Feldman procedure coincide with the quantum group expressions.

7. Comparison of the approaches

We have to mention that in the quantum case the analogy with well-developed group-theoretical Leznov–Saveliev method [1] is not completely fulfilled. First of all it remains unclear how to introduce Lax pairs for the quantum analogues of the classical exactly solvable systems containing elements of the quantum universal enveloping algebra $U_q(\mathfrak{g})$. As mentioned before the quantum flat connection elements (12) introduced in [23, 24] contain both Heisenberg field operators satisfying the quantum equations (24) and elements of affine Lie algebras. At the same time, in corresponding solutions we used group elements for ordinary groups parameterized by Heisenberg field operators. This is possibly the reason why we had to introduce an infinite constant ς in the Lax pair (let us mention that ς is absent in the final expressions for solutions). Nevertheless, as it was shown in [25], even if we use such pairs of Lax operators, a quantum group structure of the theory shows up. Moreover, an algebraic origin of the summands containing Ω remains unclear (see (12)). It is possible that the Lax operator pair (12) appears as a result of a gauge transformation of a symmetric version of Lax operator pairs $A_\pm = \sum\limits_\alpha(u_{\pm\alpha}h_\alpha + f_{\pm\alpha}x_{\pm\alpha})$ (here $u_{\pm\alpha}$, $f_{\pm\alpha}$ are operators). Nevertheless, such a transformation is not known. We hope that in the future one could construct on the base of elements of the quantum universal enveloping algebras suitable elements of quantum connections not containing auxiliary infinite constants. Then the zero curvature condition would lead to the quantum conformal and affine Toda equations.

Acknowledgements

We would like to thank the organizers of the the 9-th Biennial Conference on Classical and Quantum Relativistic Dynamics of Particles and Fields University of Connecticut, Storrs, CT, USA 9–13 June 2014.

References

[1] Leznov A N and Saveliev M V 1992 *Group–theoretical methods for integration of non–linear dynamical systems, Progress in Physics Series* **15** (Birkhauser–Verlag, Basel)
[2] Razumov A V and Saveliev M V 1997 *Lie algebras, geometry, and Toda–type systems, Cambridge lecture notes in physics* (Cambridge University Press)
[3] Dubrovin B A, Krichever I M and Novikov S P 1985 *Integrable systems* I. *Modern problems of Mathematics. Fundamental directions. Dynamical systems-4.* (VINITI, Moscow)
[4] Olshanetsky M A and Perelomov A M 1985 *Integrable systems and finite-dimensional Lie algebras. Integrable systems II. Modern problems of Mathematics. Fundamental directions. Dynamical systems-I* (VINITI, Moscow)
[5] Faddeev L D and Takhtajan L A 1987 *Hamiltonian methods in the theory of solitons* (Springer–Verlag, Berlin)
[6] Newell A C 1985 *Solitons in mathematics and physics* (Society for Industrial and Applied Mathematics, New-York)
[7] Green M B, Schwarz J H and Witten E 1988 *Superstring theory* (Cambridge University Press)
[8] Gomez C, Ruiz-Altaba M and Sierra G 1996 *Quantum groups in two–dimensional physics* (Cambridge University Press)
[9] Olive D I, Turok N and Underwood J W R 1993 *Nucl. Phys.* **B401** 663
[10] Kac V G 1990 *Infinite dimensional Lie algebras* Third edition (Cambridge university Press, Cambridge)
[11] Zhelobenko D P 1994 *Representations of reductive Lie algebras* (Moscow, "Nauka")
[12] Gervais J– L 1982 *Nucl. Phys.* **B 209** no 1 125
[13] Gervais J– L and Neveu A 1983 *Nucl. Phys.* **B 224** no 2 329
[14] Mikhailov A V, Olshanetsky M A and Perelomov A M 1981 *Comm. Math. Phys.***79** 473
[15] Toda M 1975 *Phys. Rep.* **18C** no. 1 1
[16] Hollowood T 1991 *PUPT-1286, hep-th/9110010*
[17] Olive D I, Turok N and Underwood J W R 1993*Nucl. Phys.* **B 409** 509
[18] Leznov A N and Saveliev M V 1979 *Lett. Math. Phys.* **3** 489
[19] Fedoseev I A and Leznov A N 1984 *Phys. Lett.* **141B** 1
[20] Razumov A V and Saveliev M V 1994 *Commun. Anal. Geom.***2** 461
[21] Leznov A N and Fedoseev I A 1982 *Theor. Math. Phys.* **53** 3 1175
[22] Leznov A N and Mukhtarov M A 1987 *Math. Phys.* **71** 1 46 1987;
[23] Mansfield P 1982 *Nucl. Phys.***B 208**277
[24] Mansfield P 1983 *Nucl. Phys.* **B 222** 419
[25] Hollowood T and Mansfield P *Nucl. Phys.* **B 330**
[26] Drinfeld V G 1985 *Sov. Math. Dokl.***32** 254
[27] Jimbo M 1985*Lett. Math. Phys.***10** 63
[28] Razumov A V, Saveliev M V and Zuevsky A 1999 *Nonabelian Toda equations associated with classical Lie groups.* Proc. the Memorial volume dedicated to M. V. Saveliev, (Dubna), math-ph/9909008;
[29] Reshetikhin N and Smirnov F 1990 *Comm. Math. Phys.***131** 157
[30] Schwinger J 1948 *Phys. Rev.* **74** 416
[31] Gradshtein I S and Ryzhik I M 1971 *Tables of integrals, sums, series and derivatives. Fifth editions.* (Moscow, "Nauka")

Consistent quantization of massless fields of any spin and the generalized Maxwell's equations

Alexander Gersten and Amnon Moalem
Department of Physics, Ben Gurion University of the Negev, Beer-Sheva, Israel

Abstract. A simplified formalism of first quantized massless fields of any spin is presented. The angular momentum basis for particles of zero mass and finite spin s of the $D^{(s-1/2,1/2)}$ representation of the Lorentz group is used to describe the wavefunctions. The advantage of the formalism is that by equating to zero the $s-1$ components of the wavefunctions, the $2s-1$ subsidiary conditions (needed to eliminate the non-forward and non-backward helicities) are automatically satisfied. Probability currents and Lagrangians are derived allowing a first quantized formalism. A simple procedure is derived for connecting the wavefunctions with potentials and gauge conditions. The spin 1 case is of particular interest and is described with the $D^{(1/2,1/2)}$ vector representation of the well known self-dual representation of the Maxwell's equations. This representation allows us to generalize Maxwell's equations by adding the E_0 and B_0 components to the electric and magnetic four-vectors. Restrictions on their existence are discussed.

Key words: wave equations, massless particles, any spin, generalized Maxwell's equations

1. Introduction

In a previous paper [2] consistent quantization of massless fields of any spin was developed from first principles. Probability currents and Lagrangians were derived allowing a first quantized formalism. A simple procedure was derived for connecting the wavefunctions with potentials and gauge conditions. Here we will present a simplified version of this formalism and elaborate an extension of the self-dual representation of Maxwell's equations. This representation allows us to generalize Maxwell's equations by adding non-zero time like components E_0 and B_0 to the electric and magnetic four-vectors. The existence of these components allows the formation of scalar electromagnetic waves. We will examine some results.

In our previous paper [2] we started our formalism using Dirac's [4] derivation of equations for massless particles with spin s, which in the ordinary vector notation are,

$$\frac{1}{s}\left[s\hat{p}_0 I^{(2s+1)} + S_x\hat{p}_x + S_y\hat{p}_y + S_z\hat{p}_z\right]\psi = \left[\frac{\hat{E}}{c}I^{(2s+1)} + \frac{\boldsymbol{S}}{s}\cdot\hat{\boldsymbol{p}}\right]\psi = 0, \tag{1}$$

$$\left[s\hat{p}_x I^{(s)} + S_x\hat{p}_0 - iS_y\hat{p}_z + iS_z\hat{p}_y\right]\psi = 0, \tag{2}$$

Journal of Physics: Conference Series **615** (2015) 012011 doi:10.1088/1742-6596/615/1/012011

$$\left[s\hat{p}_y I^{(s)} + S_y\hat{p}_0 - iS_z\hat{p}_x + iS_x\hat{p}_z\right]\psi = 0, \tag{3}$$

$$\left[s\hat{p}_z I^{(s)} + S_z\hat{p}_0 - iS_x\hat{p}_y + iS_y\hat{p}_x\right]\psi = 0, \tag{4}$$

where ψ is a $(2s+1)$ component wave function and S_n are the $(2s+1)\times(2s+1)$ spin matrices which satisfy,

$$[S_x, S_y] = iS_z, \quad [S_z, S_x] = iS_y, \quad [S_y, S_z] = iS_x, \quad S_x^2 + S_y^2 + S_z^2 = s(s+1)I^{(2s+1)}. \tag{5}$$

In the above, the $\hat{p}_n$ are the momentum operators, $\hat{p}_0 = \hat{E}/c$, $\hat{E}$ the energy operator, and $I^{(2s+1)}$ is a $(2s+1)\times(2s+1)$ unit matrix. Eqs. (1-4) were analyzed extensively by Bacry [5], who derived them using Wigner's condition [6] on the Pauli-Lubanski vector W^μ for massless fields,

$$W^\mu = s\hat{p}^\mu, \quad \mu = x_0, x, y, z, \tag{6}$$

$$W^\mu = \frac{-i}{2}\varepsilon^{\mu\nu\rho\lambda}S_{\rho\lambda}\hat{p}_\nu, \quad S_{\rho\lambda} = \begin{pmatrix} 0 & S_x & S_y & S_z \\ -S_x & 0 & -iS_z & iS_y \\ -S_y & iS_z & 0 & -iS_x \\ -S_z & -iS_y & iS_x & 0 \end{pmatrix} \equiv (\mathbf{S}, i\mathbf{S}). \tag{7}$$

Dirac suggested using Eq. (1) as the basic helicity equation and substitute from it the $\hat{p}_0\psi$ into the other 3 equations (2-4). Free massless particles may have only two helicity projections (forward and backward along the momentum vector). Using the Dirac procedure one obtains the $2s+1$ component Eq. (1) and $2s-1$ independent subsidiary conditions which reduce the number of helicity projections to 2,

$$\left[\hat{E}I^{(2s+1)} + \frac{c}{s}\boldsymbol{S}\cdot\hat{\boldsymbol{p}}\right]\psi^{(2s+1)} = 0, \tag{8}$$

$$(\boldsymbol{\Pi}\cdot\hat{\mathbf{p}})\,\psi^{(2s+1)} = 0, \tag{9}$$

where the components of $\boldsymbol{\Pi}$ are $(2s-1)\times(2s+1)$ matrices of the subsidiary conditions. These two sets of equations can be combined into $(2s+1)+(2s-1)=4s$ equations [1],

$$\left[\hat{E}I^{(4s)} + c\boldsymbol{\Gamma}^{(4s)}\cdot\hat{\boldsymbol{p}}\right]\Phi^{(4s)} = 0, \tag{10}$$

equivalent to the former equations, provided that the rows of the $\boldsymbol{\Pi}$ matrices are normalized so that the eigenvalues of $\boldsymbol{\Gamma}^{(4s)}\cdot\hat{\boldsymbol{p}}$ are $\pm p$ (two helicities) and,

$$\left(\hat{E}I^{(4s)} - c\boldsymbol{\Gamma}^{(4s)}\cdot\hat{\boldsymbol{p}}\right)\left(\hat{E}I^{(4s)} + c\boldsymbol{\Gamma}^{(4s)}\cdot\hat{\boldsymbol{p}}\right) = \left(\hat{E}^2 - c^2\hat{\boldsymbol{p}}^2\right)I^{(4s)}, \tag{11}$$

i.e. they factorize the d'Alembertian. The wave function $\Phi^{(4s)}$ is of the form [1] (with 2s-1 zeros),

$$\Phi^{(4s)} = \begin{pmatrix} 0 \\ \vdots \\ 0 \\ \psi_{2s+1}^{(2s+1)} \\ \vdots \\ \psi_{-2s-1}^{(2s+1)} \end{pmatrix}, \tag{12}$$

and the components of $\mathbf{\Gamma}^{(4s)}$ are $4s \times 4s$ matrices, which form a (reducible) representation of the algebra of the Pauli matrices,

$$\sigma_0 = \begin{pmatrix} 1 & 0 \\ 0 & 1 \end{pmatrix}, \quad \sigma_x = \begin{pmatrix} 0 & 1 \\ 1 & 0 \end{pmatrix}, \quad \sigma_y = \begin{pmatrix} 0 & -i \\ i & 0 \end{pmatrix}, \quad \sigma_z = \begin{pmatrix} 1 & 0 \\ 0 & -1 \end{pmatrix}. \tag{13}$$

2. Wave equations for free massless fields

We shall use the following relations [3],

$$W_\mu W^\mu = -s\,(s+1)\,\hat{p}_\mu \hat{p}^\mu I^{(s)}, \quad W_\mu \hat{p}^\mu = 0, \tag{14}$$

with the aim of obtaining the Wigner condition Eq. (6) and simultaneously to factorize the d'Alembertian (in the form $\hat{p}_\mu \hat{p}^\mu$). One can easily see that,

$$(W_\mu - s\hat{p}_\mu)\,(W^\mu + s\hat{p}^\mu) = -s\,(2s+1)\,\hat{p}_\mu \hat{p}^\mu I^{(s)}, \tag{15}$$

or

$$(W_\mu + s\hat{p}_\mu)\,(W^\mu - s\hat{p}^\mu) = -s\,(2s+1)\,\hat{p}_\mu \hat{p}^\mu I^{(s)}. \tag{16}$$

From Eqs. (15-16) one can see that if ψ, the wavefunction, satisfies,

$$(W^\mu + s\hat{p}^\mu)\,\psi = 0, \quad or \quad (W^\mu - s\hat{p}^\mu)\,\psi = 0, \quad \mu = 0,1,2,3, \tag{17}$$

then the basic massless particle requirement,

$$\hat{p}_\mu \hat{p}^\mu I^{(s)} \psi = 0, \tag{18}$$

will be satisfied. Explicitly we have,

$$W^0 = W_0 = -S_x \hat{p}_x - S_y \hat{p}_y - S_z \hat{p}_z, \tag{19}$$

$$-W^1 = W_x = -S_x\hat{p}_0 - iS_y\hat{p}_z + iS_z\hat{p}_y, \tag{20}$$

$$-W^2 = W_y = -S_y\hat{p}_0 - iS_z\hat{p}_x + iS_x\hat{p}_z, \tag{21}$$

$$-W^3 = W_z = -S_z\hat{p}_0 - iS_x\hat{p}_y + iS_y\hat{p}_x. \tag{22}$$

Now the main problem is to choose the basis of a space on which the above operators will act. In Ref. [2] we have shown that the angular momentum basis of the $D^{(s-1/2,1/2)}$ representation of the Lorentz group is the proper one to use. This basis is sum of the bases of spins s and spin $s-1$ with

$$(2s+1) + (2s-1) = 4s \text{ components.}$$

In this basis the wavefunction is,

$$\Phi^{(4s)} = \begin{pmatrix} \psi_{s-1}^{(2s-1)} \\ \vdots \\ \psi_{-s+1}^{(2s-1)} \\ \psi_{s}^{(2s+1)} \\ \vdots \\ \psi_{-s}^{(2s+1)} \end{pmatrix} \Longrightarrow \begin{pmatrix} 0 \\ \vdots \\ 0 \\ \psi_{s}^{(2s+1)} \\ \vdots \\ \psi_{-s}^{(2s+1)} \end{pmatrix}. \tag{23}$$

The advantage of the formalism, as we have shown in Ref. [2], is that by equating to zero the spin $(s-1)$ components of the wavefunction $\psi_{\mu}^{(2s-1)}$, the $2s-1$ subsidiary conditions (needed to eliminate the non-forward and non-backward helicities, according to Wigner's analysis [6] are automatically satisfied. So we are left with equations Eq. (17) with $\psi = \Phi^{(4s)}$, but not all equations are needed, the equations with $\mu = 1, 2, 3$ are the equations Eqs. $(2-4)$ from which the subsidiary conditions were derived. Now they are not needed and instead of Eqs. (17) we have,

$$\left(s\hat{p}^0 + W^0\right)\Phi^{(4s)} = 0, \quad or \quad \left(s\hat{p}^0 - W^0\right)\Phi^{(4s)} = 0, \tag{24}$$

which are the generalized helicity equations having solutions only in the forward and backward direction. Note that in Eq. (24) W^0 is a $4s \times 4s$ matrix. The explicit form of Eq. (24) is given in Refs. [1, 2] with the result,

$$\left[\hat{E}\Gamma_0^{(4s)} - c\mathbf{\Gamma}^{(4s)}\cdot\hat{\boldsymbol{p}}\right]\Phi^{(4s)} = 0, \quad or \quad \left[\hat{E}\Gamma_0^{(4s)} + c\mathbf{\Gamma}^{(4s)}\cdot\hat{\boldsymbol{p}}\right]\Phi^{(4s)} = 0, \tag{25}$$

where $\Gamma_0^{(4s)} = I^{(4s)}$ is the $4s \times 4s$ unit matrix, $\hat{E} = \hat{p}^0/c$, and the Γ matrices have the form (in the angular momentum basis),

$$\Gamma_k^{(4s)} = \frac{1}{s} \begin{pmatrix} (-1)^{k+s-1}\, S_k^{(2s-1)} & \Pi_k \\ \Pi_k^H & (-1)^{k+s}\, S_k^{(2s+1)} \end{pmatrix}, \quad k = 1, 2, 3, \tag{26}$$

where $S_k^{(2s-1)}$ are the spin $(s-1)$ matrices of Eq. (5), $S_k^{(2s+1)}$ are the spin (s) matrices of Eq. (5), Π_k is the $(2s-1) \times (2s+1)$ normalized matrix of the subsidiary conditions and Π_k^H is the Hermitian conjugate of Π_k. The matrix elements of Π_k can be evaluated directly from the normalized subsidiary conditions

$$\psi_\mu^{(2s-1)} = \sqrt{s\,(2s+1)} \sum_{m_1,m_2} \langle 1m_1; sm_2 | (1s)s-1, \mu\rangle p_{m_1} \psi_{m_2}^{(2s+1)} = 0, \tag{27}$$

where the expansion is in terms of Clebsch-Gordan coefficients and the factor $\sqrt{s\,(2s+1)}$ comes from the normalization factor of the r.h.s. of Eq. (15).

For example, for spin 1, one obtains,

$$\Pi_x = \begin{pmatrix} -1 & 0 & 0 \end{pmatrix}, \quad \Pi_y = \begin{pmatrix} 0 & -1 & 0 \end{pmatrix}, \quad \Pi_z = \begin{pmatrix} 0 & 0 & -1 \end{pmatrix}, \tag{28}$$

In a Cartesian basis, the $S_k^{(3)}$ are

$$S_x = \begin{pmatrix} 0 & 0 & 0 \\ 0 & 0 & -i \\ 0 & i & 0 \end{pmatrix}, \quad S_y = \begin{pmatrix} 0 & 0 & i \\ 0 & 0 & 0 \\ -i & 0 & 0 \end{pmatrix}, \quad S_z = \begin{pmatrix} 0 & -i & 0 \\ i & 0 & 0 \\ 0 & 0 & 0 \end{pmatrix}, \tag{29}$$

$$\Gamma_k^{(4)} = \begin{pmatrix} 0 & \Pi_k \\ \Pi_k^H & S_k^{(3)} \end{pmatrix}, \quad k = 1, 2, 3, \quad \Gamma_0^{(4)} = \text{unit matrix}, \tag{30}$$

For spin 2, one obtains,

$$\Pi_x = \begin{pmatrix} \sqrt{3} & 0 & -\frac{1}{\sqrt{2}} & 0 & 0 \\ 0 & \sqrt{\frac{3}{2}} & 0 & -\sqrt{\frac{3}{2}} & 0 \\ 0 & 0 & \frac{1}{\sqrt{2}} & 0 & -\sqrt{3} \end{pmatrix}, \tag{31}$$

$$\Pi_y = i \begin{pmatrix} -\sqrt{3} & 0 & -\frac{1}{\sqrt{2}} & 0 & 0 \\ 0 & -\sqrt{\frac{3}{2}} & 0 & -\sqrt{\frac{3}{2}} & 0 \\ 0 & 0 & -\frac{1}{\sqrt{2}} & 0 & -\sqrt{3} \end{pmatrix}, \tag{32}$$

$$\Pi_z = \begin{pmatrix} 0 & -\sqrt{3} & 0 & 0 & 0 \\ 0 & 0 & -2 & 0 & 0 \\ 0 & 0 & 0 & -\sqrt{3} & 0 \end{pmatrix}. \tag{33}$$

$$\Gamma_k^{(8)} = \frac{1}{2}\begin{pmatrix} (-1)^{k+1}\, S_k^{(3)} & \Pi_k \\ \Pi_k^H & (-1)^k\, S_k^{(5)} \end{pmatrix}, \quad k = 1, 2, 3, \tag{34}$$

More details can be found in Refs. [1, 2].

3. First quantization

As is in the case of the Klein-Gordon or Dirac equations, we substitute in Eqs. (25)

$$\hat{E} \Longrightarrow i\hbar\frac{\partial}{\partial t}, \quad \hat{\mathbf{p}} \Longrightarrow -i\hbar\nabla, \tag{35}$$

with a result

$$i\hbar\left(\Gamma_0^{(4s)}\frac{\partial}{\partial t} + c\mathbf{\Gamma}^{(4s)} \cdot \nabla\right)\Phi^{(4s)} = 0. \tag{36}$$

Similarly to the Schroedinger and Dirac equations we define the Hamiltonian $\mathcal{H}$ from Eq. (36),

$$i\hbar\frac{\partial}{\partial t}\Phi^{(4s)} = -i\hbar c\mathbf{\Gamma}^{(4s)} \cdot \nabla\Phi^{(4s)} = \mathcal{H}\Phi^{(4s)}, \tag{37}$$

and find a conserved probability current (the superscript H denotes the Hermitian conjugate),

$$\partial_t\left(\left(\Phi^{(4s)}\right)^H \Phi^{(4s)}\right) + \frac{c}{s}\nabla\cdot\left(\left(\Phi^{(4s)}\right)^H \tilde{\mathbf{S}}\Phi^{(4s)}\right) = 0. \tag{38}$$

Thus the probability density (which should be normalized) is,

$$\rho = \left(\Phi^{(4s)}\right)^H \Phi^{(4s)}, \quad \int\int\int dxdydz\rho = 1. \tag{39}$$

In the above the density is non-negative and the scalar product is positive definite. Having this result we can find a Lagrangian density,

$$\mathcal{L} = i\hbar\left(\Phi^{(4s)}\right)^H\left(\Gamma_0^{4s}\frac{\partial}{\partial t} + c\mathbf{\Gamma}^{(4s)} \cdot \nabla\right)\Phi^{(4s)}, \tag{40}$$

and using the definition of the energy momentum tensor $T^{\mu\nu}$, we find [1],

$$\int\int\int dxdydzT^{00} = \int\int\int dxdydz\left(\Phi^{(4s)}\right)^H \mathcal{H}\Phi^{(4s)} = <\mathcal{H}>, \tag{41}$$

$$\int\int\int dxdydz T^{0k} = \int\int\int dxdydz \left(\Phi^{(4s)}\right)^H (c\hat{p}_k)\, \Phi^{(4s)} =< c\hat{p}_k >, \quad k = 1, 2, 3, \tag{42}$$

i.e. consistent with the expectation values of energy and momentum.

4. Wave equations for free fields

From the Lagrangian Eq. (40) we obtain the two equations Eqs. (25), for particles of spin s ($4s$ is the dimension of the matrices),

$$i\hbar\left(\Gamma_0^{(4s)}\frac{\partial}{\partial t} + c\mathbf{\Gamma}^{(4s)}\cdot\nabla\right)\Phi^{(4s)} = 0, \tag{43}$$

$$i\hbar\left(\Gamma_0^{(4s)}\frac{\partial}{\partial t} - c\mathbf{\Gamma}^{(4s)}\cdot\nabla\right)\Phi^{(4s)} = 0. \tag{44}$$

The Γ matrices form a reducible representation of the Pauli matrices Eq. (13). For the spin $\frac{1}{2}$ Eqs. (43-44) coincide with the massless neutrino equations. The Γ matrices have the same eigenvalues as the Pauli matrices ± 1 ($2s$ degenerate). With the built-in subsidiary conditions only two solutions of each of the equations Eqs. (43-44) are possible. Thus Eqs. (43-44) are the restricted to forward or backward helicity equations for free massless particles. As the energy operator is proportional to the helicity the two solutions only differ in sign of the energy, i.e. one solution corresponds to positive energy and the second to negative energy. One can also note that the solutions of Eq. (43) are also solutions of Eq. (44) with the opposite sign of the helicity. Therefore Eq. (43) has positive energy solution in the helicity's forward direction; its negative energy solution in the helicity's backward direction is the positive energy solution of Eq. (44) for the helicity's backward direction.

In Summary Eq. (43) is the equation of the helicity in the forward direction and Eq. (44) is the equation for the helicity in the backward direction, if the physical requirement of positive energy is imposed.

4.0.1. Examples For spin 1, the photon with helicity ± 1 and total momentum $P = \sqrt{p_x^2 + p_y^2 + p_z^2}$, the normalized solution of Eq. (43) (in Cartesian basis) for positive energy and forward helicity is

$$\Phi^{(4)} = \frac{1}{\sqrt{2}P}\begin{pmatrix} 0 \\ -\frac{p_x p_z - i p_y P}{\sqrt{p_x^2+p_y^2}} \\ -\frac{i p_x P + p_y p_z}{\sqrt{p_x^2+p_y^2}} \\ \sqrt{p_x^2+p_y^2} \end{pmatrix} \exp\left(\frac{1}{i\hbar}(Et - \boldsymbol{p}\cdot\boldsymbol{x})\right), \tag{45}$$

and the normalized solution of Eq. (44) (in Cartesian basis) for positive energy and backward helicity is

$$\Phi^{(4)} = \frac{1}{\sqrt{2}P}\begin{pmatrix} 0 \\ -\frac{ip_yP+p_xp_z}{\sqrt{p_x^2+p_y^2}} \\ -\frac{p_yp_z-ip_xP}{\sqrt{p_x^2+p_y^2}} \\ \sqrt{p_x^2+p_y^2} \end{pmatrix} \exp(\frac{-1}{i\hbar}(Et - \boldsymbol{p}\cdot\boldsymbol{x})). \tag{46}$$

For spin 2, the graviton with helicity ± 2 and total momentum $P = \sqrt{p_x^2+p_y^2+p_z^2}$, the normalized solution of Eq. (43) (in angular momentum basis) for positive energy and forward helicity is

$$\Phi^{(8)} = \frac{\left(p_x^2+p_y^2\right)^2}{4P^4}\begin{pmatrix} 0 \\ 0 \\ 0 \\ \frac{1}{2}\left(\frac{p_z-P}{p_x-ip_y}\right)^2 \\ \left(\frac{p_z-P}{p_x-ip_y}\right) \\ \sqrt{\frac{3}{2}} \\ \frac{p_x-ip_y}{p_z-P} \\ \frac{1}{2}\left(\frac{p_x-ip_y}{p_z-P}\right)^2 \end{pmatrix} \exp(\frac{1}{i\hbar}(Et - \boldsymbol{p}\cdot\boldsymbol{x})), \tag{47}$$

and the normalized solution of Eq. (44) (in angular momentum basis) for positive energy and backward helicity is

$$\Phi^{(8)} = \frac{\left(p_x^2+p_y^2\right)^2}{4P^4}\begin{pmatrix} 0 \\ 0 \\ 0 \\ \frac{1}{2}\left(\frac{p_z+P}{p_x-ip_y}\right)^2 \\ \left(\frac{p_z+P}{p_x-ip_y}\right) \\ \sqrt{\frac{3}{2}} \\ \frac{p_x-ip_y}{p_z+P} \\ \frac{1}{2}\left(\frac{p_x-ip_y}{p_z+P}\right)^2 \end{pmatrix} \exp(\frac{-1}{i\hbar}(Et - \boldsymbol{p}\cdot\boldsymbol{x})). \tag{48}$$

5. Generalized Maxwell's equations

5.1. Maxwell's equations

Maxwell's equations [9] in Gaussian system units are,

$$\nabla\cdot\mathbf{E} = \alpha j_0, \tag{49}$$

$$\nabla \times \mathbf{B} - \partial_0 \boldsymbol{E} = \alpha \boldsymbol{j}, \tag{50}$$

$$\nabla \cdot \mathbf{B} = 0, \tag{51}$$

$$\nabla \times \mathbf{E} + \partial_0 \boldsymbol{B} = 0, \tag{52}$$

where $\alpha = \frac{4\pi}{c}$, $\quad j_0 = c\rho$, $\quad \partial_0 = \frac{1}{c}\frac{\partial}{\partial t}$.

Minkowski [8] found the covariant form of Maxwell's equations,

$$\partial_\mu F^{\mu\nu} = \alpha j^\nu, \quad \partial_\mu \tilde{F}^{\mu\nu} = 0, \quad \nu = 0, 1, 2, 3, \tag{53}$$

where the antisymmetric tensor $F^{\mu\nu}$ and its dual $\tilde{F}^{\mu\nu}$ are defined via the electric and magnetic fields $\mathbf{E}$ and $\mathbf{B}$ respectively as,

$$(F^{\mu\nu}) = \begin{pmatrix} 0 & -E_x & -E_y & -E_z \\ E_x & 0 & -B_z & B_y \\ E_y & B_z & 0 & -B_x \\ E_z & -B_y & B_x & 0 \end{pmatrix}, \tag{54}$$

$$\tilde{F}^{\mu\nu} = \frac{1}{2}\epsilon^{\mu\nu\alpha\beta} F_{\alpha\beta} = \begin{pmatrix} 0 & -B_x & -B_y & -B_z \\ B_x & 0 & E_z & -E_y \\ B_y & -E_z & 0 & E_x \\ B_z & E_y & -E_x & 0 \end{pmatrix}, \tag{55}$$

where $\epsilon^{\mu\nu\alpha\beta}$ is the totally antisymmetric tensor ($\epsilon^{0123} = 1, \quad \epsilon^{\mu\nu\alpha\beta} = -\epsilon_{\mu\nu\alpha\beta}$). The sum of the $F^{\mu\nu}$ and $i\tilde{F}^{\mu\nu}$ is the self-dual antisymmetric tensor, which depends only on the combination

$$\boldsymbol{\Psi} = (\mathbf{E} + i\mathbf{B}), \tag{56}$$

$$F^{\mu\nu} + i\tilde{F}^{\mu\nu} = \begin{pmatrix} 0 & -E_x - iB_x & -E_y - iB_y & -E_z - iB_z \\ E_x + iB_x & 0 & iE_z - B_z & B_y - iE_y \\ E_y + iB_y & B_z - iE_z & 0 & iE_x - B_x \\ E_z + iB_z & iE_y - B_y & B_x - iE_x & 0 \end{pmatrix}, \tag{57}$$

$$= \begin{pmatrix} 0 & -\Psi_x & -\Psi_y & -\Psi_z \\ \Psi_x & 0 & i\Psi_z & -i\Psi_y \\ \Psi_y & -i\Psi_z & 0 & i\Psi_x \\ \Psi_z & i\Psi_y & -i\Psi_x & 0 \end{pmatrix} \tag{58}$$

The self dual Maxwell's equations take the form [1],

$$\partial_\mu \left(F^{\mu\nu} + i\tilde{F}^{\mu\nu}\right) = -\left(\Gamma_\mu^{(4)}\right)^{\nu i} \partial_\mu \Psi_i = \alpha j^\nu, \tag{59}$$

where the $\Gamma_\mu^{(4)}$ matrices are the spin 1 matrices defined by Eq. (30). Thus within the $D^{\left(\frac{1}{2},\frac{1}{2}\right)}$ representation of the Lorentz group, Maxwell's equation can be written as,

$$\left(\Gamma_0^{(4)}\partial_0 + \mathbf{\Gamma}^{(4)} \cdot \nabla\right) \Phi^{(4)} = -\alpha J, \quad \Phi^{(4)} = \begin{pmatrix} 0 \\ \Psi_x \\ \Psi_y \\ \Psi_z \end{pmatrix}, \quad J = \begin{pmatrix} c\rho \\ j_x \\ j_y \\ j_z \end{pmatrix}. \tag{60}$$

Eq. (60) is the same as Eq. (59) which is Lorentz covariant, thus Eq. (60) is also Lorentz invariant. This invariance also applies to the zero component of $\Phi^{(4)}$, which was proven explicitly by Lomont [7]. For $J = 0$, the free Maxwell equations takes the form,

$$\left(\Gamma_0^{(4)}\partial_0 + \mathbf{\Gamma}^{(4)} \cdot \nabla\right) \Phi^{(4)} = 0, \tag{61}$$

which is exactly Eq. (43). The zero component is the perpendicularity condition for the photon equation and it ensures that the helicity can be only in the forward or backward direction. It also insures that the photon will be a particle of spin 1. This condition also remains for the inhomogenious Maxwell's equations Eq. (60). But in the presence of interactions it is not obvious that the spin 1 property should remain. Therefore we will now generalize the non-homogenious Maxwell equations and replace the zero component with $\Psi_0 = E_0 + iB_0$.

5.2. The generalized equations

The generalized equations are,

$$\left(\Gamma_0^{(4)}\partial_0 + \mathbf{\Gamma}^{(4)} \cdot \nabla\right) \Phi^{(4)} = -\alpha J, \quad \Phi^{(4)} = \begin{pmatrix} \Psi_0 \\ \Psi_x \\ \Psi_y \\ \Psi_z \end{pmatrix}, \quad J = \begin{pmatrix} c\rho \\ j_x \\ j_y \\ j_z \end{pmatrix}. \tag{62}$$

from which one can extract the generalized Maxwell's equations in the form

$$\partial_0 E_0 + \boldsymbol{\nabla} \cdot \boldsymbol{E} = \alpha j_0, \tag{63}$$

$$\partial_0 \boldsymbol{E} + \nabla E_0 - \nabla \times \mathbf{B} = -\alpha \boldsymbol{j}, \tag{64}$$

$$\partial_0 B_0 + \nabla \cdot \boldsymbol{B} = 0, \tag{65}$$

$$\boldsymbol{\nabla} \times \boldsymbol{E} + \boldsymbol{\partial_0 B} + \boldsymbol{\nabla} B_0 = 0. \tag{66}$$

Similar equations were derived by Dvoeglazov [10]. We will now look for new features of these equations.

5.3. Energy density

According to Eqs. (37) and (41) the energy density is proportional to,

$$\begin{pmatrix} \Psi_0^* & \Psi_x^* & \Psi_y^* & \Psi_z^* \end{pmatrix} (i\hbar c\partial_0) \begin{pmatrix} \Psi_0 \\ \Psi_x \\ \Psi_y \\ \Psi_z \end{pmatrix} = i\hbar c \left(\Psi_0^* \partial_0 \Psi_0 + \mathbf{\Psi}^* \cdot (\partial_0 \mathbf{\Psi})\right) \tag{67}$$

$$= i\hbar c \left[(E_0 - iB_0)\, \partial_0 \,(E_0 + iB_0) + (\mathbf{E} - i\mathbf{B}) \cdot (\partial_0 \,(\mathbf{E} + i\mathbf{B}))\right], \tag{68}$$

which is different from the conventional form proportional to

$$\mathbf{\Psi}^* \cdot \mathbf{\Psi} = (\mathbf{E} - i\mathbf{B}) \cdot (\mathbf{E} + i\mathbf{B}) = \mathbf{E}^2 + \mathbf{B}^2$$

5.4. The potentials

Let us introduce the potentials,

$$A^{(4)} = \begin{pmatrix} A_0 \\ A_x \\ A_y \\ A_z \end{pmatrix}, \tag{69}$$

where the scalar potential $V = cA_0$. Their relation to the wavefunction $\Phi^{(4)}$ can be obtained by requiring that in the Lorenz gauge they satisfy the wave equation,

$$\partial_\mu \partial^\mu A^{(4)} = \alpha J. \tag{70}$$

The Γ matrices, which are a representation of the Pauli matrices, will factorize the d'Alembertian the same way as the Pauli matrices factorize it, i.e.,

$$\left(\Gamma_0^{(4)} \partial_0 - \mathbf{\Gamma}^{(4)} \cdot \nabla\right) \left(\Gamma_0^{(4)} \partial_0 + \mathbf{\Gamma}^{(4)} \cdot \nabla\right) = \Gamma_0^{(4)} \partial_\mu \partial^\mu \tag{71}$$

In order to get Eq. (70), using Eqs. (62) and (71) the relation between the $\Phi^{(4)}$ and $A^{(4)}$ has to be

Journal of Physics: Conference Series **615** (2015) 012011 doi:10.1088/1742-6596/615/1/012011

$$\Phi^{(4)} = \begin{pmatrix} E_0 + iB_0 \\ E_x + iB_x \\ E_y + iB_y \\ E_z + iB_z \end{pmatrix} = -\left(\Gamma_0^{(4)}\partial_0 - \mathbf{\Gamma}^{(4)} \cdot \nabla\right) \begin{pmatrix} A_0 \\ A_x \\ A_y \\ A_z \end{pmatrix} \tag{72}$$

$$= -\begin{pmatrix} \partial_0 A_0 + \partial_1 A_x + \partial_2 A_y + \partial_3 A_z \\ \partial_0 A_x + \partial_1 A_0 + i\left(\partial_2 A_z - \partial_3 A_y\right) \\ \partial_0 A_y + \partial_2 A_0 - i\left(\partial_1 A_z - \partial_3 A_x\right) \\ \partial_0 A_z + \partial_3 A_0 + i\left(\partial_1 A_y - \partial_2 A_x\right) \end{pmatrix}, \tag{73}$$

from which the relation between the fields and the potentials (in the Lorenz gauge) is,

$$E_0 = -\partial_0 A_0 - \nabla \cdot \mathbf{A}, \tag{74}$$

$$B_0 = 0, \tag{75}$$

$$\mathbf{E} = -\nabla A_0 - \partial_0 \mathbf{A}, \tag{76}$$

$$\mathbf{B} = \nabla \times \mathbf{A}. \tag{77}$$

Eqs. (74) and (75) are new relations of the generalized Maxwell's equations, but the Lorenz condition,

$$\partial_0 A_0 + \nabla \cdot \mathbf{A} = 0, \tag{78}$$

enforces the spin 0 component E_0 to be zero, this result should be gauge invariant as the electric and magnetic fields are. Thus only when the Lorenz condition is not satisfied (while Eq. (70) is valid) the E_0 can be different from zero. This result indicates that spin 1 is conserved in Maxwell's equations. In order to have $E_0 \neq 0$ a spin changing interaction must be added to the 4-current, as the E_0 component belongs to the spin 0 part of the four vector E_μ which belongs to the $D^{\left(\frac{1}{2},\frac{1}{2}\right)}$ representation of the Lorentz group with an angular momentum basis of spin 0 and spin 1.

5.5. Wave equations

Let us multiply Eq. (62) by $\left(\Gamma_0^{(4)}\partial_0 - \mathbf{\Gamma}^{(4)} \cdot \nabla\right)$,

$$\Gamma_0^{(4)}\partial_\mu\partial^\mu\Phi^{(4)} = \left(\Gamma_0^{(4)}\partial_0 - \mathbf{\Gamma}^{(4)} \cdot \nabla\right)\left(\Gamma_0^{(4)}\partial_0 + \mathbf{\Gamma}^{(4)} \cdot \nabla\right)\Phi^{(4)} = -\alpha\left(\Gamma_0^{(4)}\partial_0 - \mathbf{\Gamma}^{(4)} \cdot \nabla\right)J, \tag{79}$$

from which we obtain,

$$\Box E_0 = \alpha\left(\partial_0 j_0 + \nabla \cdot \mathbf{j}\right), \tag{80}$$

$$\Box B_0 = 0, \tag{81}$$

$$\Box \mathbf{E} = \alpha\left(\partial_0 \mathbf{j} + \nabla j_0\right), \tag{82}$$

$$\Box \mathbf{B} = -\alpha \nabla \times \mathbf{j}, \tag{83}$$

where $\Box = \partial_\mu \partial^\mu$.

Current conservation requires,

$$\partial_0 j_0 + \nabla \cdot \mathbf{j} = \mathbf{0}, \tag{84}$$

therefore when the current is conserved

$$\Box E_0 = 0, \tag{85}$$

and only if the current is not conserved the possibility of $\Box E_0 \neq 0$ will exist. As indicated before, a spin changing interaction can induce the E_0 component.

5.6. Scalar electrodynamic waves

There are claims that scalar electrodynamic waves have been detected [11]. As we have shown, within the framework of electrodynamics with conserved currents the scalar waves cannot exist. There exists a misconception that longitudinal waves are the scalar waves. Longitudinal waves can exist as solutions of Maxwell's equations when the perpendicularity condition $\nabla \cdot \mathbf{E} = \nabla \cdot \mathbf{B} = \mathbf{0}$ is violated, for instance when $\nabla \cdot \mathbf{E} = \alpha j_0$, i.e. in the presence of a charge distribution. Therefore in Ref. [11] the longitudinal waves were actually vector waves. Other claims of scalar electromagnetic waves come from K. Meyl [12], which are based on a theory different from Maxwell's. A modification of Maxwell's equation to allow for current non-conservation was proposed by Hively and Giacos [13] with the aim to experimentaly detect scalar electromagnetic waves.

5.7. Summary and conclusions

A simplified formalism of first quantized massless fields of any spin is presented. Angular momentum basis for particles of zero mass and finite spin s of the $D^{(s-1/2,1/2)}$ representation of the Lorentz group is used to describe the wavefunctions. The advantage of the formalism is that by equating to zero the $s-1$ components of the wavefunctions, the $2s-1$ subsidiary conditions (needed to eliminate the non-forward and non-backward helicities) are automatically satisfied.

Probability currents and Lagrangians are derived allowing a first quantized formalism. A simple procedure is derived of connecting the wavefunctions with potentials and gauge conditions.

The spin 1 case is of particular interest and is described with the $D^{(1/2,1/2)}$ vector representation of the well known self-dual representation of the Maxwell's equations Eq. (60). The wave function is the 4-vector

$$\Phi^{(4)} = \begin{pmatrix} 0 \\ \Psi_x \\ \Psi_y \\ \Psi_z \end{pmatrix} = \begin{pmatrix} 0 \\ E_x + iB_x \\ E_y + iB_y \\ E_z + iB_z \end{pmatrix}. \tag{86}$$

The zero component remains unchanged under Lorentz transformations [7, 1]. It is well known that the vector $D^{(1/2,1/2)}$ representation has an angular momentum basis of spin 1 plus spin 0. The meaning of it is that the wave function Eq. (86)contains only the spin 1 contribution. Thus Maxwell's equations contain an intrinsic spin 1. This representation allows to generalize Maxwell's equations by adding the E_0 and B_0 timelike components to create the electric and magnetic four-vectors,

$$\Phi^{(4)} = \begin{pmatrix} E_0 + iB_0 \\ E_x + iB_x \\ E_y + iB_y \\ E_z + iB_z \end{pmatrix}. \tag{87}$$

We found that as long as the electromagnetic current is conserved, E_0 and B_0 have to be zero. Thus scalar electromagnetic waves can be created only if charged conservation is violated, or if a spin changing interaction is added to the conserved current.

References

[1] A. Gersten and A. Moalem, Journal of Physics: Conference Series **330** (2011), 012010.
[2] A. Gersten and A. Moalem, Journal of Physics: Conference Series **437** (2013), 012019.
[3] A.Gersten, *Foundations of Physics Letters*, **13**, 185-192, (2000)
[4] P.A.M. Dirac, *Proc. Roy. Soc.* **A155**, 447-59 (1936).
[5] H. Bacry, *Nuovo Cimento*, **A32**, 448-60 (1976).
[6] E.P. Wigner, *Ann. Math.*, **40**, 149 (1939).
[7] J.S. Lomont, *Phys. Rev.* **111**, 1710-1716 (1958)
[8] H. Minkowski, "Die Grundgleichungen fiir die elektromagnetischen Vorgiinge in bewegten Körpern", *Nachrichten von der Gesellschaft der Wissenschaften zu Gottingen, Mathematisch-Physikalische Klasse*, pp. 53-111, (1908). Presented in the session of December 21, 1907. Published in 1908.
English translation "The Fundamental Equations for Electromagnetic Processes in Moving Bodies" in M. Saha, *The Principle of Relativity*, Calcutta: University Press, pp. 1-69, (1920), and by Wikisource: Translation:The Fundamental Equations for Electromagnetic Processes in Moving Bodies http://en.wikisource.org/?curid=674267
[9] J.D. Jackson, *Classical Electrodynamics, Third Edition*, John Wiley & sons, New York 1999, p.781
[10] V.V. Dvoeglazov, *J. Phys. A: Math. Gen.* **33** , pp 5011-5016, (2000).
[11] C. Monstein and J.P. Wesley, Observation of scalar longitudinal electrodynamic waves, *Europhys. Lett.* **59** (4), pp. 514-520 (2002)
[12] K. Meyl, *Self-Consistent Electrodynamics, The unified theory is evolving, if the discovered potential vortex replaces the vector potential in the dielectric*, ISBN 978-3-940 03-15-6, (2010).
[13] Hively, L.M. and Giakos, G.C. (2012), 'Toward a more complete electrodynamic theory', *Int. J. Signal and Imaging Systems Engineering*, Vol. 5, No. 1, pp.3–10.

Energy accumulation in relativistic sub-Bohr orbitals, Franklin's relativistic rotation of quarks and gravitational field bounceback as processes relevant to explosion of supernovae

N. Ben-Amots

P.O.Box 3193, Haifa 3103201, Israel

E-mail: benamots@alumni.technion.ac.il

Abstract. We discuss exploding stars exceeding 8 $M_{\rm Sun}$. Supernova processes start with implosion before the explosion. The processes causing implosion are relatively well understood. Not so the processes that reverse the implosion into explosion. We propose and discuss three possible processes that can reverse implosion of a star into an explosion, causing supernova: a) energy accumulation in sub-Bohr electron orbitals populated by an electron moving at high relativistic velocities [1], b) Franklin's relativistic rotation of quarks [2], [3] and c) Einstein's gravitational field bounceback [4]. The relativistic sub-Bohr orbitals are derived by using an analogy to exponential gravitation (recent refs [1], [4], [5], [6], [7] and [8]).

1. Supernova

Supernova type II is a kind of exploding star, in which the whole star first collapses and then bounces back into an explosion in a short time. A supernova can outshine its entire galaxy during the first days after the explosion. Matter is ejected at velocities up to thousands of km/sec [9], [10], creating a propagating cloud of supernova remnants, and usually also a neutron star remnant is inferred. The remnants are not investigated here. One of the possibilities that initiate a supernova explosion is core collapse when the core exceeds Chandrasekhar mass of 1.4 $M_{\rm Sun}$. A neutron star may be thus created.

Thousands of observations of supernovae led to a classification of many types of supernovae. See detailed reviews [11] and [12] Chapter 8. We are dealing here with type II supernovae caused by explosion of stars exceeding 8 $M_{\rm Sun}$.

In a star the gravitational attraction, which tends to cause the collapse of the star, is balanced by the pressure caused by exothermic nuclear reactions, in this case fusion. When the exothermic nuclear fuel in the core of the star is exhausted, the star core collapses causing collapse of the outer shells of the star as well.

Main heat processes in stars usually considered in simulations of supernovae include fusion of:

Journal of Physics: Conference Series **615** (2015) 012012 doi:10.1088/1742-6596/615/1/012012

(i) 4^1H, which join to form ^4He (by what is called CNO process), while releasing energy that works against gravitational implosion,
(ii) 3^4He, which join to ^{12}C while releasing energy that works against gravitational implosion.

Other exothermic reactions produce nuclei up to ^{56}Fe if the star is sufficiently massive and if processes that initiate earlier implosion are avoided, namely

(i) Electron-positron pair production,
(ii) Electron capture (relevant to nuclei like ^{16}O, ^{32}Si or few isotopes of Mg or Fe [12], [13], [14]).

If one of these processes occurs before ^{56}Fe production, a supernova can occur earlier. The conditions for these two processes to occur and trigger supernova are known ([13], [14]; [12] for review).

If the evolution of the star avoids these two processes then at temperature approaching $10^{10}K^o$ corresponding to 0.85 MeV, ^{56}Fe disintegrates to $13^4He + 4$ neutrons, which is an endothermic process that contributes to collapse and implosion because it *demands* 2.22 MeV per nucleon.

At temperature approaching $2{\times}10^{10}K^o$ corresponding to 1.7 MeV, ^4He disintegrates endothermically to 4 nucleons, enhancing implosion because this demands 8 MeV per nucleon.

Computer simulations under many different assumptions (star rotation, neutrino pressure, different state equations, magnetic field, different nuclear reactions, electron capture, pair production, general relativity effects, convection) show implosion, followed by weak bounceback, but not bounce back to an explosion so violent that it blows up almost the entire star, causing the supernova [11], [12], [15], [16]. Podsiadlowski [11] wrote: "But... this energy [of the outward propagating shock] is quickly consumed; a prompt shock is always found to stall and is unable to drive an explosion." and "It has remained one of the most enduring unsolved problems in supernova physics, how a fraction of the implosion energy can be deposited just below the accretion shock [during implosion] and be allowed to accumulate till enough energy is available to drive an explosion."

Shaviv [12] notes that "once the collapse starts, the basic problem is how the infall of the outer layers onto the core is transformed into an explosion which ejects the envelope at speeds of thousands of kilometer per second. This problem is very complex, probably involves not yet known physics, and has not yet been completely solved."

Woosley and Heger [16] concluded that "to this day, a first-principle model of how the explosion is energized is lacking."

A "List of unsolved problems in physics" [17] poses the problem: What is the exact mechanism by which an implosion of a dying star becomes an explosion?

Figure 1 illustrates the problem schematically.

Without sufficient storage of energy for explosion, the result is implosion to what is considered as black hole, but we observe supernova explosions!

Following are three proposals for energy accumulation and release. One of them includes also deployment of significant additional energy.

2. Exponential gravitation and exponential electric charge: Sub-Bohr orbitals

Our first proposal is based on variable electric charge [1], analogous to variable mass gravitation [1], [4], [5], [6], [7] and [8], which we summarize in short below.

First about variable mass gravitation.

The force between a small and a large mass m and M interacting gravitationally is

$$F = -m(r)MG/r^2 \tag{1}$$

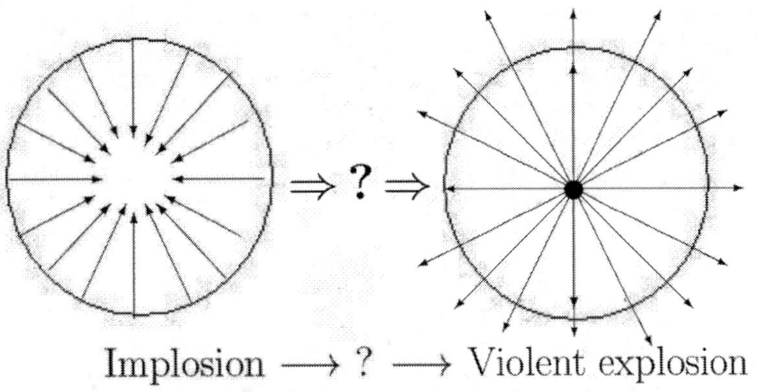

Figure 1. Schematic chain of events in the creation of supernova. The intermediate stage inverting the implosion to explosion is the subject of this paper

We let m to vary with r (the distance between point masses m and M). The variation of r is quasistatic. The force is the negative of the derivative of the energy $E(r)$:

$$F = -d[E(r)]/dr = -d[m(r)c^2]/dr = -c^2 d[m(r)]/dr \tag{2}$$

Equating (1) to (2):

$$m(r)MG/r^2 = c^2 d[m(r)]/dr \tag{3}$$

Solving, we obtain variable exponential mass

$$m(r) = m_\infty \exp(-MG/c^2 r) \tag{4}$$

for masses $m(r)$ and M *with zero velocity*. m_∞ is the value of the mass $m(r)$ at a distance (from M) $r \longrightarrow \infty$. Others have also arrived at the same solution (4) (Milne [18], [19], [20], [21] and [22], Majerník [23], Yilmaz [24], [25], Kiesslinger [26], Marmet [27], and more authors [28], [29], [30], [31], [32], [33], [34] and [35], and recently Majerník [5], [6] and Walker [7], [8]). This variable mass theory lays beyond the Einstein's Equations.

Because of analogy between the law of gravitational attraction force and Coulomb forces, we can consider a variable electric charge analogous to the variable rest mass of equation (4) (We deal with only one electron orbiting a proton):

$$q(r) = q_\infty \exp(-QY/r) \tag{5}$$

where Q is the electric charge of the central mass and Y is a constant (we skip here its explanation). (In contrast, usually it is assumed that the elementary electric charge is invariable).

Ben-Amots [1] used variable $q(r)$ of equation (5) in the Bohr atom model and found an additional *set* of sub-Bohr circular orbitals of the electron orbiting around a proton with relativistic velocity and with radiuses orders of magnitude smaller than the Bohr radius. The sub-Bohr orbitals found by Ben-Amots have positive orbital energy, the first orbital energy being 51 MeV.

Journal of Physics: Conference Series **615** (2015) 012012 doi:10.1088/1742-6596/615/1/012012

By means of a different theory Milne ([19], [20], [21] and [22]) found just one additional sub-Bohr orbital of the electron (and not a set of sub-Bohr orbitals) and calculated its energy as 50 MeV. Later Corben (see Schild[36]) independently arrived at the same sub-Bohr orbital as Milne's sub-Bohr orbital, with the same orbital energy. Except for Ben-Amots ([1] and here), variable electric charge was not used for the calculation of sub-Bohr orbitals, and in fact exponential electric charge was never used for elementary charge.

The radiuses of the sub-Bohr set of electron orbitals $r_{sub-Bohr}$ are much smaller than the radiuses of the Bohr electron orbitals r_{Bohr}. They are of the order of 10^{-13}cm, but larger than the accepted proton radius. Our derivation (Ben-Amots [1]) leads to a second order algebraic equation for r that gives approximations to both solutions:

(i) The Bohr set of circular electron orbitals, and

(ii) A *set* of *sub-Bohr* orbitals, which have radius of about 2.8×10^{-13}cm, which is slightly more than twice of the accepted proton radius of about 1.2×10^{-13}cm.

The two solutions of the second order algebraic equations are:

$$[1/2 + (1/2)\sqrt{1-(4q_{e\infty}^2\pi/nch)^2}](nh/2\pi q_{e\infty})^2/m_{e\infty} \approx r_{\rm Bohr} \tag{6}$$

and

$$[1/2 - (1/2)\sqrt{1-(4q_{e\infty}^2\pi/nch)^2}](nh/2\pi q_{e\infty})^2/m_{e\infty} = r_{\rm sub-Bohr} \tag{7}$$

where n = 1,2,3 . . .

Notice that (6) and (7) of the same n are two solutions of the same second order algebraic equation, the difference being the plus or minus sign in the two solutions.

In Equation (6), noting that $(4q_{e\infty}^2\pi/nch)^2 <<1$ we get for r_{Bohr} the value

$$r_{\rm Bohr} = (nh/2\pi q_{e\infty})^2/m_{e\infty}, \quad n = 1, 2, 3..., \tag{8}$$

which is the well known Bohr set of radiuses.

(In Equation (7) we cannot neglect $(4q_{e\infty}^2\pi/nch)^2$: Such neglect leads to absence of sub-Bohr orbital(s), characteristic to the assumption of constant rest electric charge).

From Equation (7) by using the expansion $\sqrt{1-x^2} = 1 - x^2/2 - ...$ for small x we get in *first* approximation of sub-Bohr radius:
$r_{\rm sub-Bohr} \approx (2q_{e\infty}^2\pi/nch)^2(nh/2\pi q_{e\infty})^2/m_{e\infty} = q_{e\infty}^2/m_{e\infty}c^2$ equal for all n (found by Milne [20], [21], [22] by other means).

By *second* order approximation using the expansion $\sqrt{1-x^2} = 1 - x^2/2 - x^4/8 - ...$ for small x we get a *set* of very close sub-Bohr radiuses for different n=1,2,3 . . .

$$r_{\rm sub-Bohr} = q_{e\infty}^2/m_{e\infty}c^2 + [2\pi q_{e\infty}^3/(nc^2h)]^2/m_{e\infty} \tag{9}$$

or,

$$r_{\rm sub-Bohr} = [(q_{e\infty}/c)^2/m_{e\infty}][1 + 1/(137.036n)^2] \tag{10}$$

So, for larger n we get *smaller* $r_{\rm sub-Bohr}$.

The electron in our sub-Bohr orbitals has *positive* energy, which for small quantum number n is about $n\times51.5$ MeV. See Figure 2. This high positive energy is a consequence of the centrifugal forces that in combination with the exponential forces create a narrow but high barrier peaked at about 2.8×10^{-13}cm, which prevents electron capture by the proton. Because of the high positive energy, a sub-Bohr orbital is unstable at pressures lower than in supernova core during implosion and may be populated by an electron only under these enormous pressures.

Journal of Physics: Conference Series **615** (2015) 012012 doi:10.1088/1742-6596/615/1/012012

For the corresponding sub-Bohr orbitals of hydrogen our theory has the lowest orbital energy of 51 MeV implying relativistic electron velocity of approximately $v = c - 8\,\text{km/sec}$ or

$$v = c\cos(1/137.036) \tag{11}$$

and in general orbitals of energy

$$E_n = [51{+}51.5(n{-}1)]\ \text{MeV}, \qquad n{=}\ 1, 2, 3 \ldots n_{max} \tag{12}$$

after subtracting the rest mass of the electron 0.5 MeV. The radiuses approach a limit close to $2.8{\times}10^{-13}$cm.

In other words this is a whole *set* of orbitals with energies given by (12)

$$E_{\text{n}}[\text{MeV}] = 51\ ;\ 102.5\ ;\ 154\ ;\ \ldots E_{\max},\quad n = 1, 2, 3\ \ldots\ n_{\max} \tag{13}$$

(while Milne [20], [21], [22] obtained only *one* 50 MeV orbital).

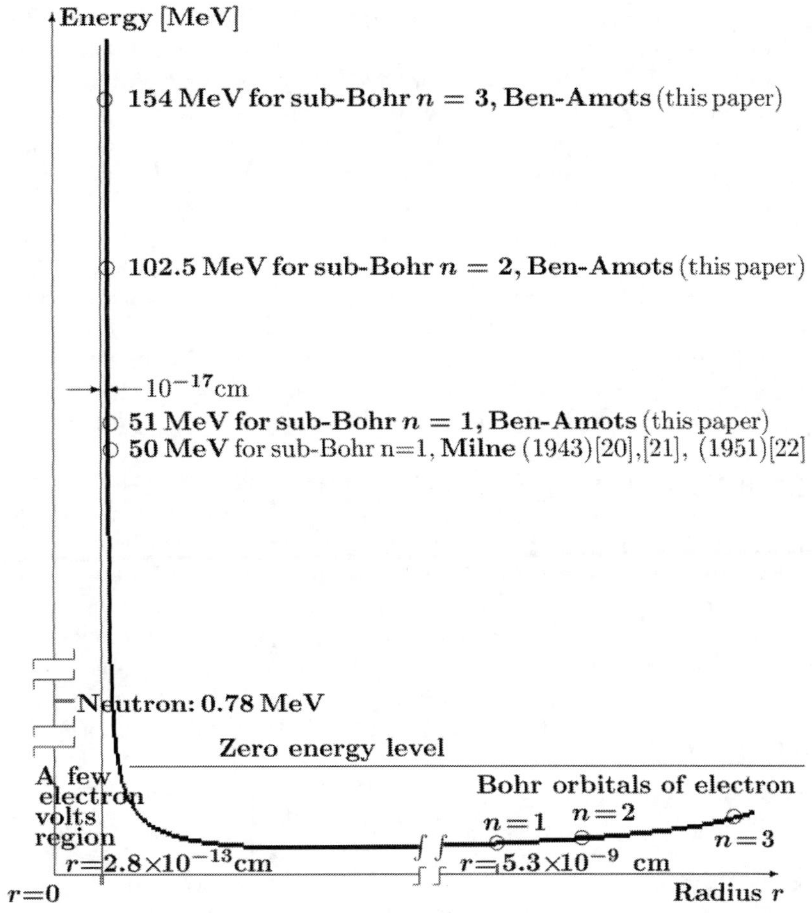

Figure 2. Energy dependence on the distance r between a proton and an electron orbiting the proton in sub-Bohr and Bohr orbits with quantum number n=1, 2, 3 (not to scale)

Journal of Physics: Conference Series **615** (2015) 012012 doi:10.1088/1742-6596/615/1/012012

3. Sub-Bohr orbitals as accumulator of energy in supernova

When the core of the star collapses its temperature increases. When the temperature approaches 6×10^{11}K° (corresponding to 51 MeV) electrons will start to populate the first sub-Bohr orbital around protons in the core.

An electron that populates the first sub-Bohr orbital

(i) takes an energy of 51.5 MeV from the available thermal energy, and

(ii) does not contribute anymore to the total number of free particles in the core of the star.

Both 1 and 2 enhance implosion. But populated sub-Bohr orbitals *accumulate* significant energy. As the pressure increases further with collapse, the temperature rises and the same electron can jump to inner, higher, more energetic orbitals, accumulating additional 51.5 MeV for each ascending step.

We think that in reality there is a limit to reachable sub-Bohr orbital, the highest having an energy $E_{\max}$, beyond which no higher-energy sub-Bohr orbital can be populated, mainly because the narrow barrier separating the sub-Bohr energy branch from the neutron energy (Figure 2) becomes easily penetrable by the electron at high E_n (sub-Bohr). Thus, if an electron in the $n_{\max}$ orbital gains more energy because of increased pressure, its probability to populate a higher orbital becomes small. So, when this electron goes inside to be captured by the proton to form a neutron, it suddenly releases accumulated energy of $[50.22+(n_{\max}-1)\times51.5]$ MeV. This is actually a new type of electron capture (from a sub-Bohr orbital).

In the electron capture by the proton from a sub-Bohr orbital a neutrino is emitted too.

To get an idea of the released energy, for example if $n_{\max}\geq19$ then the maximum accumulated energy is larger than the rest energy of the proton, i.e. $M_P c^2$!

This can be part of the bounceback energy we are searching for to solve the enigma of the energy needed for the supernova explosion. As the star collapses the pressure in the core increases, so the energy of the core increases at the expense of gravitational energy. Then, there is a sudden release of energy when the sub-Bohr electrons in the core of the star are captured by the protons to form neutrons, or separate from the protons, releasing the sub-Bohr orbital huge accumulated energy, which stops the implosion and causes explosion of the star.

Note that models based on relativity alone, in which constant electric charge is assumed, miss the set of sub-Bohr electron orbitals.

Ordinary electron capture is a weak interaction reaction that happens slowly at ordinary pressures, but faster with increased pressure. During implosion of the core the process of populating the sub-Bohr orbitals may or may not occur faster than ordinary electron capture, depending on how fast the pressure rises during the implosion of the core. The conditions in stars under pressure at which ordinary electron capture occurs are known [13], [14]. Under certain conditions during the implosion ordinary electron capture may occur before the electron populates a sub-Bohr orbital, while under other conditions populating sub-Bohr orbitals happens before ordinary electron capture rate increases, depending on the rate of the raise of pressure during implosion of the star. Apparently this determines the final result of an implosion of a star, that is the star's fate - supernova explosion or no external explosion, (leading to what we think could be yet unknown type of star of high density and mass above $5\,M_{Sun}$).

In section 1 above we presented the yet unsolved problem, of where and how the kinetic energy of implosion of the star is accumulated and reused to be transformed into kinetic energy of bounceback explosion.

We showed a first possible way to accumulate energy sufficient to create a significantly strong bounceback, but not yet sufficient for a full supernova explosion.

The sub-Bohr electrons, while being energy accumulators, do not *supply* more energy than they accumulate. Alone, they enable a significant bounceback, but a limited one, with no

Journal of Physics: Conference Series **615** (2015) 012012 doi:10.1088/1742-6596/615/1/012012

larger kinetic energy than the gravitational energy given by the infalling matter to the sub-Bohr electrons.

What supplies the additional energy that results in so strong explosion that increases the velocity of the ejected matter of the entire or almost the entire star up to thousands km/sec?

In the next section we suggest another bounceback process that may be involved too and is capable of producing the additional necessary energy.

4. Quenching the rotation of quarks as source of energy in supernova

A possible mechanism that can explain the source of energy during collapse of the supernova involves quarks. The high pressure in the collapsed core of the star produced by sudden stopping of the fast collapse may cause sudden close contact between the spinning quarks of neighbor nucleons. The friction between these close contact quarks causes fast quenching of their rotation (we presume that after the quarks of neighbor nucleons touch each other, the circumferential layers of the quarks interpenetrate, because rolling of all touching neighboring quarks is not possible where four or more quarks are involved) (Ben-Amots [3]). (Rolling here means revolving of touching spheres without slipping at the point of the touching surfaces).

The high speed rotation of the quark means highest circumferential velocity and highest density at the equator of the quark [3]). The quark density at its equator where velocity is at maximum is about 4.3×10^{15} gram/cm^3. As with known dense matter the velocity of light in it should be significantly smaller than c. Yet the relative velocity between two quarks in the interpenetrating layers is about $0.99999999999984\,c$ when two quarks are revolving with parallel angular momentum vectors (calculated according to Ben-Amots [3]).

The penetration causes intense Cherenkov radiation in this case, which, at stellar scale, becomes important. The energy needed for Cherenkov radiation within the star is taken from the kinetic energy of the rotation of the quarks. The rotation energy of the quark spinning with Franklin's relativistic rotation [2] constitutes more than 99% of the quark mass (Ben-Amots [3]). High energy Cherenkov photons are created. Quenching quarks rotation between compressed neutrons *supplies* energy.

The kinetic energy of a mass m_0 moving at a tenth of the speed of light is

$$E_k = m_0c^2(1/\sqrt{1-0.1^2}-1) \approx 0.005\,m_0c^2 \tag{14}$$

In the extreme case of a supernova model where the entire mass of the star is ejected at a tenth of the speed of light c, the kinetic energy is 0.5% of M_*c^2. The necessary mass of quarks whose rotation should be fully quenched for supplying this maximal kinetic energy is only about 0.5% of the mass of the star. Additional energy is necessary for thermal heating and mostly for neutrino ejection. In all, quenching of the rotation energy of the quarks of a few percents of the mass of the star is necessary for all forms of energy of supernova explosion.

The quenching of the quark rotation causing associated photon pressure may be the much looked after source of additional energy needed in models of supernovae. Together with the energy accumulated in the sub-Bohr orbitals it may stop the implosion of a star and reverse it to an explosion with energy beyond the gravitational potential energy gained during the implosion.

5. Gravitational field as accumulator of energy in supernova

Einstein and Infeld in their book "The evolution of physics" [37] argued: "The mass of the matter, and mass of the gravitational field, can be indistinguishable" and later Einstein [38] wrote: "Matter cannot be concentrated arbitrarily."

We deduce that gravitational field also cannot be concentrated infinitely. Ben-Amots [4] went into some details:

> The gravitational field of a spherically symmetric body has mass by itself and is attracted inward towards the central mass. The central attraction creates an inward flux of mass of the gravitational field. Yet this cannot be the whole picture. The [pressure of] gravitational field fall should be almost balanced by radiation outward by the central mass.

The residue of these two (the inward flux of "falling field" and the radiation outward by the central mass) is the origin of the gravitational field ([4] and section 2 above).

In the event of implosion of a celestial spherical body of large mass (above $8\,M_{Sun}$) in the earlier stage of a supernova, the inner core is compressed, causing an increase of the gravitational radiation outward, and with it an increase of the repelling pressure of this radiation, in accord with the above mentioned Einstein's claim [38] that "matter cannot be concentrated arbitrarily." The increase of the repelling pressure of the radiation together with the pressure of the neutrinos, stop the implosion of upper layers and contributes to an eventual bounceback outward in the process of supernovae explosion.

Under the present assumptions, the mass/field increased radiation outward by the inner core during the implosion (of the outer layers) stage of a supernova, causes the core to expand outward. It stops the outer layers implosion inward, and redirects them to bounce outward. The falling outer layers collide with the expanding core and are bounced back outward. The sharply increased pressure caused by collision, raises the temperature that ignites the fusion of the helium, carbon, oxygen etc. of the falling outer layers, and causes the final explosion of the supernova.

Summing up: the implosion compresses the gravitational field beyond a maximal balanced density. The overcompressed gravitational field causes outward pressure, which stops the implosion and triggers outward explosion.

6. Remarks

(i) In some cases the implosion might be stopped without producing supernova explosion, but instead by forming a massive celestial object stabilized by continuous energy production of quenching the rotation of the quarks, which produces electromagnetic radiation pressure. This radiation opposes gravitational collapse and supplies radiation energy of the massive celestial body as suggested in Ben-Amots [3]).

(ii) We are aware of the drawbacks of the approximations in our derivation of the sub-Bohr orbitals, but still our results are similar to Milne's result [18], [19], [20], [21] and [22], derived by independent method not involving variable electric charge. We expect that in the future more precise solutions of Milne's type calculations will give his sub-Bohr orbital, and in addition our sub-Bohr orbitals of $102.5, 154\ldots$ [MeV] orbitals.

7. Summary

We suggested various processes that alone or together may stop implosion of a star and cause its explosion,

(i) First possibility: The $[51+(n-1)\times 51.5]\,\mathrm{MeV}$ set of sub-Bohr relativistic electron orbitals around a proton accumulates and then releases energy. The accumulated energy is significant at temperature greater than $6\times 10^{11}\mathrm{K}^{\circ}$ (corresponding to $51\,\mathrm{MeV}$.)

(ii) The compressed massive gravitational field during implosion is an additional accumulator of energy, which is then released causing bounceback.

(iii) Quenching of quark rotation of nucleons as a *source* of energy additional to the *accumulated* energies of 1 and 2.

Acknowledgements
The author is indebted to Dan Igner, who helped find the appropriate wording for a clearer presentation. Thanks to IARD for allowing me to present this work in the IARD2014 conference.

References

[1] Ben-Amots N 2007 *Found. Phys.* **37** 773
[2] Franklin P 1922 *Proc. Nat. Acad. Sci. USA* **8** 265
[3] Ben-Amots N 2003 *Found. Phys.* **33** 1369
[4] Ben-Amots N 2011 *J. Phys.: Conf. Series* **330** 012017
[5] Majerník V 2006 Energy conservation at the gravitational collapse *Preprint* arXiv:astro-ph/0609313
[6] Majerník V 2009 *Phys. Essays* **22** 413
[7] Walker C 2010 *Calculation of gravitational potential by the method of multiple redshifts* http://sites.google.com/site/revisingnewton
[8] Walker C 2012 *Newtonian potential and the Einstein equivalence principle*, Foundational Questions Context http://www.fqxi.org/data/essay-contest-files/Walker_shells.pdf
[9] Wikipedia (Retrieved 2014) *Supernova* http://en.wikipedia.org/wiki/Supernova
[10] Sadler E M, Stathakis R A, Boyle B J and Eckers R D 1998 *Int. Astr. Union Circular* 6901
[11] Podsiadlowski P 2012 *Supernovae and Gamma-ray bursts*, Chapter of course, Oxford University http://www-astro.physics.ox.ac.uk/~podsi/sn_podsi.pdf
[12] Shaviv G 2009 *The Life of Stars: The Controversial Inception and Emergence of the Theory of Stellar Structure* (Berlin: Springer). Chapter 8
[13] Shaviv G and Kovetz A 1970 *Ast. Space Sci.* **7** 416
[14] Rakavy G and Shaviv G 1968 *Ast. Space Sci.* **1** 429
[15] Liebendörfer M, Fisher T, Fröhlich C, Thielemann F-K and Whitehouse S 2008 *J. Phys. G: Nucl. Part. Phys.* **35** 014056
[16] Woosley S E and Heger A 2007 *Phys. Rep.* **442** 269
[17] Wikipedia (Retrieved 2013) *List of unsolved problems in physics* http://en.wikipedia.org/wiki/List_of_unsolved_problem_in_physics
[18] Milne E A 1938 *Proc. Roy. Soc. London* p 333. See pp 351-355
[19] Milne E A 1943 *Phil. Mag.* **34** 235. See pp 239-245
[20] Milne E A 1943 *Phil. Mag.* **34** 246. See pp 246-250
[21] Milne E A 1943 *Phil. Mag.* **34** 712
[22] Milne E A 1951 *Kinematic Relativity* (Oxford: Clarendon Press) pp 203–210, 214–220. (1st edn 1948)
[23] Majerník V 1985 *Ast. Space Sci.* **113** 199
[24] Yılmaz H 1972 *Nuo. Cim.* B **10** 79. See p 87
[25] Yılmaz H 1973 *Nuo. Cim. Lett.* **7** 337
[26] Kiesslinger R *Gravitation Theory on the Testbench* (Nürnberg, Germany)
[27] Marmet P 1997 *Einstein's Theory of Relativity versus Classical Mechanics* (Gloucester, Ontario, Canada: Newton Physics Books). See Chapter 12
[28] Richterek L and Majerník V 1999 *Acta Univ. Palacki. Olomuc., Fac. Rer. Nat., Physica* **38** 47
[29] Hatch R R 1999 *Galilean electrodynamics* **10(4)** 69
[30] Vankov A A 2000 Does light gravitates? (Proposal on a new test of equivalence principle) *Preprint* arXiv/0012050
[31] Vankov A A 2001 Proposal of experimental test of general relativity theory *Preprint* arXiv/0105057
[32] Vankov A A 2003 On problem of mass origin and self-energy divergence in relativistic mechanics and gravitational physics *Preprint* gr-qc/0311063
[33] Vankov A A 2004 Elimination of high-energy divergence in relativistic Lagrangean formulation of gravitating particle dynamics *Preprint* arXiv/physics/0402117
[34] Turanyanin D 2004 *On interaction of motional masses. J. Theoretics*. See addendum. http://www.journaloftheoretics.com/Links/Papers/Turan.pdf
[35] Ibison M 2006 *Class. Quant. Grav.* **23** 577 (*Preprint* arXiv/0705.0080v1)
[36] Schild A 1963 *Phys. Rev.* **131** 2762. See p 2762
[37] Einstein A and Infeld M 1938 *The Evolution of Physics* (New York: Simon and Schuster) pp 256-258, 260
[38] Einstein A 1939 *Ann. Math.* **40** 922. See p 936

Foundations for proper-time relativistic quantum theory

Tepper L. Gill [1], Trey Morris [2] and Stewart K. Kurtz [3]

[1] Department of Mathematics, Physics and E&CE, Howard University Washington DC 20059 USA, *E-mail :* tgill@howard.edu

[2] Department of E&CE, Howard University Washington DC 20059 USA, *E-mail :* Morris.Trey.J@gmail.com

[3] Department of Electrical Engineering, Penn. State University University Park PA 16802-2703 USA, *E-mail :* skk1@psu.edu

Abstract. This paper is a progress report on the foundations for the canonical proper-time approach to relativistic quantum theory. We first review the the standard square-root equation of relativistic quantum theory, followed by a review of the Dirac equation, providing new insights into the physical properties of both. We then introduce the canonical proper-time theory. For completeness, we give a brief outline of the canonical proper-time approach to electrodynamics and mechanics, and then introduce the canonical proper-time approach to relativistic quantum theory. This theory leads to three new relativistic wave equations. In each case, the canonical generator of proper-time translations is strictly positive definite, so that it represents a particle. We show that the canonical proper-time extension of the Dirac equation for Hydrogen gives results that are consistently closer to the experimental data, when compared to the Dirac equation. However, these results are not sufficient to account for either the Lamb shift or the anomalous magnetic moment.

Introduction

Following Dirac's quantization of the electromagnetic field in 1927, and his relativistic electron theory in 1928, the equations for quantum electrodynamics QED were developed by Heisenberg and Pauli in the years 1929-30. From the beginning, when researchers attempted to use the straightforward and physically intuitive time-dependent perturbation expansion to compute physical observables, a number of divergent expressions appeared. Although it was known that the same problems also existed in classical electrodynamics, Dirac had shown that, in this case, one could account for the problem of radiation reaction without directly dealing with the self-energy divergence by using both advanced and retarded fields and a particular limiting procedure. Early attempts to develop subtraction procedures for the divergent expressions were very discouraging because they depended on both the gauge and the Lorentz frame, making them appear ambiguous. These problems were solved via the fundamental work of Feynman, Schwinger, and Tomonaga. In recent times, it is generally agreed that quantum electrodynamics (QED) is an almost perfect theory, which is in excellent agreement with experiment.

The fact that QED is very successful is without doubt. However, there are still some foundational and technical issues, which require clarification and which leaves the thoughtful student with a sense of unease in taking this as the final answer. In light of the tremendous historical success of eigenvalue analysis in physics and engineering, it is not inappropriate to reinvestigate the foundations with an eye towards clearly identifying the physical and mathematical limitations to our understanding of the hydrogen spectrum as an eigenvalue problem.

In the first section of this review we take a new look at the square-root operator and show that it has an analytic representation as a nonlocal composite of three singularities. The particle component has two negative parts and one (hard core) positive part, while the antiparticle component has two positive parts and one (hard core) negative part. This effect is confined within a Compton wavelength such that, at the point of singularity, they cancel each other providing a finite result. Furthermore, the operator looks like the identity outside a Compton wavelength.

In the second section, we provide an analytic diagonalization of the Dirac operator. Our approach leads to a complete split of the particle and antiparticle parts into two non-hermitian components, which are mapped into each other by the charge conjugation transformation. Thus, the full matrix-valued operator is hermitian and shows (as is explained in the text) that the spinor representation in the Dirac equation hides its time nonlocal property. We conclude that the Dirac and square-root operator do not represent the same physics, despite that fact that they are related by a unitary transformation.

In the third section, we introduce the canonical proper-time approach to electrodynamics and mechanics. This approach fixes the proper-time of the observed system as the clock of choice for all observers and explicitly shows that the question of simultaneity is actually a question of clock conventions. The change in convention produces a new symmetry group which is distinct from, but closely related to the Lorentz group, but has a Euclidean representation space. Thus, the new convention also replaces the standard form of Lorentz covariance by a new one. This advantage allows us to construct a parallel image of the conventional Maxwell theory for a charged particle, which is mathematically, but not physically, equivalent to the conventional form. The new wave equation contains a gauge independent term, which appears instantaneously along the direction of motion, but opposing any applied force and is zero otherwise. This is the near field (i.e., the field at the site of the charged particle). This shows that the origin of radiation reaction is not the action of a charge on itself but arises from inertial resistance to changes in motion. We show that the dissipative term is equivalent to an effective mass so that classical radiation has both a massless and a massive part. We also discuss solutions to a number of other problems that are solved with our new clock convention, which are either impossible or problematic within the standard framework.

In the forth section, we describe the canonical quantized proper-time theory. We obtain three possible relativistic wave equations, because of new possibilities, for the manner in which the potential energy may be introduced into the theory. Each new equation is generated by a strictly positive definite canonical Hamiltonian, so that it represents a consistent particle. We focus on the proper-time extension of the Dirac equation. A basic test of our proper-time theory is the extent that it compares to the Dirac theory in accounting for the hydrogen spectra. We show that our canonical proper-time version of the Dirac equation gives results which are consistently closer to the experimental data, when compared to the Dirac equation. The present theory has not yet accounted for the Lamb shift or the anomalous magnetic moment. However, the analysis in sections one, two and three support our contention that the electron is not a point particle. This non-point nature is only expected to be important in s-states, where there is a finite probability of the electron being at the center of the proton. This aspect of our research is still in progress and will be reported on at a later time.

1. The square-root equation

In the transition to relativistic mechanics, the equation $E^2 = c^2\mathbf{p}^2 + m^2c^4$ leads to the quantum Hamiltonian

$$H = \sqrt{c^2\mathbf{p}^2 + m^2c^4}.$$

Thus, it is quite natural to expect that the first choice for a relativistic wave equation would be:

$$i\hbar\frac{\partial\psi}{\partial t} = \left[\sqrt{c^2[\mathbf{p} - (e/c)\mathbf{A}]^2 + m^2c^4} + V\right]\psi,$$

where $\mathbf{p} = -i\hbar\nabla$. However, no one knew how to directly relate this equation to physically important problems. Furthermore, this equation is nonlocal, meaning, in the terminology of the times (1920-30), that it is represented by a power series in the momentum operator. One was led in this way to the Gordon-Klein and Dirac equations.

1.1. Background

Since the early work, many investigators have studied the square-root equation. It is not our intention to provide a detailed history or to identify the many important contributors to the study of this problem. In recent times, the works of Silenko (see [1], [2]) are well worth reading. They also provide a very good list of the important historical studies. In addition, he has made a number of interesting investigations into the transformational relationship between the square-root and Dirac equation ([3] is a good starting point). The recent paper by Simulik and Krivsky [4] offers another interesting approach to the square-root equation and its relationship to that of Dirac. Closer to our investigation of the square-root equation is the study by Kowalski and Rembieliński [5] (also known as the Salpeter equation). They have used it as an alternative of the Klein-Gordon equation.

In this section, we take a new look at the square-root equation. First, we investigate the extent that the non-commutativity of $\mathbf{p}$ and $\mathbf{A}$ affect our ability to give an unambiguous meaning to the square-root operator. We show that a unique analytic representation is well defined for suitable time-independent $\mathbf{A}$ provided we can solve a corresponding equation of the Schrödinger type. We then investigate a few simple cases of solvable models in order to get a feeling for the physical interpretation of this operator.

To begin, we start with the equation:

$$S[\psi] = H_s\psi = \left\{\beta\sqrt{c^2\left(\mathbf{p} - \tfrac{e}{c}\mathbf{A}\right)^2 - e\hbar c\Sigma\cdot\mathbf{B} + m^2c^4}\right\}\psi. \tag{1.1}$$

Where β and Σ are the Dirac matrices

$$\beta = \begin{bmatrix} \mathbf{I} & \mathbf{0} \\ \mathbf{0} & -\mathbf{I} \end{bmatrix}, \quad \Sigma = \begin{bmatrix} \boldsymbol{\sigma} & \mathbf{0} \\ \mathbf{0} & \boldsymbol{\sigma} \end{bmatrix};$$

$\mathbf{I}$ and $\boldsymbol{\sigma}$ are the identity and Pauli matrices respectively. Under physically reasonable mathematical conditions, the following operator is a well defined self-adjoint generator of a strongly continuous unitary group:

$$H_s^2 = c^2[\mathbf{p} - (e/c)\mathbf{A}]^2 - e\hbar c\Sigma\cdot\mathbf{B} + m^2c^4.$$

From the basic theory of fractional powers of closed linear operators, it can be shown that

$$\sqrt{H_s^2} = \left(\sqrt{H_s^2}\right)^{-1}H_s^2 = H_s^2\left(\sqrt{H_s^2}\right)^{-1}. \tag{1.2}$$

In order to construct an analytic representation for equation (1.1), we assume that $\mathbf{B}$ is constant. The general case can be found in [6]. Let $\mathbf{G} = -c^2\left(\mathbf{p} - \frac{e}{c}\mathbf{A}\right)^2$ and $\omega^2 = m^2c^4 - e\hbar c\Sigma \cdot \mathbf{B}$, so that ω is also constant. Using this notation, we can write (1.1) as

$$S[\psi] = \left\{\beta\sqrt{-\mathbf{G} + \omega^2}\right\}\psi.$$

Using the analytic theory of fractional powers of closed linear operators and equation (1.2), we can represent $S[\psi]$ as

$$S[\psi] = \frac{\beta}{\pi}\int_0^\infty \left[(\lambda + \omega^2) - \mathbf{G}\right]^{-1}(-\mathbf{G} + \omega^2)\frac{d\lambda}{\sqrt{\lambda}}[\psi], \tag{1.3}$$

where $\left[(\lambda + \omega^2) - \mathbf{G}\right]^{-1}$ is the resolvent associated with the operator $-\mathbf{G} + \omega^2$. The resolvent can be computed directly if we can find the fundamental solution to the equation:

$$\partial Q(\mathbf{x}, \mathbf{y}; t)/\partial t + (\mathbf{G} - \omega^2)Q(\mathbf{x}, \mathbf{y}; t) = \delta(\mathbf{x} - \mathbf{y}).$$

Schulman [7] has shown that the solution to the above equation is

$$Q = \int_{\mathbf{x}(0)}^{\mathbf{x}(t)} \mathcal{D}[\mathbf{x}(s)] \exp\left\{\int_0^t V[\mathbf{x}(s)]ds + \tfrac{ie}{\hbar c}\int_{\mathbf{y}}^{\mathbf{x}} \mathbf{A}[\mathbf{x}(s)] \cdot d\mathbf{x}(s)\right\}$$

where $V = c^2\omega^2/i\hbar$ and

$$\int_{\mathbf{x}(0)=\mathbf{y}}^{\mathbf{x}(t)=\mathbf{x}} \mathcal{D}[\mathbf{x}(s)] = \int_{\mathbf{x}(0)=\mathbf{y}}^{\mathbf{x}(t)=\mathbf{x}} \mathcal{D}[\mathbf{x}(s)] \exp\left\{-\tfrac{1}{4}\int_0^t \left|\frac{d\mathbf{x}(s)}{ds}\right|^2 ds\right\}$$
$$= \lim_{N\to\infty}\left[\tfrac{1}{4\pi\varepsilon_N}\right]^{nN/2} \textstyle\int_{\mathbb{R}^n} \displaystyle\prod_{k=1}^{N} dx_j \exp\left\{-\sum_{j=1}^{N}\left[\tfrac{1}{4\varepsilon_N}(x_j - x_{j-1})^2\right]\right\},$$

and $e_N = t/N$. A rigorous justification for the path integral can be found in Gill and Zachary [8]. We assume that $\int_{\mathbf{y}}^{\mathbf{x}} \mathbf{A}[\mathbf{x}(s)] \cdot d\mathbf{x}(s) = \bar{\mathbf{A}} \cdot (\mathbf{x} - \mathbf{y})$, where $\bar{\mathbf{A}}$ is the mean value of $\mathbf{A}$. Using this, we have:

$$\left[(\lambda + \omega^2) - \mathbf{G}\right]^{-1} f(\mathbf{x}) = \int_0^\infty e^{-\lambda t}\left[\int_{\mathbb{R}^3} Q(\mathbf{x}, t; \mathbf{y}, 0)f(\mathbf{y})d\mathbf{y}\right]dt$$

and

$$\left[(\lambda + \omega^2) - \mathbf{G}\right]^{-1} f(\mathbf{x})$$
$$= \int_{\mathbb{R}^3} e^{\left\{\frac{ie}{\hbar c}\bar{\mathbf{A}}\cdot(\mathbf{x}-\mathbf{y})\right\}}\left\{\int_0^\infty \exp\left[-\tfrac{(\mathbf{x}-\mathbf{y})^2}{4t} - \tfrac{\omega^2 t}{\hbar^2} - \lambda t\right]\tfrac{dt}{(4\pi t)^{3/2}}\right\} f(\mathbf{y})d\mathbf{y}$$

Using a table of Laplace transforms, the inner integral can be computed to get

$$\int_0^\infty \exp\left[-\frac{(\mathbf{x}-\mathbf{y})^2}{4t} - \frac{\omega^2 t}{\hbar^2} - \lambda t\right]\frac{dt}{(4\pi t)^{3/2}}$$
$$= \frac{1}{4\pi}\frac{\exp\left[-\sqrt{(\lambda + \mu^2)}\,\|\mathbf{x} - \mathbf{y}\|\right]}{\|\mathbf{x} - \mathbf{y}\|}$$

where $\mu^2 = \omega^2/\hbar^2$. Equation (1.3) now becomes

$$S[\psi](\mathbf{x}) = \tfrac{c\beta}{4\pi^2}\int_0^\infty \left\{ \int_{\mathbb{R}^3} \mathrm{e}^{\left\{\frac{ie}{\hbar c}\bar{\mathbf{A}}\cdot(\mathbf{x}-\mathbf{y})\right\}} \mathrm{e}^{-\left[\sqrt{(\lambda+\mu^2)}\|\mathbf{x}-\mathbf{y}\|\right]} \tfrac{(-\mathbf{G}+\omega^2)}{\|\mathbf{x}-\mathbf{y}\|}\psi(\mathbf{y})d\mathbf{y} \right\} \frac{d\lambda}{\sqrt{\lambda}}.$$

Once again, we interchange the order of integration and perform the computations to get

$$\int_0^\infty \left\{ \frac{\exp\left[-\sqrt{(\lambda+\mu^2)}\,\|\mathbf{x}-\mathbf{y}\|\right]}{\|\mathbf{x}-\mathbf{y}\|} \right\} \frac{d\lambda}{\sqrt{\lambda}} = \frac{4\mu\Gamma(\frac{3}{2})}{\pi^{1/2}} \frac{\mathbf{K}_1\left[\mu\,\|\mathbf{x}-\mathbf{y}\|\right]}{\|\mathbf{x}-\mathbf{y}\|}.$$

where $\mathbf{K}_1[\mathbf{z}]$ is the modified Bessel function of the third kind and first order. Thus, if we set, $\mathbf{a} = \frac{e}{\hbar c}\mathbf{A}$ and $\bar{\mathbf{a}} = \frac{e}{\hbar c}\bar{\mathbf{A}}$ we get

$$\begin{aligned} S[\psi](\mathbf{x}) &= \tfrac{c\beta}{2\pi^2}\int_{\mathbf{R}^3} \mathrm{e}^{[i\bar{\mathbf{a}}\cdot(\mathbf{x}-\mathbf{y})]} \frac{\mu\mathbf{K}_1\left[\mu\,\|\mathbf{x}-\mathbf{y}\|\right]}{\|\mathbf{x}-\mathbf{y}\|}(-\mathbf{G}+\omega^2)\psi(\mathbf{y})d\mathbf{y} \\ &= \tfrac{c\beta}{2\pi^2}(-\mathbf{G}+\omega^2)\int_{\mathbf{R}^3} \mathrm{e}^{[i\bar{\mathbf{a}}\cdot(\mathbf{x}-\mathbf{y})]} \frac{\mu\mathbf{K}_1\left[\mu\,\|\mathbf{x}-\mathbf{y}\|\right]}{\|\mathbf{x}-\mathbf{y}\|}\psi(\mathbf{y})d\mathbf{y}. \end{aligned} \tag{1.4}$$

Since $\nabla\cdot\mathbf{a} = 0$, we have

$$-\mathbf{G}+\omega^2 = \hbar^2\left(-\Delta + 2i\mathbf{a}\cdot\nabla + \mathbf{a}^2 + \mu^2\right),$$

so that so that (1.4) becomes

$$S[\psi](\mathbf{x}) = \tfrac{\hbar^2 c\beta}{2\pi^2}\left(-\Delta + 2i\mathbf{a}\cdot\nabla + \mathbf{a}^2 + \mu^2\right)\int_{\mathbf{R}^3} e^{i\bar{\mathbf{a}}\cdot(\mathbf{x}-\mathbf{y})} \tfrac{\mu\mathbf{K}_1[\mu\|\mathbf{x}-\mathbf{y}\|]}{\|\mathbf{x}-\mathbf{y}\|}\psi(\mathbf{y})d\mathbf{y}.$$

The operator $\left(-\Delta + 2i\mathbf{a}\cdot\nabla + \mathbf{a}^2 + \mu^2\right)$ acts on $\mathbf{x}$, making the integral singular. However, this singular representation constructed below, equation (1.5) has many of the properties observed in experiments. As will be seen, it represents the confinement of three singularities within a Compton wavelength. (A full discussion is delayed to the end of this section.)

We omit many of the computational details, which can be found in [9], but the idea is to consider a ball $\mathbf{B}_\rho(\mathbf{x})$ of radius ρ about $\mathbf{x}$, so that $\mathbb{R}^3 = \mathbb{R}^3_\rho \cup \mathbf{B}_\rho(\mathbf{x})$, where $\mathbb{R}^3_\rho = \left(\mathbb{R}^3 \backslash \mathbf{B}_\rho(\mathbf{x})\right)$ and $\partial\mathbb{R}^3_\rho = \left(\partial\mathbb{R}^3 \backslash \partial\mathbf{B}_\rho(\mathbf{x})\right)$. We then restrict all operations to $\mathbb{R}^3_\rho$ and only let $\rho \to 0$ at the end.

1.2. Free Case

The free particle case is the simplest (but still interesting), with $\mathbf{A} = \mathbf{0}$, so that

$$S[\psi](\mathbf{x}) = -\tfrac{\mu^2\hbar^2 c\beta}{\pi^2}\int_{\mathbf{R}^3}\left[\tfrac{1}{\|\mathbf{x}-\mathbf{y}\|} - 4\pi\delta\left(\mathbf{x}-\mathbf{y}\right)\right]\left\{\tfrac{\mathbf{K}_0[\mu\|\mathbf{x}-\mathbf{y}\|]}{\|\mathbf{x}-\mathbf{y}\|} + \tfrac{2\mathbf{K}_1[\mu\|\mathbf{x}-\mathbf{y}\|]}{\mu\,\|\mathbf{x}-\mathbf{y}\|^2}\right\}\psi(\mathbf{y})d\mathbf{y}. \tag{1.5}$$

If $\mathbf{x} \neq \mathbf{y}$, the effective kernel of equation (1.5) is

$$\frac{K_0\left[\mu\,\|\mathbf{x}-\mathbf{y}\|\right]}{\|\mathbf{x}-\mathbf{y}\|^2} + \frac{2K_1\left[\mu\,\|\mathbf{x}-\mathbf{y}\|\right]}{\mu\|\mathbf{x}-\mathbf{y}\|^3}.$$

Recall that the integral of $\|\mathbf{x}-\mathbf{y}\|^{-2}$ over $\mathbb{R}^3$ is finite. In order to understand the physical interpretation of equation (1.5), it will be helpful to review some properties of the modified

Bessel functions $K_0[u]$, $u^{-\frac{1}{2}}K_{1/2}[u]$ and $u^{-1}K_1[u]$. We follow Gradshteyn and Ryzhik [10], for $0 < u \ll 1$, we have that:

$$\begin{aligned}\frac{K_1[u]}{u} &= [1+\theta_1(u)]\,u^{-2}\\ \frac{K_{1/2}[u]}{u^{1/2}} &= \sqrt{\tfrac{\pi}{2}}u^{-1}\\ K_0[u] &= [1+\theta_0(u)]\ln u^{-1},\end{aligned}$$

where θ_0, $\theta_1 \to 0$ as $u \to 0$. We note that, up to a multiplicative constant, $u^{-\frac{1}{2}}K_{1/2}[u]$ is the well-known Yukawa potential [11], conjectured in 1935 to account for the short range of the nuclear interaction. From here, we see that, near $u = 0$, the singular term $u^{-1}K_1[u]$ is twice as strong as the Yukawa potential. The singular term $K_0[u]$ is actually integrable and so does not contribute at $u = 0$. Looking at equation (1.5), we see that the singular term $-8\pi K_1(u)\delta(u)$ acts to cancel the the singular term $u^{-1}K_1[u]$ at $u = 0$, so that the total integral is well defined.

The behavior of these functions is quite different, when $u \gg 1$. In this case, we have:

$$\begin{aligned}\frac{K_1[u]}{u} &= [1+\theta_1'(u)]\,\frac{\exp\{-u\}}{u^{3/2}}\\ \frac{K_{1/2}[u]}{u^{1/2}} &= \sqrt{\tfrac{\pi}{2}}\frac{\exp\{-u\}}{u}\\ K_0[u] &= [1+\theta_0'(u)]\,\frac{\exp\{-u\}}{u^{1/2}}.\end{aligned}$$

We see that each term has a exponential cutoff. However, now $K_0[u]$ has the longest range, while $u^{-1}K_1[u]$ has the shortest range. Furthermore, inspection shows that $u^{-1}K_1[u]$ is multiplied by the reduced Compton wavelength, which further shortens its range.

It is clear that our square-root operator represents an extended object with an effective extent of about a Compton wavelength.

1.3. Constant **A** *Case*

When $\mathbf{A} \neq \mathbf{0}$ is constant, $\nabla \cdot \mathbf{A} = 0$ and $\mathbf{B} = \nabla \times \mathbf{A} = 0$, so that we get ($\bar{\mathbf{a}} = \mathbf{a}$):

$$S[\psi](\mathbf{x}) = \frac{-\mu^2\hbar^2 c\beta}{\pi^2}\int_{\mathbf{R}^3} e^{i\mathbf{a}\cdot(\mathbf{x}-\mathbf{y})}\left[\frac{1}{\|\mathbf{x}-\mathbf{y}\|} - 4\pi\delta(\mathbf{x}-\mathbf{y})\right]\left\{\frac{\mathbf{K}_0[\mu\|\mathbf{x}-\mathbf{y}\|]}{\|\mathbf{x}-\mathbf{y}\|} + \frac{2\mathbf{K}_1[\mu\|\mathbf{x}-\mathbf{y}\|]}{\mu\|\mathbf{x}-\mathbf{y}\|^2}\right\}\psi(\mathbf{y})d\mathbf{y}.$$

1.4. The Constant Field Case

If $\mathbf{B} \neq \mathbf{0}$ is constant, then $\mathbf{A}(\mathbf{z}) = \frac{1}{2}\mathbf{z} \times \mathbf{B}$. Let $\mathbf{a}(\mathbf{z}) = \frac{e}{2\hbar c}\mathbf{z} \times \mathbf{B}$ and $F = -\mathbf{a} \cdot (\mathbf{x} - \mathbf{y})$. In this case $\mathbf{A}(\mathbf{z}) \cdot \mathbf{dz} = \mathbf{0}$, so we can write the final result as:

$$\begin{aligned}\mathbf{S}[\psi] = &-\frac{\hbar^2\mu^2 c\beta}{\pi^2}\left\{\int_{\mathbf{R}^3}\left[\frac{1}{\|\mathbf{x}-\mathbf{y}\|} - \frac{4\pi\delta(\mathbf{x}-\mathbf{y})}{1+iF}\right][1+iF]\,\frac{\mathbf{K}_2[\mu\|\mathbf{x}-\mathbf{y}\|]}{\|\mathbf{x}-\mathbf{y}\|}\psi(\mathbf{y})d\mathbf{y}\right\}\\ &+\frac{\hbar^2\mu^2 c\beta}{\pi^2}\int_{\mathbf{R}^3}\mathbf{a}^2\frac{\mathbf{K}_1[\mu\|\mathbf{x}-\mathbf{y}\|]}{\|\mathbf{x}-\mathbf{y}\|}\psi(\mathbf{y})d\mathbf{y},\end{aligned} \tag{1.6}$$

where

$$\frac{\mathbf{K}_2[\mu\|\mathbf{x}-\mathbf{y}\|]}{\|\mathbf{x}-\mathbf{y}\|} = \frac{\mathbf{K}_0[\mu\|\mathbf{x}-\mathbf{y}\|]}{\|\mathbf{x}-\mathbf{y}\|} + \frac{2\mathbf{K}_1[\mu\|\mathbf{x}-\mathbf{y}\|]}{\mu\|\mathbf{x}-\mathbf{y}\|^2}. \tag{1.7}$$

From equations (1.6) and (1.7), we see that a constant magnetic field makes a real difference compared to either the $\mathbf{A} = \mathbf{0}$ or $\mathbf{A} \neq \mathbf{0}$ cases, producing two extra terms, in addition to the free particle term. The first new term is purely imaginary and singular at $\mathbf{x} = \mathbf{y}$ (like the Yukawa term). Physically, we interpret this term as representing particle absorption and emission (see Mott and Massey [12]) . The second term is real, repulsive and nonsingular. In addition, the effective mass μ is constant but matrix-valued with complex components, $\mu^2 = m^2c^2/\hbar^2 - \frac{e}{\hbar c}\Sigma\cdot\mathbf{B}$. Since

$$\Sigma = \begin{pmatrix} \sigma & \mathbf{0} \\ \mathbf{0} & \sigma \end{pmatrix};\ \sigma_1 = \begin{pmatrix} 0 & 1 \\ 1 & 0 \end{pmatrix},\ \sigma_2 = \begin{pmatrix} 0 & -i \\ i & 0 \end{pmatrix},\ \sigma_3 = \begin{pmatrix} 1 & 0 \\ 0 & -1 \end{pmatrix},$$

$$\mu^2 = \begin{bmatrix} (\frac{m^2c^2}{\hbar^2} - \frac{e}{\hbar c}B_3)\mathbf{I}_2 & \frac{ie}{\hbar c}(B_2 - iB_1)\mathbf{I}_2 \\ \frac{-ie}{\hbar c}(B_2 - iB_1)\mathbf{I}_2 & (\frac{m^2c^2}{\hbar^2} + \frac{e}{\hbar c}B_3)\mathbf{I}_2 \end{bmatrix}.$$

From known properties of Bessel functions for nonintegral ν, we can represent $\mathbf{K}_\nu[u]$ as

$$(2/\pi)\mathbf{K}_\nu[u] = \frac{\mathbf{I}_{-\nu}(u) - \mathbf{I}_\nu(u)}{\sin\pi\nu} = \frac{e^{i/2(\pi\nu)}\mathbf{J}_{-\nu}(iu) - e^{-i/2(\pi\nu)}\mathbf{J}_\nu(iu)}{\sin\pi\nu}.$$

In the limit as ν approaches an integer, the above takes the indeterminate form 0/0 and is defined via L'Hôpital's rule. However, for our purposes, we assume that ν is close to an integer and $u = u_1 + iu_2,\ u_2 \neq 0$. In this case, $\mathbf{K}_\nu[u]$ acquires some of the oscillatory behavior of $\mathbf{J}_\nu[u]$. Thus, we can interpret equation (1.6) as representing a pulsating mass (extended object of variable mass) with mean value $\hbar/c\,\|\mu\|$, where $\|\mu\| = [\mu^*\mu]^{1/2}$ and μ^* is the Hermitian conjugate of μ, with the square root being computed using elementary spectral theory. If $\mathbf{B}$ is very large, we see that the effective mass can also be large. However, the operator still looks (almost) like the identity outside a Compton wavelength.

In closing, we should say a few additional words about the interesting work of Kowalski and Rembieliński [5]. They solve the Salpeter equation ($\beta = I$) and construct a number of examples. This work represents an original contribution to our understanding of the square-root equation. They approach the problem using the method of Fourier transforms and get the correct solution for $\mathbf{x} \neq \mathbf{y}$. However, this approach misses the the $\mathbf{x} = \mathbf{y}$ term, giving the impression that their equation is not defined at that point. This minor defect can be easily fixed, by adding our delta term, which makes their solutions well-defined for all $\mathbf{x}$. More important, is to note that replacing the indentity operator by the (general) β matrix provides a generalization of their solutions for all spin-values.

1.5. Conclusions

From our analysis, we have the following conclusions concerning the square-root operator:

(i) In the simplest case, $\mathbf{A} = \mathbf{0}$, the square-root operator has a representation as a nonlocal composite of three singularities. The particle component has two negative parts and one (hard core) positive part, while the antiparticle component has two positive parts and one (hard core) negative part. This effect is confined within a Compton wavelength such that, at the point of singularity, they cancel each other providing a finite result. Furthermore, the operator looks like the identity outside a Compton wavelength. (Recall that the experimental observation of three singularities in proton and neutron scattering experiments led to the quark model.)

(ii) A constant magnetic field induces changes in both the mass and the shape of this extended object. It also increases the number of singularities. This suggests that the square-root operator represents a charge/mass density, for otherwise it could not be affected by a constant magnetic field.

Journal of Physics: Conference Series **615** (2015) 012013 doi:10.1088/1742-6596/615/1/012013

(iii) The square-root operator is not physically the same as the Dirac operator despite the fact that they are related by a unitary transformation. (We will discuss this point further in the next section.)

2. The Dirac Equation

The first successful attempt to resolve the question of how best to handle the square-root equation was made by Dirac in 1926 [13]. Dirac noticed that the Pauli matrices could be used to write $c^2\mathbf{p}^2 + m^2c^4$ as $\left[c\alpha \cdot \mathbf{p} + mc^2\beta\right]^2$. The matrix α is defined by $\alpha = (\alpha_1, \alpha_2, \alpha_3)$, where

$$\alpha_i = \begin{pmatrix} \mathbf{0} & \sigma_i \\ \sigma_i & \mathbf{0} \end{pmatrix}, \quad \sigma_1 = \begin{pmatrix} 0 & 1 \\ 1 & 0 \end{pmatrix}, \sigma_2 = \begin{pmatrix} 0 & -i \\ i & 0 \end{pmatrix}, \sigma_3 = \begin{pmatrix} \mathbf{1} & \mathbf{0} \\ \mathbf{0} & -\mathbf{1} \end{pmatrix}.$$

Thus, Dirac [13] showed that an alternative representation of the square-root equation could be taken as:

$$i\hbar\frac{\partial \Psi}{\partial t} = \left[c\alpha \cdot \mathbf{p} + mc^2\beta\right]\Psi. \tag{2.1}$$

In this case, Ψ must be viewed as a vector-valued function or spinor. To be more precise, $\Psi \in L^2\left(\mathbb{R}^3, \mathbb{C}^4\right) = L^2\left(\mathbb{R}^3\right) \otimes \mathbb{C}^4$ is a four-component column vector $\Psi = (\psi_1, \psi_2, \varphi_1, \varphi_2)^t$. In this representation, $\psi = (\psi_1, \psi_2)^t$ represents the particle (positive energy) component, and $\varphi = (\varphi_1, \varphi_2)$ represents the antiparticle (negative energy) component of the theory (for details, see Thaller [14]).

A fair understanding of the Dirac equation can only be claimed in recent times, and, as pointed out by D. Finkelstein, "Dirac introduced a Lorentz-invariant Clifford algebra into the complex algebra of observables of the electron". (See, in particular, Biedenharn [15] or deVries [16] and Hestenes [17].) Despite successes, both practical and theoretical, there still remain a number of conceptual, interpretational, and technical misunderstandings about this equation. It is generally believed that it is not possible to separate the particle and antiparticle components directly without approximations (when interactions are present). The various approximations found in the literature may have led to this belief. In addition, the historically important algebraic approaches of Foldy-Wouthuysen [18], Pauli [19], and Feynman and Gell-Mann [20] have no doubt further supported such ideas.

In this section we show that it is possible to directly separate the particle and antiparticle components of the Dirac equation without approximations, even when scalar and vector potentials of quite general character are present (see [9]). We show that the square root operator cannot be considered physically equivalent to the Dirac operator from another point of view. In addition, we offer another interpretation of the zitterbewegung and the fact that the expected value of a velocity measurement of a Dirac particle at any instant of time is $\pm c$.

2.1. Complete Separation

It turns out that a direct analytic separation is actually quite simple and provides additional insight into the particle and antiparticle components. In order to see this, let $\mathbf{A}(\mathbf{x}, t)$ and $V(\mathbf{x})$ be given vector and scalar potentials and, after adding $V(\mathbf{x})$ and making the transformation $\mathbf{p} \to \pi = \mathbf{p} - e/c\mathbf{A}$, write (2.1) in two-component form as:

$$\begin{aligned} i\hbar\frac{\partial \psi}{\partial t} &= (V + mc^2)\psi + c(\sigma \cdot \pi)\varphi \\ i\hbar\frac{\partial \varphi}{\partial t} &= (V - mc^2)\varphi + c(\sigma \cdot \pi)\psi. \end{aligned} \tag{2.2}$$

We write the second equation as:

$$\left[\frac{\partial}{\partial t}+iB_1\right]\varphi=-iD\psi,\quad B_1=\tfrac{1}{\hbar}(V-mc^2)\text{ and }D=\tfrac{1}{\hbar}c(\sigma\cdot\pi).$$

In this form, we see that from an analytical point of view equation (2.2) is a first order inhomogeneous partial differential equation. This equation can be solved via the Green's function method if we first solve

$$\left[\frac{\partial}{\partial t}+iB_1\right]u(t)=\delta(t).$$

It is easy to see that the solution to this equation is

$$u(t)=\theta(t)\exp\{-iB_1t\},\ \theta(t)=\left\{\begin{array}{l}1,\ t>0\\0,\ t<0\end{array}\right.,$$

so that

$$\varphi(t)=\int_{-\infty}^{t}c\exp\{-iB_1(t-\tau)\}\left[(\sigma\cdot\pi)/i\hbar\right]\psi(\tau)d\tau.$$

It now follows via convolution that:

$$\begin{aligned}i\hbar\frac{\partial\psi}{\partial t}&=(V+mc^2)\psi\\&+\left[c^2(\sigma\cdot\pi)/i\hbar\right]\int_{-\infty}^{t}\exp\{-iB_1(t-\tau)\}(\sigma\cdot\pi)\psi(\tau)d\tau.\end{aligned}\tag{2.3}$$

In a similar manner, we obtain the complete equation for φ:

$$\begin{aligned}\hbar\frac{\partial\varphi}{\partial t}&=(V-mc^2)\varphi\\&+\left[c^2(\sigma\cdot\pi)/i\hbar\right]\int_{-\infty}^{t}\exp\{-iB_2(t-\tau)\}(\sigma\cdot\pi)\varphi(\tau)d\tau,\end{aligned}\tag{2.4}$$

where

$$v(t)=\theta(t)\exp\{-iB_2t\},\ \theta(t)=\left\{\begin{array}{l}1,\ t>0\\0,\ t<0\end{array}\right.,$$

$B_2=\frac{1}{\hbar}(V+mc^2)$.

Thus, we have decomposed $L^2(\mathbb{R}^3,\mathbb{C}^4)$ as $L^2(\mathbb{R}^3,\mathbb{C}^4)=L^2(\mathbb{R}^3,\mathbb{C}^2)\oplus L^2(\mathbb{R}^3,\mathbb{C}^2)$. One copy of $L^2(\mathbb{R}^3,\mathbb{C}^2)$ contains the particle (positive energy) wave component, while the other copy contains the antiparticle (negative energy) wave component. Which of these copies corresponds to the components $\psi=(\psi_1,\psi_2)^t$ and which to the components $\varphi=(\varphi_1,\varphi_2)^t$ depends, to some extent, on the properties of the scalar potential V. It may have been noticed that equations (2.3) and (2.4) are non-hermitian. It is shown in [9] that they are mapped into each other by the charge conjugation operator, so that the full matrix representation is hermitian. An unsettled issue is the definition of the appropriate inner product for the two subspaces, which will account for the quantum constraint that the total probability integral is normalized. We can satisfy this requirement if we set $(\psi,\,\chi)=\psi_1\bar{\chi}_1+\psi_2\bar{\chi}_2$, $(\psi,\,\chi)_1=(A_1\psi,\,A_1\chi)$ and $(\varphi,\,\eta)_2=(A_2\varphi,\,A_2\eta)$, where $A_1\psi=cu(t)*\left[(\sigma\cdot\pi)/i\hbar\right]\psi(t)$ and $A_2\varphi=cv(t)*\left[(\sigma\cdot\pi)/i\hbar\right]\varphi(t)$. We can now define the particle and antiparticle inner products by:

$$\begin{aligned}\langle\psi\,,\chi\rangle_p&=\int_{\mathbf{R}^3}\left[(\psi,\,\chi)+(\psi,\,\chi)_1\right]d\mathbf{x}\\\langle\varphi\,,\eta\rangle_{ap}&=\int_{\mathbf{R}^3}\left[(\varphi,\,\eta)+(\varphi,\,\eta)_2\right]d\mathbf{x},\end{aligned}\tag{2.5}$$

so that the normalized probability densities satisfy:

$$\begin{aligned} \rho_\psi &= |\psi|^2 + \left| \int_{-\infty}^{t} c\exp\{-iB(t-\tau)\} \left[(\sigma\cdot\pi)/i\hbar\right] \psi(\tau) d\tau \right|^2 \\ \rho_\varphi &= |\varphi|^2 + \left| \int_{-\infty}^{t} c\exp\{-iB'(t-\tau)\} \left[(\sigma\cdot\pi)/i\hbar\right] \varphi(\tau) d\tau \right|^2 . \end{aligned} \tag{2.6}$$

2.1.1. Interpretations Writing the Dirac equation and the direct separation in two-component matrix form, we have:

$$i\hbar\frac{\partial}{\partial t}\begin{bmatrix}\psi\\ \varphi\end{bmatrix} = \begin{bmatrix}(V+mc^2) & c(\sigma\cdot\pi)\\ c(\sigma\cdot\pi) & (V-mc^2)\end{bmatrix}\begin{bmatrix}\psi\\ \varphi\end{bmatrix}$$

and

$$i\hbar\frac{\partial}{\partial t}\begin{bmatrix}\psi\\ \varphi\end{bmatrix} = \begin{bmatrix}\begin{array}{l}(V+mc^2)\\ +\left[c^2(\sigma\cdot\pi)/i\hbar\right]\left[u*(\sigma\cdot\pi)\right]\end{array} & \mathbf{0}\\ \mathbf{0} & \begin{array}{l}(V-mc^2)\\ +\left[c^2(\sigma\cdot\pi)/i\hbar\right]\left[v*(\sigma\cdot\pi)\right]\end{array}\end{bmatrix}\begin{bmatrix}\psi\\ \varphi\end{bmatrix}$$

We call the latter equation the analytic diagonalization of the Dirac equation because the wave function has not changed.

The standard approach to the diagonalization of the Dirac equation (without an external potential V) is via the Foldy-Wouthuysen representation [18]. Assuming that $\mathbf{A}$ does not depend on t, the following generalization can be found in deVries [16]:

$$i\hbar\frac{\partial}{\partial t}\begin{bmatrix}\Phi_1\\ \Phi_2\end{bmatrix} = \begin{bmatrix}\sqrt{c^2\pi^2 - ec\hbar(\Sigma\cdot\mathrm{B}) + m^2c^4} & 0\\ 0 & -\sqrt{c^2\pi^2 - ec\hbar(\Sigma\cdot\mathrm{B}) + m^2c^4}\end{bmatrix}\begin{bmatrix}\Phi_1\\ \Phi_2\end{bmatrix}$$

where

$$\Sigma = \begin{pmatrix}\boldsymbol{\sigma} & \mathbf{0}\\ \mathbf{0} & \boldsymbol{\sigma}\end{pmatrix}.$$

In this case, $\begin{bmatrix}\Phi_1 & \Phi_2\end{bmatrix}^t = U_{FW}\begin{bmatrix}\psi & \varphi\end{bmatrix}^t$ and our square-root operator $S = U_{FW}\mathbf{H}_{\mathrm{D}}U_{FW}^{-1}$.

From equation (2.2), we conclude that the coupling of the particle and antiparticle wave functions in the first-order form of the Dirac equation hides the second order nonlocal time nature of the equation. We know that the square-root operator is nonlocal in space. Thus, the implicit time nonlocality of the Dirac equation is mapped into the explicit spatial nonlocality of the square-root equation by the Foldy-Wouthuysen transformation. This is a mathematical relationship, which is not physically equivalent.

The time nonlocal behavior raises questions about the zitterbewegung. The physically reasonable interpretation of the zitterbewegung and the fact that the expected value of a velocity measurement (of a Dirac particle) at any instant in time $\pm c$ are reflections of the fact that the Dirac equation makes a spatially extended particle appear as a point in the present by forcing it to oscillate between the past and future at speed $\pm c$.

3. Classical Proper-time Theory

In this section, we briefly review the classical theory. The theory was first introduced in 2001 [21] and further discussed in [22]. However, the theory has its roots in the foundations of quantum electrodynamics as developed by Feynman and Dyson.

3.1. Background

Following the suggestions of Feynman and Dyson, our program began with the development of a mathematical theory for Feynman's time-ordered operator calculus, where time is accorded its natural role as the director of physical processes. Briefly, our theory is constructive in that operators acting at different times actually commute (in the mathematical sense). This approach allows us to develop a general perturbation theory for all theories generated by unitary evolutions. We are also able to reformulate our theory as a physically motivated sum over paths as suggested by Feynman. Our purpose was to prove the last two remaining conjectures of Dyson concerning the mathematical foundations for QED (see [23]). (A. Salam confirmed Dyson's first conjecture [24], while S. Weinberg [25] confirmed his second one.) In particular, we showed that:

(i) The renormalized perturbation series of quantum electrodynamics is at most asymptotic. (We also provided the remainder so that, in the mathematical sense, the expansion is exact.)

(ii) The ultraviolet divergence of quantum electrodynamics is caused by a violation of the time-energy uncertainty relations (at each point in time).

As a special case, our approach also provided the first rigorous mathematical foundation for the Feynman path integral formulation of quantum mechanics (see [8]).

In the Feynman world-view the universe is a three-dimensional motion picture in which more and more of the future appears as time evolves. Time is a physically defined variable with properties distinct from those of the three spacial variables. This view is inconsistent with the Minkowski world-view, in which time is an additional coordinate for space-time geometry.

With this inconsistency in mind, we began to investigate the possibility that an alternative formulation of both classical and quantum theory could exist, which encodes the Feynman world-view. We discovered the canonical proper-time approach to classical electrodynamics, in which the proper-time of the observed system is used as opposed to the proper-time of the observer.

3.2. Maxwell's equations

For the local-time version of Maxwell's equations, it is convenient to start with the standard definition of proper-time:

$$d\tau^2 = dt^2 - \frac{1}{c^2}d\mathbf{x}^2 = dt^2\left[1 - \frac{\mathbf{w}^2}{c^2}\right], \quad \mathbf{w} = \frac{d\mathbf{x}}{dt}. \tag{3.1}$$

Motivated by geometry and the philosophy of the times, Minkowski suggested that we use the proper-time to define a metric for the space-time implementation of the special theory of relativity. Physically, it is well-known that $d\tau$ is not an exact one-form because a particle can traverse many different paths (in space) during any given τ interval. This reflects the fact that the distance traveled in a given τ interval depends on the forces acting on the particle. This also implies that the clock of the source carries additional physical information about the acting forces. In order to see this, rewrite equation (3.1) as:

$$dt^2 = d\tau^2 + \frac{1}{c^2}d\mathbf{x}^2 = d\tau^2\left[1 + \frac{\mathbf{u}^2}{c^2}\right], \quad \mathbf{u} = \frac{d\mathbf{x}}{d\tau}. \tag{3.2}$$

For any other observer, we have:

$$dt'^2 = d\tau^2 + \frac{1}{c^2}d\mathbf{x}'^2 = d\tau^2\left[1 + \frac{\mathbf{u}'^2}{c^2}\right], \quad \mathbf{u}' = \frac{d\mathbf{x}'}{d\tau}. \tag{3.3}$$

It follows that observers can use one unique clock to discuss all events associated with the source (simultaneity). We also note that, the phase space variables remain unchanged because the momentum $\mathbf{p} = m\mathbf{w} = m_0\mathbf{u}$, where $m = \gamma m_0$.

From equations (3.2) and (3.3) we see explicitly that, the new metric for each observer is exact, while the representation space is now Euclidean. To see that this is a change in the clock convention, assume that we are observing a particle moving with constant velocity relative to the unprimed (inertial) frame. In this case, we can integrate equation (3.2) obtaining:

$$t = \left(\sqrt{1+\tfrac{\mathbf{u}^2}{c^2}}\right)\tau.$$

The inverse relationship is

$$\tau = \left(\sqrt{1+\tfrac{\mathbf{u}^2}{c^2}}\right)^{-1} t = \left(\sqrt{1-\tfrac{\mathbf{w}^2}{c^2}}\right) t,$$

where $\mathbf{w} = \frac{\mathbf{dx}}{\mathbf{dt}}$. Now, t and τ differ by a scale factor, so that either may be used (a convention). The advantage of the τ representation is that the same τ is also available to our prime observer:

$$\tau = \left(\sqrt{1+\tfrac{\mathbf{u}'^2}{c^2}}\right)^{-1} t' = \left(\sqrt{1-\tfrac{\mathbf{w}'^2}{c^2}}\right) t'.$$

The important point of our theory is that, this convention is also available in the general case. (In order to show that our approach does not complicate matters, in Section 3.3 we construct the general transformation group.)

In the new formalism, the natural definition of velocity is no longer $\mathbf{w} = d\mathbf{x}/dt$ but $\mathbf{u} = d\mathbf{x}/d\tau$. This suggests that there may be a certain duality in the relationship between t, τ and $\mathbf{w}$, $\mathbf{u}$. To see that this is indeed the case, recall that $\mathbf{u} = \mathbf{w}\Big/\sqrt{1-(\mathbf{w}^2/c^2)}$. Solving for $\mathbf{w}$, we get that $\mathbf{w} = \mathbf{u}\Big/\sqrt{1+(\mathbf{u}^2/c^2)}$. If we set $b = \sqrt{c^2+\mathbf{u}^2}$, this relationship can be written as

$$\frac{\mathbf{w}}{c} = \frac{\mathbf{u}}{b}. \tag{3.4}$$

For reasons to be clear momentarily, we call b the collaborative speed of light. Indeed, we see that

$$\frac{1}{c}\frac{\partial}{\partial t} = \frac{1}{c}\frac{\partial \tau}{\partial t}\frac{\partial}{\partial \tau} = \frac{1}{c}\frac{1}{\sqrt{1+(\mathbf{u}^2/c^2)}}\frac{\partial}{\partial \tau} = \frac{1}{b}\frac{\partial}{\partial \tau}. \tag{3.5}$$

For our prime observer, it is easy to see that the corresponding result is:

$$\frac{\mathbf{w}'}{c} = \frac{\mathbf{u}'}{b'}, \qquad \frac{1}{c}\frac{\partial}{\partial t'} = \frac{1}{b'}\frac{\partial}{\partial \tau}. \tag{3.6}$$

From equations (3.5) (and (3.6)) we see that the non-invariance of $t, (t')$ and the invariance of c on the left is replaced by the non-invariance of $b, (b')$ and the invariance of τ on the right. These equations represent mathematically equivalent relations. Thus, wherever they are used consistently as replacements for each other, they can't change the mathematical relationships. In order to see their impact on Maxwell's equations, in c.g.s. units, we have:

$$\begin{gathered}
\nabla\cdot\mathbf{B} = 0, \qquad \nabla\cdot\mathbf{E} = 4\pi\rho, \\
\nabla\times\mathbf{E} = -\frac{1}{c}\frac{\partial\mathbf{B}}{\partial t}, \quad \nabla\times\mathbf{B} = \frac{1}{c}\left[\frac{\partial\mathbf{E}}{\partial t} + 4\pi\rho\mathbf{w}\right].
\end{gathered} \tag{3.7}$$

Using equations (3.1) and (3.2) in (3.3), we have *the mathematically identical representation for Maxwell's equations*:

$$\nabla \cdot \mathbf{B} = 0, \qquad \nabla \cdot \mathbf{E} = 4\pi\rho,$$
$$\nabla \times \mathbf{E} = -\frac{1}{b}\frac{\partial \mathbf{B}}{\partial \tau}, \quad \nabla \times \mathbf{B} = \frac{1}{b}\left[\frac{\partial \mathbf{E}}{\partial \tau} + 4\pi\rho\mathbf{u}\right].$$

Thus, Maxwell's equations are equally valid when the local time of the particle is used to describe the fields. This leads to the following conclusions:

(i) There are two distinct clocks to use in the representation of Maxwell's equations. (The choice of clocks is a convention.)

(ii) Since the two representations are mathematically equivalent, mathematical equivalence is not the same as physical equivalence.

(iii) When the proper-time is used, the constant speed of light c is replaced by the effective speed of light b, which depends on the motion of the system (i.e., $b = \sqrt{c^2 + \mathbf{u}^2}$). Thus, we have a natural varying speed of light theory (VSL), as opposed to a postulated one (see Magueijo [26] or Moffat [27]).

Let us now derive the corresponding wave equations in the local-time variable. Taking the curl of the last two equations Maxwell equations (above), and using standard vector identities, we get:

$$\begin{aligned} &\frac{1}{b^2}\frac{\partial^2 \mathbf{B}}{\partial \tau^2} - \frac{\mathbf{u}\cdot\mathbf{a}}{b^4}\left[\frac{\partial \mathbf{B}}{\partial \tau}\right] - \nabla^2 \cdot \mathbf{B} = \frac{1}{b}\left[4\pi\nabla \times (\rho\mathbf{u})\right], \\ &\frac{1}{b^2}\frac{\partial^2 \mathbf{E}}{\partial \tau^2} - \frac{\mathbf{u}\cdot\mathbf{a}}{b^4}\left[\frac{\partial \mathbf{E}}{\partial \tau}\right] - \nabla^2 \cdot \mathbf{E} = -\nabla(4\pi\rho) - \frac{1}{b}\frac{\partial}{\partial \tau}\left[\frac{4\pi(\rho\mathbf{u})}{b}\right]. \end{aligned} \tag{3.8}$$

where $\mathbf{a} = d\mathbf{u}/d\tau$ is the effective acceleration. The new (gauge independent) term appears instantaneously along the direction of motion, but opposing any applied force and is zero otherwise. This is the near field (i.e., the field at the site of the charged particle). This is exactly what one expects of the back reaction caused by the inertial resistance of a particle to accelerated motion and, according to Wheeler and Feynman [28], is precisely what is meant by radiation reaction. Thus, the collaborative use of the observer's coordinate system and the local clock of the observed system provides intrinsic information about the local field dynamics not available in the conventional formulation of Maxwell's theory. It is shown in [22], that the theory does not require point particles, self-energy divergence, mass renormalization or the Lorentz Dirac equation.

It is also shown in [21] that, for a closed system of interacting charged particles, the proper-time of the center of mass corresponds to the historical clock of Horwitz, Piron, and Fanchi (see [29] and [30]). In this case, $b = c$ and the corresponding Maxwell equations represent the far field (only retarded potentials). It was further shown that, from this vantage point, the particle interactions appear as the delayed action-at-a-distance type. This verifies the Wheeler-Feynman conjecture that field theory and delayed action-at-a-distance are complimentary manifestations of the same physics (see [28]). The requirement of total conservation of momentum, angular momentum and energy allowed us to prove the complete absorption of radiation by all particles in the system. (Recall that, this was an assumption in the Wheeler and Feynman approach *and the center-fold of their theory.*)

If we make a scale transformation (at fixed position) with $\mathbf{E} \to (b/c)^{1/2}\mathbf{E}$ and $\mathbf{B} \to (b/c)^{1/2}\mathbf{B}$, the equations in (3.4) transform to

$$\begin{aligned}\frac{1}{b^2}\frac{\partial^2\mathbf{B}}{\partial\tau^2} - \nabla^2\cdot\mathbf{B} + \left[\frac{\ddot{b}}{2b^3} - \frac{3\dot{b}^2}{4b^4}\right]\mathbf{B} &= \frac{c^{1/2}}{b^{3/2}}\left[4\pi\nabla\times(\rho\mathbf{u})\right],\\ \frac{1}{b^2}\frac{\partial^2\mathbf{E}}{\partial\tau^2} - \nabla^2\cdot\mathbf{E} + \left[\frac{\ddot{b}}{2b^3} - \frac{3\dot{b}^2}{4b^4}\right]\mathbf{E} &= -\frac{c^{1/2}}{b^{1/2}}\nabla(4\pi\rho) - \frac{c^{1/2}}{b^{3/2}}\frac{\partial}{\partial\tau}\left[\frac{4\pi(\rho\mathbf{u})}{b}\right].\end{aligned} \tag{3.9}$$

This is the Klein-Gordon equation with an effective mass μ given by

$$\mu = \left\{\frac{\hbar^2}{c^2}\left[\frac{\ddot{b}}{2b^3} - \frac{3\dot{b}^2}{4b^4}\right]\right\}^{1/2} = \left\{\frac{\hbar^2}{c^2}\left[\frac{\mathbf{u}\cdot\ddot{\mathbf{u}} + \dot{\mathbf{u}}^2}{2b^4} - \frac{5\left(\mathbf{u}\cdot\dot{\mathbf{u}}\right)^2}{4b^6}\right]\right\}^{1/2}. \tag{3.10}$$

Remark 3.1 *We note that, when b is constant, $\mathbf{a} = \mathbf{0}$, $\mu = \mathbf{0}$ and $t = \frac{b}{c}\tau$ (also $t' = \frac{b'}{c}\tau$), so that the local time theory is both mathematically and physically equivalent to the standard theory. However, when b is not constant $\mu \neq 0$ and the two approaches are not physically equivalent.*

For additional insight, let $(\mathbf{x}(\tau), \tau)$ represent the field position and $(\bar{\mathbf{x}}(\tau'), \tau')$ the retarded position of a source charge e, with $\mathbf{r} = \mathbf{x} - \bar{\mathbf{x}}$. If we set $r = |\mathbf{x} - \bar{\mathbf{x}}|$, $s = r - (\frac{(\mathbf{r}\cdot\mathbf{u})}{b})$, and $\mathbf{r_u} = \mathbf{r} - \frac{r}{b}\mathbf{u}$, then we were able to compute the $\mathbf{E}$ and $\mathbf{B}$ fields directly in [21] to obtain:

$$\mathbf{E}(\mathbf{x},\tau) = \frac{e\left[\mathbf{r_u}(1 - \mathbf{u}^2/b^2)\right]}{s^3} + \frac{e\left[\mathbf{r}\times(\mathbf{r_u}\times\mathbf{a})\right]}{b^2s^3} + \frac{e(\mathbf{u}\cdot\mathbf{a})\left[\mathbf{r}\times(\mathbf{u}\times\mathbf{r})\right]}{b^4s^3}$$

and

$$\mathbf{B}(\mathbf{x},\tau) = \frac{e\left[(\mathbf{r}\times\mathbf{r_u})(1 - \mathbf{u}^2/b^2)\right]}{rs^3} + \frac{e\mathbf{r}\times\left[\mathbf{r}\times(\mathbf{r_u}\times\mathbf{a})\right]}{rb^2s^3} + \frac{er(\mathbf{u}\cdot\mathbf{a})(\mathbf{r}\times\mathbf{u})}{b^4s^3}.$$

(It is easy to see that $\mathbf{B}$ is orthogonal to $\mathbf{E}$.) The first two terms in the above equations are standard, in the $(\mathbf{x}(t), \mathbf{w}(t))$ variables. The third part of both equations is new and arises because of the dissipative term in our wave equation. (Once again, this term is zero when b is constant.) It is easy to see that $\mathbf{r}\times(\mathbf{u}\times\mathbf{r}) = r^2\mathbf{u} - (\mathbf{u}\cdot\mathbf{r})\mathbf{r}$, so we get a component along the direction of motion. (Thus, the $\mathbf{E}$ field has a longitudinal part.) This confirms our claim that the new dissipative term is equivalent to an effective mass that arises due to the collaborative acceleration of the particle. This means that the cause for radiation reaction comes directly from the use of the local clock to formulate Maxwell's equations. It follows that, in this approach there is no need to assume advanced potentials, self-interaction, mass renormalization and the Lorentz-Dirac equation in order to account for it (radiation reaction), as is required when the observer clock is used. Furthermore, no assumptions about the structure of the source are needed.

3.3. Proper-time Lorentz Group

We now identify the new transformation group that preserves the first postulate of the special theory. The standard (Lorentz) time transformations between two inertial observers can be written as

$$t' = \gamma(\mathbf{v})\left[t - \mathbf{x}\cdot\mathbf{v}/c^2\right], \qquad t = \gamma(\mathbf{v})\left[t' + \mathbf{x}'\cdot\mathbf{v}/c^2\right]. \tag{3.11}$$

We want to replace t, t' by τ. To do this, use the relationship between dt and $d\tau$ to get:

$$t = \tfrac{1}{c}\int_0^\tau b(s)ds = \tfrac{1}{c}\bar{b}\tau, \quad t' = \tfrac{1}{c}\int_0^\tau b'(s)ds = \tfrac{1}{c}\bar{b}'\tau, \tag{3.12}$$

where we have used the mean value theorem of calculus to obtain the end result, so that both $\bar{b}$ and $\bar{b}'$ represent an earlier τ-value of b and b' respectively. Note that, as b and b' depend on τ, the transformations (3.12) represent explicit nonlinear relationships between t, t' and τ (during interaction). This is to be expected in the general case when the system is acted on by external forces. However, if b is constant (so is b'), then t, t' and τ differ by a scale transformation, which means they are physically equivalent, in addition to their natural mathematical equivalence.

If we set

$$\mathbf{d}^* = \mathbf{d}/\gamma(\mathbf{v}) - (1-\gamma(\mathbf{v}))\left[(\mathbf{v}\cdot\mathbf{d})/(\gamma(\mathbf{v})\mathbf{v}^2)\right]\mathbf{v},$$

we can write the transformations that fix τ as:

$$\begin{aligned}
\mathbf{x}' &= \gamma(\mathbf{v})\left[\mathbf{x}^* - (\mathbf{v}/c)\bar{b}\tau\right], & \mathbf{x} &= \gamma(\mathbf{v})\left[\mathbf{x}'^* + (\mathbf{v}/c)\bar{b}'\tau\right], \\
\mathbf{u}' &= \gamma(\mathbf{v})\left[\mathbf{u}^* - (\mathbf{v}/c)b\right], & \mathbf{u} &= \gamma(\mathbf{v})\left[\mathbf{u}'^* + (\mathbf{v}/c)b'\right], \\
\mathbf{a}' &= \gamma(\mathbf{v})\left\{\mathbf{a}^* - \mathbf{v}\left[(\mathbf{u}\cdot\mathbf{a})/(bc)\right]\right\}, & \mathbf{a} &= \gamma(\mathbf{v})\left\{\mathbf{a}'^* + \mathbf{v}\left[(\mathbf{u}'\cdot\mathbf{a}')/(b'c)\right]\right\}.
\end{aligned} \tag{3.13}$$

If we put equation (3.12) in (3.11), differentiate with respect to τ and cancel the extra factor of c, we get the transformations between b and b':

$$b'(\tau) = \gamma(\mathbf{v})\left[b(\tau) - \mathbf{u}\cdot\mathbf{v}/c\right], \qquad b(\tau) = \gamma(\mathbf{v})\left[b'(\tau) + \mathbf{u}'\cdot\mathbf{v}/c\right]. \tag{3.14}$$

From these results, it follows that, at the local level, during interaction, equations (3.13) and (3.14) provide a nonlinear and nonlocal representation of the Lorentz group. We call it the proper-time Lorentz group.

3.4. Proper-Time Particle Theory

We now investigate the corresponding particle theory. The key concept to our approach may be seen by examining the time evolution of a dynamical parameter $W(\mathbf{x},\mathbf{p})$, via the standard formulation of classical mechanics, described in terms of the Poisson brackets:

$$\frac{dW}{dt} = \{H, W\}. \tag{3.15}$$

We can also represent the dynamics via the proper time by using the representation $d\tau = (1/\gamma)dt = (mc^2/H)dt$, so that:

$$\frac{dW}{d\tau} = \frac{dt}{d\tau}\frac{dW}{dt} = \frac{H}{mc^2}\{H, W\}.$$

Assuming a well-defined (invariant) rest energy (mc^2) for the particle, we determine the canonical proper-time Hamiltonian K such that:

$$\{K, W\} = \frac{H}{mc^2}\{H, W\}, \quad K|_{\mathbf{p}=0} = H|_{\mathbf{p}=0} = mc^2.$$

Using

$$\begin{aligned}
\{K, W\} &= \left[\frac{H}{mc^2}\frac{\partial H}{\partial \mathbf{p}}\right]\frac{\partial W}{\partial \mathbf{x}} - \left[\frac{H}{mc^2}\frac{\partial H}{\partial \mathbf{x}}\right]\frac{\partial W}{\partial \mathbf{p}} \\
&= \frac{\partial}{\partial \mathbf{p}}\left[\frac{H^2}{2mc^2} + a\right]\frac{\partial W}{\partial \mathbf{x}} - \frac{\partial}{\partial \mathbf{x}}\left[\frac{H^2}{2mc^2} + a'\right]\frac{\partial W}{\partial \mathbf{p}},
\end{aligned}$$

we get that $a = a' = \frac{1}{2}mc^2$, so that (assuming no explicit time dependence)

$$K = \frac{H^2}{2mc^2} + \frac{mc^2}{2}, \quad \text{and} \quad \frac{dW}{d\tau} = \{K, W\}.$$

Journal of Physics: Conference Series **615** (2015) 012013 doi:10.1088/1742-6596/615/1/012013

Since τ is invariant during interaction (minimal coupling), we make the natural assumption that (the form of) K also remains invariant. Thus, if $\sqrt{c^2\mathbf{p}^2 + m^2c^4} \to \sqrt{c^2\pi^2 + m^2c^4} + V$, where $\mathbf{A}$ a vector potential, V is a potential energy term and $\pi = \mathbf{p} - \frac{e}{c}\mathbf{A}$. In this case, K becomes:

$$K = \frac{\pi^2}{2m} + mc^2 + \frac{V^2}{2mc^2} + \frac{V\sqrt{c^2\pi^2 + m^2c^4}}{mc^2}.$$

If we set $H_0 = \sqrt{c^2\pi^2 + m^2c^4}$, use standard vector identities with $H_0 = mcb - V$, $\nabla \times \pi = -\frac{\mathbf{e}}{\mathbf{c}}\mathbf{B}$, and compute Hamilton's equations, we get:

$$\mathbf{u} = \frac{d\mathbf{x}}{d\tau} = \left[1 + \frac{V}{H_0}\right]\frac{\pi}{m} = \left[\frac{mbc}{mbc - V}\right]\frac{\pi}{m} \Rightarrow \pi = m\mathbf{u} - \frac{V}{bc}\mathbf{u}$$

and

$$\begin{aligned} \frac{d\mathbf{p}}{d\tau} &= -\frac{\left[(\pi\cdot\nabla)\,\pi + \frac{\mathbf{e}}{\mathbf{c}}\pi \times \mathbf{B}\right]}{m}\left[1 + \frac{V}{H_0}\right] - \nabla V \frac{H_0}{mc^2}\left[1 + \frac{V}{H_0}\right] \\ &= \tfrac{e}{c}\left(\mathbf{u}\cdot\nabla\right)\mathbf{A} + \tfrac{e}{c}\mathbf{u}\times\mathbf{B} - \nabla V\frac{b}{c}\left[1 + \frac{V}{mcb}\right]. \end{aligned} \tag{3.16}$$

Further reduction, using the definition of $\mathbf{E}$, with $V = e\Phi$, we have:

$$\begin{aligned} \frac{c}{b}\left[\frac{d\mathbf{p}}{d\tau} - \frac{e}{c}\frac{d\mathbf{A}}{d\tau}\right] &= -\frac{e}{b}\frac{\partial\mathbf{A}}{\partial\tau} + \tfrac{e}{b}\mathbf{u}\times\mathbf{B} - e\nabla\Phi\left[1 + \frac{V}{mcb}\right] \\ &= e\mathbf{E} + \tfrac{e}{b}\mathbf{u}\times\mathbf{B} - e\nabla\Phi\frac{V}{mcb}. \end{aligned} \tag{3.17}$$

It is clear that the additional term in equation (3.17) acts to oppose the force imposed by the charged particle part of the $\mathbf{E}$ field (i.e., $-\nabla V$). In order to see the physical meaning of the term, assume an interaction between a proton and an electron, where $\mathbf{A} = \mathbf{0}$ and V is the Coulomb interaction, so that (3.17) becomes:

$$\frac{c}{b}\frac{d\mathbf{p}}{d\tau} = -\nabla V - \nabla V\frac{V}{mcb}. \tag{3.18}$$

Using $H_0 \approx mc^2$, we see that $\lim_{r\to r_0}\mathbf{u} = \mathbf{0}$ and $\lim_{r\to r_0}\mathbf{a} = \mathbf{0}$, so that:

$$0 = -\nabla V - \nabla V\frac{V}{mc^2} \tag{3.19}$$

and the classical electron radius, r_0, is a critical point (i.e., $-\nabla V - \nabla V(V/mc^2) = 0$). Thus, for $0 < r < r_0$, the force becomes repulsive. We interpret this as a fixed region of repulsion, so that the singularity $r = 0$ is impossible to reach at the classical level. The neglected terms are attractive but of lower order. This makes the critical point less than r_0. Thus, in general, the electron experiences a strongly repulsive force when it gets too close to the proton. This means that the classical principle of impenetrability, namely that no two particles can occupy the same space at the same time occurs naturally. It is this additional term that leads us to suspect that the electron may not act like a point particle in the s-states of hydrogen, where it has a finite probability of being at the center of the proton.

The above observation also implies that, two electrons will experience an attraction if they can come close enough together as for example, at very low energies (temperatures).

The Lagrangian representation reveals the close relationship to the non-relativistic case. If we solve for $\mathbf{p}$, we get

$$\mathbf{p} = m\mathbf{u} - \frac{V\mathbf{u}}{cb} + \frac{e}{c}\mathbf{A}.$$

Using this in K along with $b^2 = u^2 + c^2$, we have

$$\begin{aligned} K &= \frac{\left(m\mathbf{u} - \frac{V\mathbf{u}}{bc}\right)^2}{2m} + mc^2 + \frac{V^2}{2mc^2} + \frac{V\left(mcb - V\right)}{mc^2} \\ &= \tfrac{1}{2}mu^2 - \frac{Vu^2}{bc} + \frac{V^2u^2}{2mb^2c^2} + mc^2 - \frac{V^2}{2mc^2} + \frac{Vb}{c}. \end{aligned}$$

From $\mathcal{L}d\tau = \mathbf{p}\cdot d\mathbf{x} - Kd\tau$, we can write $\mathcal{L}$ as

$$\begin{aligned} \mathcal{L} &= \left[m\mathbf{u} - \frac{V\mathbf{u}}{bc}\right]\cdot\mathbf{u} + \tfrac{e}{c}\mathbf{A}\cdot\mathbf{u} \\ &- \left\{\tfrac{1}{2}mu^2 - \frac{Vu^2}{bc} + \frac{V^2u^2}{2mb^2c^2} + mc^2 + \frac{Vb}{c} - \frac{V^2}{2mc^2}\right\} \\ &= \tfrac{1}{2}mu^2 + \tfrac{e}{c}\mathbf{A}\cdot\mathbf{u} - mc^2 - \frac{Vb}{c} + \frac{V^2}{2mc^2}\left[1 - \frac{u^2}{b^2}\right]. \end{aligned}$$

From this representation, it is clear that the neglect of second order terms gives us the non-relativistic theory.

4. Relativistic Quantum Theory

The Klein-Gordon and Dirac equations were first discovered in early attempts to make quantum mechanics compatible with the Minkowski formulation of special theory of relativity. Both were partially successful but could no longer be interpreted as particle equations and a complete theory required quantum fields and the associated problems. For a recent discussion of other problems, one can consult [9] (see also [6]).

In this section we introduce the canonical extension of the Dirac and square-root equations. Let $\mathbf{A}(\mathbf{x},t)$ and $V(\mathbf{x})$ be given vector and scalar potentials and, after adding $V(\mathbf{x})$ and making the transformation $\mathbf{p} \to \pi = \mathbf{p} - \frac{e}{c}\mathbf{A}$.

To quantize our theory, we follow the standard procedure leading to the equation:

$$i\hbar\frac{\partial\Phi}{\partial\tau} = K\Phi = \left[\frac{H^2}{2mc^2} + \frac{mc^2}{2}\right]\Phi.$$

However, in addition to the Dirac Hamiltonian, there are two other possible Hamiltonians, depending on the way the potential appears with the square-root operator:

$$\beta\sqrt{c^2\pi^2 - \mathbf{e}c\hbar\boldsymbol{\Sigma}\cdot\mathbf{B} + \mathbf{m^2c^4}} + V$$

and

$$\beta\sqrt{c^2\pi^2 - \mathbf{e}c\hbar\boldsymbol{\Sigma}\cdot\mathbf{B} + (\mathbf{mc^2} + \beta\mathbf{V})^2}.$$

We have identified three possible canonical proper-time particle equations for spin-$\frac{1}{2}$ particles. (We also note that, these equations can be modified to apply to particles of any spin, by a minor change in the β matrix.)

(i) The canonical proper-time version of the Dirac equation:

$$i\hbar\frac{\partial\Psi}{\partial\tau}=\left\{\frac{\boldsymbol{\pi}^2}{2m}+\beta V+mc^2-\frac{e\hbar\Sigma\cdot\mathbf{B}}{2mc}+\frac{V\alpha\cdot\pi}{mc}-\frac{i\hbar\alpha\cdot\nabla V}{2mc}+\frac{V^2}{2mc^2}\right\}\Psi. \tag{4.1}$$

(ii) The canonical proper-time version of the square-root equation, using the first possibility:

$$i\hbar\frac{\partial\Psi}{\partial\tau}=\left\{\frac{\boldsymbol{\pi}^2}{2m}-\frac{e\hbar\Sigma\cdot\mathbf{B}}{2mc}+mc^2+\frac{V^2}{2mc^2}\right\}\Psi+\frac{V\beta\sqrt{c^2\boldsymbol{\pi}^2-\mathbf{ec\hbar\Sigma\cdot B+m^2c^4}}}{2mc^2}\Psi+\frac{\beta\sqrt{c^2\boldsymbol{\pi}^2-\mathbf{ec\hbar\Sigma\cdot B+m^2c^4}}}{2mc^2}V\Psi. \tag{4.2}$$

(iii) The canonical proper-time version of the square-root equation, using the second possibility:

$$i\hbar\frac{\partial\Psi}{\partial\tau}=\frac{\boldsymbol{\pi}^2}{2m}+\beta V+mc^2-\frac{e\hbar\Sigma\cdot\mathbf{B}}{2mc}+\frac{V^2}{2mc^2}. \tag{4.3}$$

If $V=0$, all equations reduce to:

$$i\hbar\frac{\partial\Psi}{\partial\tau}=\left\{\frac{\boldsymbol{\pi}^2}{2m}+mc^2-\frac{e\hbar\Sigma\cdot\mathbf{B}}{2mc}\right\}\Psi.$$

The close relationship to the Schrödinger operator, makes it easy to see that, in all cases, K is positive definite. In mathematical terms, the lower order terms are relatively bounded with respect to $\boldsymbol{\pi}^2/2m$. It follows that, unlike the Dirac and Klein-Gordon approach, we can interpret (4.1)-(4.3) as representations for particles. In the above equations, we have assumed that V is time independent. (However, since $\mathbf{A}(\mathbf{x},t)$ can have general time-dependence, $\sqrt{c^2\boldsymbol{\pi}^2-\mathbf{ec\hbar\Sigma\cdot B+m^2c^4}}$ need not be related to the Dirac operator by a Foldy-Wouthuysen type transformation.)

We plan to investigate the last two equations at a later time. In the next section, we focus on the canonical proper-time Dirac extension.

4.1. The Dirac Theory

Since the Dirac equation forms the basis for QED, an important test of our proper-time extension is how well it compares to the Dirac equation in its description of the hydrogen spectrum. In this section, we compare the Dirac equation with the canonical proper-time extension for the Hydrogen atom problem.

If we let $\mathbf{A}=\mathbf{0}$, $V_0=-\frac{e^2}{r}$ and consider the standard Dirac Hydrogen atom eigenvalue problem,

$$\lambda_n\Psi_n=H_D\Psi_n,$$

where λ_n is the n-th eigenvalue and Ψ_n is the corresponding eigenfunction. For this case, if j is the total angular momentum and α is the fine structure constant, we have

$$\lambda_n=mc^2\left[1+\frac{\alpha^2}{\left[n-\left|j+\frac{1}{2}\right|+\sqrt{\left(j+\frac{1}{2}\right)^2-\alpha^2}\right]^2}\right]^{-1/2}.$$

For the proper-time extension, with the same eigenfunction, we have

$$E_n\Psi_n = \left[\frac{H_D^2}{2mc^2} + \frac{mc^2}{2}\right]\Psi_n = \left[\frac{\lambda_n^2}{2mc^2} + \frac{mc^2}{2}\right]\Psi_n.$$

In order to compare the two expressions to order $O(\alpha^8)$, we begin with the following approximations:

$$\begin{aligned}
&(1)\quad (1+x)^{-1} \simeq 1 + x + x^2 + O(x^3)\\
&(2)\quad (1+x)^{-1/2} \simeq 1 - \tfrac{1}{2}x + \tfrac{3}{8}x^2 + O(x^3)\\
&(3)\quad (1-x)^{1/2} \simeq 1 - \tfrac{1}{2}x - \tfrac{1}{8}x^2 + O(x^3).
\end{aligned}$$

Let $\kappa = j + \frac{1}{2}$, then our proper time extension can be written as:

$$\begin{aligned}
E_n &= \left[\frac{\lambda_n^2}{2mc^2} + \frac{mc^2}{2}\right] = \frac{mc^2}{2}\left\{1 + \left[1 + \frac{\alpha^2}{\left(n-|\kappa|+\sqrt{\kappa^2-\alpha^2}\right)^2}\right]^{-1}\right\}\\
&\simeq \frac{mc^2}{2}\left\{1 + \left[1 - \frac{\alpha^2}{\left(n-|\kappa|+\sqrt{\kappa^2-\alpha^2}\right)^2} + \frac{\alpha^4}{\left(n-|\kappa|+\sqrt{\kappa^2-\alpha^2}\right)^4}\right]\right\}\\
&= \frac{mc^2}{2}\left[2 - \frac{\alpha^2}{\left(n-|\kappa|+\sqrt{\kappa^2-\alpha^2}\right)^2} + \frac{\alpha^4}{\left(n-|\kappa|+\sqrt{\kappa^2-\alpha^2}\right)^4}\right].
\end{aligned} \tag{4.4}$$

Using (2), we have:

$$\begin{aligned}
\lambda_n &\simeq mc^2\left[1 + \frac{\alpha^2}{\left(n-|\kappa|+\sqrt{\kappa^2-\alpha^2}\right)^2}\right]^{-1/2}\\
&\simeq mc^2\left[1 - \frac{\alpha^2}{2\left(n-|\kappa|+\sqrt{\kappa^2-\alpha^2}\right)^2} + \frac{3\alpha^4}{8\left(n-|\kappa|+\sqrt{\kappa^2-\alpha^2}\right)^4}\right].
\end{aligned}$$

Using (3), we can approximate $\sqrt{\kappa^2-\alpha^2}$ to get

$$\begin{aligned}
&\sqrt{\kappa^2-\alpha^2} \simeq |\kappa|\left(1 - \tfrac{\alpha^2}{2\kappa^2} - \tfrac{\alpha^4}{8\kappa^4}\right) = |\kappa| - \tfrac{\alpha^2}{2|\kappa|} \Rightarrow\\
&n - |\kappa| + \sqrt{\kappa^2-\alpha^2} \simeq n - \tfrac{\alpha^2}{2|\kappa|} \Rightarrow\\
&\alpha^2\left[n - |\kappa| + \sqrt{\kappa^2-\alpha^2}\right]^{-2} \simeq \frac{\alpha^2}{n^2}\left[\left(1 - \tfrac{\alpha^2}{2n|\kappa|}\right)^2\right]^{-1} \simeq \frac{\alpha^2}{n^2}\left[1 + \tfrac{\alpha^2}{n|\kappa|} - \tfrac{\alpha^4}{4n^2|\kappa|^3}\right]\\
&= \frac{\alpha^2}{n^2} + \frac{\alpha^4}{n^3|\kappa|} - \frac{\alpha^6}{4n^4|\kappa|^3}
\end{aligned}$$

and

$$\alpha^4\left[n - |\kappa| + \sqrt{\kappa^2-\alpha^2}\right]^{-4} \simeq \frac{\alpha^4}{n^4}\left[\left(1 - \tfrac{\alpha^2}{2n|\kappa|}\right)^4\right]^{-1} \simeq \frac{\alpha^4}{n^4}\left[1 + \tfrac{2\alpha^2}{n|\kappa|}\right].$$

With the last result, we now have:

$$\begin{aligned}
\lambda_n &\simeq mc^2\left\{1 - \frac{\alpha^2}{2n^2}\left[1 + \frac{\alpha^2}{n|\kappa|} - \frac{\alpha^4}{4n|\kappa|^3}\right] + \frac{3\alpha^4}{8n^4}\left[1 + \frac{\alpha^2}{n|\kappa|}\right]\right\}\\
&= mc^2\left\{\left[1 - \frac{\alpha^2}{2n^2} - \frac{\alpha^4}{2n^4}\left(\frac{n}{|\kappa|} - \frac{3}{4}\right)\right] + \frac{\alpha^6}{8n^5|\kappa|}\left(\frac{n^2}{|\kappa|^2} + 3\right)\right\}.
\end{aligned} \tag{4.5}$$

For E_n, we have

$$\begin{aligned} E_n &\simeq \frac{mc^2}{2}\left\{2-\left[\frac{\alpha^2}{n^2}+\frac{\alpha^4}{n^3\left|\kappa\right|}-\frac{\alpha^6}{4n^4\left|\kappa\right|^2}\right]+\frac{\alpha^4}{n^4}\left[1+\frac{2\alpha^2}{n\left|\kappa\right|}\right]\right\} \\ &= mc^2\left\{\left[1-\frac{\alpha^2}{2n^2}-\frac{\alpha^4}{2n^4}\left(\frac{n}{\left|\kappa\right|}-1\right)\right]+\frac{\alpha^6}{4n^5\left|\kappa\right|}\left(\frac{n}{\left|\kappa\right|}+8\right)\right\}. \end{aligned} \tag{4.6}$$

It is now easy to see that, to order α^4, $\lambda_n - E_n = \frac{-\alpha^4}{8n^4}$, so that the E_n values are systematically lower than the λ_n values.

Table 1 below provides a relative comparison between the Dirac and proper-time extension compared with the experimental data for s-states, compiled by National Institute for Standards and Technology (NIST) of the US government.

Table 1: Comparison with NIST data for s-states

State	Dirac	Proper-time	Nist	Δ-DNIST	Δ-PTNIST
2s	10.20439429	10.20422448	10.19881008	.00558421	.00541440
3s	12.09411035	12.09393146	12.08749443	.00661592	.00643603
4s	12.75550914	12.75532871	12.74853244	.00697670	.00679627
5s	13.06164150	13.06146066	13.05449789	.00714361	.00696277

As can be seen from the last two columns, the proper-time extension consistently provides results that a closer to the experimental data for all cases. (We have omitted the 1s-state comparison because of the NIST normalization condition.)

In Table 2, we see the same comparative results for the p, d and f-states.

Table 2: Comparison with NIST data for p, d and f-states

State	Dirac	Proper-time	Nist	Δ-DNIST	Δ-PTNIST
2p (j=1/2)	10.20439429	10.20422448	10.19880553	0.005588760	0.005418952
2p (j=3/2)	10.20443957	10.20426976	10.19885089	0.005588681	0.005418870
3p (j=1/2)	12.09411035	12.09393146	12.08749292	0.006617431	0.006438537
3p (j=3/2)	12.09412377	12.09394488	12.08750636	0.006617407	0.006438512
3d (j=3/2)	12.09412377	12.09394488	12.08750634	0.006617430	0.006438535
3d (j=5/2)	12.09412824	12.09394935	12.08751082	0.006617422	0.006438528
4p (j=1/2)	12.75550914	12.75532871	12.74853167	0.006977467	0.006797044
4f (j=7/2)	12.75551763	12.75533720	12.74854038	0.006976250	0.006796820

Thus, in all cases, the canonical proper-time extension of the Dirac equation provides a closer approximation to the known experimental data for the Hydrogen spectra compared to the Dirac equation. In all cases, the changes are in the forth decimal place. This is insufficient to account for either the Lamb shift or the anomalous magnetic moment.

4.2. Future Direction

In what follows, let $V = V_0 = \frac{-e^2}{r}$. Based on our analysis of the square-root operator in the first section and the Dirac operator in the second section, we are in the process of investigating the possibility that in s-states, the potential energy takes on the form:

$$\begin{aligned} V &= -\frac{e^2\sqrt{M^2c^4 - ec\hbar\Sigma\cdot\mathbf{B} + c^2\pi^2}}{Mc^2 r} = -\frac{e^2}{r}\frac{\left(1-\frac{e\hbar\Sigma\cdot\mathbf{B}}{M^2c^3}+\frac{\pi^2}{M^2c^2}\right)}{\sqrt{1-\frac{e\hbar\Sigma\cdot\mathbf{B}}{M^2c^3}+\frac{\pi^2}{M^2c^2}}} \\ &\simeq V_0 + \frac{r_0}{r}\frac{e\hbar\Sigma\cdot\mathbf{B}}{2Mc} - \frac{r_0}{r}\frac{\pi^2}{2M}, \end{aligned} \tag{4.7}$$

where $r_0 = e^2/Mc^2$. There are three possible choices for M:

(i) The electron cannot be treated as a point particle in s-states of hydrogen, so that $M = m$, the mass of the electron and r_0 is the classical electron radius.

(ii) Neither the electron nor the proton can be treated as point particles in s-states of hydrogen, so that $M = \mu$, the reduced mass and r_0 is the classical mixed reduced radius.

(iii) The electron can be treated as a point particle in s-states of hydrogen, but the proton cannot so that $M = m_p$, the mass of the proton and r_0 is the classical reduced proton radius.

It is clear that, at the zero-th order, we recover the Coulomb potential and the second term in (4.7) is a first order approximation. Assuming the first case, our eigenvalue problem becomes:

$$\begin{aligned} E\Psi = \Bigg\{ &\frac{\boldsymbol{\pi}^2}{2m} + \beta V + mc^2 - \frac{e\hbar\Sigma\cdot\mathbf{B}}{2mc} \\ &+ \frac{V\alpha\cdot\pi}{mc} - \frac{i\hbar\alpha\cdot\nabla V}{2mc} + \frac{V^2}{2mc^2} \Bigg\}\Psi. \end{aligned} \tag{4.8}$$

As a first try, we set $V = V_0$ for the terms containing α, use our first order approximation in the second term and V itself in the last term, so that

$$\beta V \simeq \beta V_0 + \beta\frac{r_0}{r}\frac{e\hbar\Sigma\cdot\mathbf{B}}{2mc} - \beta\frac{r_0}{r}\frac{\pi^2}{2m}.$$

For the last term, we use the approximation:

$$\begin{aligned} \frac{V^2}{2mc^2} &= \frac{1}{2}\left\{\frac{V_0}{mc^2}\sqrt{1-\frac{e\hbar\Sigma\cdot\mathbf{B}}{m^2c^3}+\frac{\pi^2}{m^2c^2}}\frac{V_0}{mc^2}\sqrt{1-\frac{e\hbar\Sigma\cdot\mathbf{B}}{m^2c^3}+\frac{\pi^2}{m^2c^2}}\right\} \\ &\simeq \frac{1}{2}\left\{\left[\frac{V_0}{mc^2}\right]^2\left[1-\frac{e\hbar\Sigma\cdot\mathbf{B}}{m^2c^3}+\frac{\pi^2}{m^2c^2}\right]\right\} + \frac{V_0}{mc^2}\frac{\mathbf{p}^2[V_0]}{2m^2c^2}\sqrt{1-\frac{e\hbar\Sigma\cdot\mathbf{B}}{m^2c^3}+\frac{\pi^2}{m^2c^2}} \\ &\simeq \frac{1}{2}\left\{\left[\frac{V_0}{mc^2}\right]^2\left[1-\frac{e\hbar\Sigma\cdot\mathbf{B}}{m^2c^3}+\frac{\pi^2}{m^2c^2}\right]\right\} + \frac{V_0}{mc^2}\frac{\mathbf{p}^2[V_0]}{2m^2c^2} \\ &= \frac{1}{2}\left\{\left[\frac{V_0}{mc^2}\right]^2\left[1-\frac{e\hbar\Sigma\cdot\mathbf{B}}{m^2c^3}+\frac{\pi^2}{m^2c^2}\right]\right\} - \frac{V_0}{mc^2}\frac{2\pi\hbar^2}{m^2c^2}\delta(r). \end{aligned}$$

Using these terms, we have:

$$\begin{aligned} E\Psi &= \left\{1-\beta\frac{r_0}{r}+\frac{r_0^2}{r^2}\right\}\frac{\pi^2}{2m}\Psi + \left[1+\beta\frac{r_0}{r}-\frac{r_0^2}{r^2}\right]\frac{e\hbar\Sigma\cdot\mathbf{B}}{2mc}\Psi + mc^2\Psi \\ &+\beta V_0 + \frac{V_0\alpha\cdot\pi}{mc}\Psi - \frac{i\hbar\alpha\cdot\nabla V_0}{2mc}\Psi + \frac{V_0^2}{2mc^2}\Psi + \frac{r_0}{r}\frac{2\pi\hbar^2}{mc^2}\delta(r)\Psi \\ &= \left\{\frac{\pi^2}{2m}+\beta V_0 - \frac{e\hbar\Sigma\cdot\mathbf{B}}{2mc}+mc^2+\frac{V_0\alpha\cdot\pi}{mc}-\frac{i\hbar\alpha\cdot\nabla V_0}{2mc}+\frac{V_0^2}{2mc^2}\right\}\Psi \\ &+\left\{\left[\frac{r_0^2}{r^2}-\beta\frac{r_0}{r}\right]\frac{\pi^2}{2m} - \left[\frac{r_0^2}{r^2}-\beta\frac{r_0}{r}\right]\frac{e\hbar\Sigma\cdot\mathbf{B}}{2mc} + \frac{r_0}{r}\frac{2\pi\hbar^2}{m^2c^2}\delta(r)\right\}\Psi \\ &= E_0\Psi + K'\Psi, \end{aligned}$$

where

$$K' = \left[\frac{r_0^2}{r^2}-\beta\frac{r_0}{r}\right]\frac{\pi^2}{2m} - \left[\frac{r_0^2}{r^2}-\beta\frac{r_0}{r}\right]\frac{e\hbar\Sigma\cdot\mathbf{B}}{2mc} + \frac{r_0}{r}\frac{2\pi\hbar^2}{m^2c^2}\delta(r).$$

In conclusion, if this approach is as successful as we believe, we will still need to justify our approximation methods.

Acknowledgments

The work reported here represents the outcome of a program begun by the author, J. Lindesay and W. W. Zachary at Howard University over thirty years ago. The work in section two is joint with M. Alfred (Howard) and that in section three is joint with Gonzalo Ares de Parga (Mexico).

This paper is an extended version of a talk given at the The 9th Biennial Conference on Classical and Quantum Relativistic Dynamics of Particles and Fields held at the University of Connecticut in June 2014.

References

[1] A. J. Silenko Phys. Rev. D **89** (2014).
[2] A. J. Silenko Phys. Rev. A **77** **44** (2008).
[3] A. J. Silenko Eur. Phys. Journal C, **57** (2008), 595.
[4] V.M. Simulik and I.Yu. Krivsky Universal Journal of Physics and Application **2** (2014), 115-128.
[5] K. Kowalski and J. Rembieliński, arXiv.1110.5146v1 (2011).
[6] T.L. Gill and W.W. Zachary, *Analytic Representation of the square-root operator*, J. Phys. A: Math. Gen. **38** (2005) 2479-2496.
[7] L. S. Schulman, *Techniques and Applications of Path Integration*, Wiley, New York, (1981).
[8] T.L. Gill and W.W. Zachary, *A new class of Banach spaces*, J. Phys. A: Math. Theor. **41** (2008) 1-15, doi:10.1088/1751-8113/41/49/495206.
[9] T. L. Gill, W. W. Zachary and M. Alfred, *J. Phys. A: Math.. and Gen.* **38** (2005) 6955-6976.
[10] I. S. Gradshteyn and I. M. Ryzhik, *Tables of Integrals, Series and Products*, Academic Press, New York, (1980).
[11] H. Yukawa, Proc. Phys. Math. Soc. Japan **17** (1935) 48.
[12] N. F. Mott and H. S. Massey, *Theory of Atomic Collisions Clarendon*, Oxford, 1965.
[13] P. A. M. Dirac, Proc. Roy. Soc (London) **A117** (1928) 610, **A118** (1928) 351.
[14] B. Thaller, *The Dirac Equation*, Springer-Verlag, New York, (1992).
[15] L. C. Biedenharn, Phys. Rev. **126** (1962) 845.
[16] E. deVries, Fortsch. d. Physik **18** (1970) 149.
[17] D. Hestenes, *Space Time Algebra*, Gordon and Breach, New York, (1966).
[18] L. L. Foldy and S. A. Wouthuysen, Phys. Rev. **78**, (1950) 29.
[19] W. Pauli, *Handbuch der Physik, 2d ed.*, **24**, (1933).
[20] R. P. Feynman and M. Gell-Mann, Phys. Rev. **109**, (1958) 193.
[21] T.L. Gill, W.W. Zachary, and J. Lindesay, *The Classical Electron Problem*, Foundations of Phys. **31** (2001) 1299-1354.
[22] T. L. Gill and W. W. Zachary, *Two Mathematically Equivalent Versions of Maxwell's Equations*, Foundations of Phys. **41** (2011) 99-128.
[23] T. L. Gill and W. W. Zachary, *Foundations for relativistic quantum theory I: Feynman's operator calculus and the Dyson conjectures*, Journal of Mathematical Physics **43** (2002), 69-93.
[24] A. Salam, *Overlapping divergence and the S-matrix*, Phys. Rev. **82**, (1951) 217-227.
[25] S. Weinberg, *High energy behavior in quantum field theory*, Phys. Rev. **118**, (1960) 838-849.
[26] J. Magueijo , *New varying speed of light theories* Rept. Prog. Phys. **66** (11): (2003) 351-366.
[27] J. Moffat, *Superluminary Universe: A Possible Solution to the Initial Value Problem in Cosmology* , Int. J. Mod. Phys. D **2** (3): (1993) 351-366.
[28] J.A. Wheeler, R.P. Feynman, *Rev. Mod. Phys.* **21** (1949) 425.
[29] L. P. Horwitz and C. Piron, Helv. Phys. Acta **46**, (1981) 316.
[30] J. R. Fanchi, *Parametrized Relativistic Quantum Theory* Kluwer Academic, Dordrecht, (1993).

Canonical proper time quantum gravitation

James Lindesay

Computational Physics Laboratory, Department of Physics, Howard University, Washington, DC 20059

E-mail: jlindesay@howard.edu

Abstract. At the root of the tensions involved in modeling the quantum dynamics of gravitating systems are the subtleties of quantum locality. Quantum mechanics describes physical phenomena using a theory of non-local phase relationships (non-local in the sense that quantum states maintain a space-like coherence that is acausal). However, the principle of equivalence in general relativity asserts that freely falling frames are locally inertial frames of reference. Thus, gravitating systems are often described using constituents that are freely falling, undergoing geodesic motion defining well localized trajectories. The canonical proper time formulation of relativistic dynamics is particularly useful for describing such inertial constituents using the coordinates of non-inertial observers. The physics of the simplest of gravitating inertial quantum systems, consistent with presented experimental evidence, will be examined. Subsequently, descriptions of both weakly and strongly gravitating quantum systems will be developed using canonical proper gravitation.

1. Introduction

Quantum mechanics remains an enigmatic formulation when examined through intuitions framed in classical mechanics. Some of the fundamentals of quantum mechanics are itemized as follows:

- Measurements cannot be objectified in quantum physics. Certain physical parameters (like position and momentum) cannot be simultaneously measured in a single measurement.
- Quantum formulations of physical systems inherently incorporate unknown and unknowable quantities (as demonstrated in EPR entanglements[1], and Bell's hidden variable analysis[2]).
- What *can* be known about a quantum coherent system is contained in its quantum state vector $|\Psi\rangle$, state components Ψ_m, or wave function $\Psi(x,t)$, and probabilities $\Psi_m^* \Psi_m$ or $\Psi^*(x,t)\Psi(x,t)dx$.
- A given system is not the wavefunction itself, since wavefunctions can only be used to describe outcome likelihoods consistent with physical conservation laws. The system will be measured to be in one state or another, not spread amongst a set of states. This is even true of classical distributions of unmeasured states. However, for classical distributions the constituent states of an ensemble can in principle be known on some level.

As an illustration of the difference between a classically coherent system and a quantum coherent system, consider a wave-front propagating through space. For a classical ocean wave breaking upon a beach, each floating plankton and grain of sand on the beach is a detector of the classically coherent wave. However, for a quantum wave breaking upon detectors, there is no intermediate detection, and only one detector will measure the particle, resulting in a "collapse of the wavefunction" describing its propagation. Subsequence to that detection, the description of the quantum system provided by the prior wavefunction becomes obsolete. Any measurement of a quantum system fixes the immediate past of that system to be consistent with that particular measurement, but leaves any future measurements uncertain while the system remains coherent.

The double slit experiment provides a good demonstration of inherent quantum behaviors. In the double slit experiment illustrated in Figures 1, particles are incident from the left upon a solid wall that has only two slits through which any particle can pass without being absorbed. Classical particles either hit the barrier or hit the screen on the right, clustering just behind the

Figure 1. *Illustration of the result of 100 particles incident from the left upon a barrier with two horizontal slits for classical particles (left) vs quantum particles (right).*

slits. However, quantum particles are singly detected on the screen on the right in a distribution that is widely spread over the screen in a manner characterized by a wave that coherently passes through both slits. Notice that if the quantum particles are continuously observed, they behave like the classical particles. Furthermore, if any quantum particle is detected going through either slit, the particle is later detected as if it were just a classical particle (or perhaps diffracted wave) going only through that slit. Particles that are not detected going through either slit maintain space-like quantum coherence throughout the propagation and self-interfere going through both slits. Any measurement that localizes the propagation of the particle through one slit or the other breaks its coherence. The phase relationships in this experiment are determined purely by the classical geometry, and remain unchanged irregardless of any motions of the observer.

There are general constraints upon measurements of quantum systems. If one makes many measurements upon a set of identically prepared quantum coherent systems, the average of those measurements satisfy the uncertainty principle $\Delta x \, \Delta p \geq \frac{\hbar}{2}$, where ΔQ is the root mean squared deviation of measurements of quantity Q from its average value. This principle follows from the triangle identity for the magnitudes of two state vectors and their resultant, and the non-commutivity of the momentum and position operators[3]. This non-commutivity prevents *any* single measurement of both the momentum and position of a quantum state.

The incorporation of the non-local behaviors of quantum systems into gravitation is not trivial. If the gravitational interaction itself is to be quantized, the non-linear effects become particularly difficult to understand, especially for strongly gravitating systems. Also, experimental evidence presented in the next section demonstrates that gravitational quanta *cannot* be the mechanism by which space-time localization occurs during the gravitation of quantum coherent systems. In other words, space-time coordinates cannot "bubble up" during the quasi-stationary gravitation of coherent systems in a manner that localizes regions of those systems. The coordinates themselves cannot be entangled with the systems.

Classical gravitation is unique amongst interactions in that vastly differing objects with identical geometric initial conditions will follow identical trajectories, allowing the geometrization of gravitational dynamics. Locally, the motions of objects are essentially the same as those of inertial objects as viewed by accelerating observers, allowing descriptions of the dynamics in terms of various coordinate representations of those motions. The *principle of equivalence* effectively equates the inertial mass of an object with the mass determining the strength of gravitational coupling. This characteristic of gravity allows classical motions of arbitrary local objects in terms of unique *geodesics* independent of gravitational couplings. Covariance then allows one to use *any* convenient coordinates to describe those motions.

What follows will explore the consequences of the principle of equivalence on the gravitation, co-gravitation, and self-gravitation of quantum coherent systems. Section 2 summarizes some of the experiments involving quantum gravity, illustrates inertial systems as viewed by accelerating observers, and develops canonical proper gravitation. Section 3 explores some properties of a non-singular black hole, in particular demonstrating that quantum non-locality can in principle prevent gravitational collapse to a singularity. Finally, section 4 will briefly discuss the modeling of gravitating quantum fields.

2. Gravitation and Quantum Coherence

2.1. Experiments in Quantum Gravity

During the 1970's, experiments[4] were performed on quantum coherent neutrons undergoing self-interference in the Earth's gravitational field. Those experiments involved a form of double slit interference illustrated in Figure 2. In the experiment, singly traversing neutrons are counted

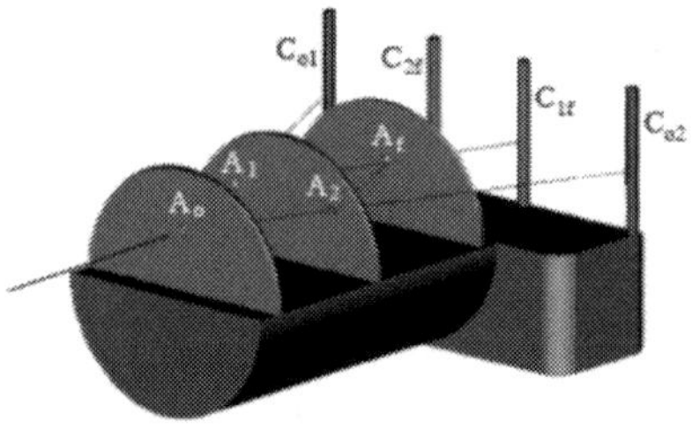

Figure 2. *Diagram of apparatus used to demonstrate quantum coherence of gravitating neutrons. Neutrons incident upon aperture A_o self-interfere through apertures A_1 and A_2, and are detected via counters C_{1f} and C_{2f} for which no classical trajectory for the neutrons exists.*

at the two counters C_{1f} and C_{2f} for which no classical trajectory exist for the incident neutrons. The apparatus is then rotated about the axis of incidence of the neutrons, changing the relative gravitational potential (and therefore the relative phase) between A_1 and A_2. The formulation of the interference pattern in the expected difference in counts between the counters requires use of both Newton's constant G_N and Planck's constant $\hbar$ in same formula, which was verified by the experiment. One could therefore conclude that gravitating quantum systems maintain spatial coherence. This means that gravitational interactions cannot break quantum coherence in order to establish spatial and temporal relationships, despite common interpretations that gravity *is* geometry. The establishment of the space-time geometry that localizes the propagation of each neutron through the slits cannot also localize the gravitating neutron at either slit. The experiment verifies the principle of equivalence for a quantum system, which will here be stated as follows: *The coherence state of an inertial system is not affected by the motions of the observer.* The result from the Earth-based laboratory should be indistinguishable from the result on an accelerating spaceship and apparatus with incident inertial neutrons.

Other experimental observations have demonstrated the maintenance of quantum coherence during gravitation:

- Interference in ticks of quantum atomic clocks superposed across differing gravitational potentials due to Compton frequencies (temporal coherence)[5].
- Wannier-Stark ladders (interference between fermions on rungs of an optical ladder[6]), which can be used as precise gravimeters[7].
- CMB photons maintain coherence for 14 billion years over a change in cosmological scale of 1100 (very dynamic!), while sometimes undergoing gravitational lensings.
- Red-shift / time dilation effects are important even for everyday behaviors of Global Positioning Satellites (GPS), resulting in a shift of +46ms/day from gravity and -7ms/day from motions (the GPS system requires 50 nanoseconds/day accuracy)[8].
- Precession of superfluid vortex arrays in Earth's gravitational field.

One can interpret these observations as demonstrating that the establishment of geometric relationships in a quantum coherent gravitating system does not break quantum coherence.

2.2. Uniform Proper Acceleration

Examine the differential Lorentz transformation from the (momentary) rest frame of a system with constant proper acceleration $(c\tau, z_P)$ to an inertial Minkowski frame (ct, z) expressed in terms of the rapidity ζ:

$$\begin{aligned} dct &= \cosh\zeta\, dc\tau + \sinh\zeta\, dz_P, \\ dz &= \sinh\zeta\, dc\tau + \cosh\zeta\, dz_P. \end{aligned} \tag{2.1}$$

For an observer at rest in the proper frame of reference, (2.1) implies that

$$v_P = \frac{dz_P}{dc\tau} = 0 \quad \Rightarrow \quad \frac{dct}{dc\tau} = \cosh\zeta\,, \ \frac{dz}{dct} = \tanh\zeta. \tag{2.2}$$

Since the direction of acceleration remains unchanged, the rapidities characterizing the Lorentz velocity transformations add, implying that

$$\frac{v'}{c} = \tanh(\zeta + \zeta') \quad \Rightarrow \quad \frac{\delta v_P}{c} = \tanh(\delta\zeta) = \delta\zeta. \tag{2.3}$$

This directly relates the proper acceleration a to the rapidity ζ:

$$\frac{dv_P}{d\tau} \equiv a, \ \frac{d\zeta}{d\tau} = \frac{a}{c} \rightarrow \zeta = \frac{a\tau}{c}. \tag{2.4}$$

Therefore, the solution for the trajectory undergoing uniformly accelerating motion using special relativity is given by

$$\begin{aligned} z(\tau) &= z_P + \tfrac{c^2}{a}\left(\cosh\tfrac{a\tau}{c} - 1\right), \\ t(\tau) &= \tfrac{c}{a}\sinh\tfrac{a\tau}{c}, \\ \tfrac{v(\tau)}{c} &= \tanh\tfrac{a\tau}{c} \end{aligned} \tag{2.5}$$

where z_P is the proper coordinate of the accelerating system, a is the proper acceleration of that system, and τ is the proper time coordinate for that system. The system is seen to be undergoing hyperbolic motion.

2.2.1. Clocks and redshifts

A particularly useful coordinate system can be developed using a spatially dependent form for the local proper acceleration[9]:

$$a = \frac{c^2}{z_P}. \tag{2.6}$$

Using this identification for the proper accelerations in (2.5), all of the trajectories with proper locations z_P approach a single outgoing light-like asymptote passing through the origin, which is a (shared) horizon with proper coordinate $z_P = 0$. The coordinates for an inertial observer (ct, z) corresponding to accelerating coordinates $(c\tau, z_P)$ satisfy

$$ct = z_P \sinh \frac{c\tau}{z_P}, \quad z = z_P \cosh \frac{c\tau}{z_P}. \tag{2.7}$$

Direct coordinate transformation properties can immediately be developed between the inertial observer and the accelerating observers within the shared causal patch of space-time.

A depiction of two such accelerating trajectories (2.7) is shown in Figure 3. In the figure, all

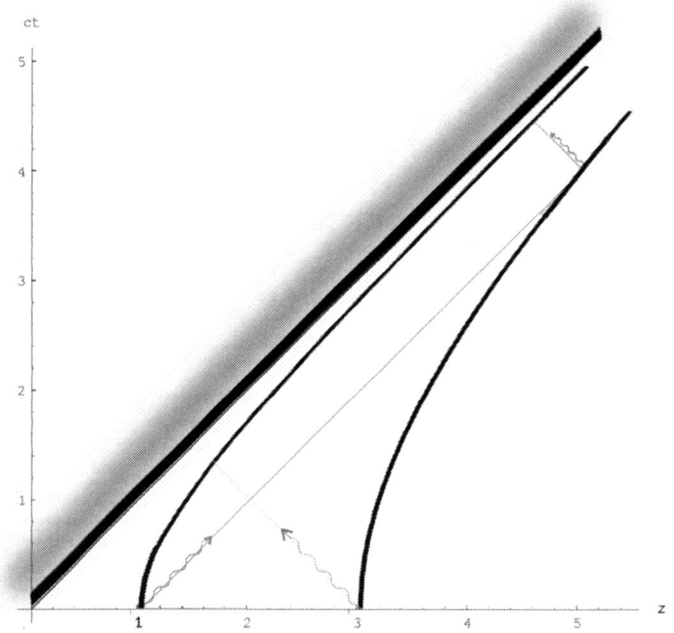

Figure 3. *The traversal of light rays as detected by uniformly accelerating observers, and their usefulness in constructing a standard clock.*

observers with accelerating coordinates are momentarily at rest relative to the inertial observers at $t = 0 = \tau$, with momentarily corresponding proper locations $z = z_P$. This surface represents a surface of synchronicity for the accelerating clocks with the inertial clocks. The trajectories of $z_P = 1$ and $z_P = 3$ are depicted by bold hyperbolas, and the shared horizon of the accelerating observers $z_P = 0$ is depicted as the outgoing light-like bold line through the origin of the Minkowski coordinates.

One can compare the temporal dilations for the different accelerating observers by examining the detection of two light beams emitted simultaneously by two observers towards each other at

Journal of Physics: Conference Series **615** (2015) 012014 doi:10.1088/1742-6596/615/1/012014

$\tau = 0 = t$, as demonstrated in Figure 4. In the first (bottom) frame, the clocks at two separate proper coordinates are synchronized at $\tau = 0$ as the rod is momentarily at rest with respect to the Minkowski frame, and the photons are emitted at each location towards the other. In the second frame, the ingoing photon has traversed along a greater proportion of the proper intervals than has the outgoing photon. By the third frame, the ingoing photon has reached the innermost observer at the proper time demonstrated on the left clock, as the outgoing photon continues towards the outermost observer. The outgoing photon finally reaches the outermost observer in the fifth frame at the proper time indicated. The innermost observer detects a blue-shifted photon, and has a slower moving clock, relative to the outermost observer.

A standard clock can be constructed by the observers in Figure 3 using a photon emitter/detector facing outward at the innermost observer, and a mirror facing inward held by the outermost observer. One tick of such a clock is depicted by the emission of an outgoing photon and the detection of it's reflection. Such a standard clock will tick in the uniformly accelerating system with uniform ticks $\Delta\tau$.

2.3. An Inertial Wavefront Described Using Accelerating Coordinates

Next, the propagation of an inertial neutron as it passes through a uniformly accelerating double slit will be examined. The neutron will be assumed to move with a uniform momentum in the x-direction, while the double slit barrier and detector will be assumed to uniformly accelerate in the z-direction. A wavefront of the neutron is depicted from various perspectives in the illustrations in Figure 5. In the leftmost set of sequential frames, the wavefronts, slits, and detector are displayed on the Minkowski ct-z plane, with the x-direction of propagation suppressed, in sequential snapshots as the proper time of each of the various observers progresses. The incident particle wavefront is displayed in the lower frames as a gray plane progressing upward with fixed z coordinate, becoming cylindrical surfaces after traversing the slits in the upper frames. The asymptotes of the accelerating observer are drawn as the gray light cone with apex at the origin, the double slit barriers are accelerating black solid lines, and the detector is an accelerating black rectangle.

The middle frames in Figure 5 demonstrate the perspective of an inertial observer as the wavefront approaches the uniformly accelerating barrier in the lower frames, becoming interfering wavefronts in the upper frames prior to encountering the accelerating detector. This represents the perspective from which the dynamics can be calculated most directly. The wavefront is a vertical plane prior to the barrier, and interfering cylindrical surfaces subsequent to the barrier. The frames on the right are from the perspective of the accelerating barrier and detector. From this perspective, the approaching wavefront is no longer a vertical planar surface, and the interfering surfaces droop due to the acceleration prior to detection.

2.4. Canonical Proper Gravitation

The canonical proper time formulation for relativistic systems[10, 11] can be quite useful in providing insights into the dynamics of inertial systems using the coordinates of accelerating observers. The canonical proper time generator K is defined to generate translations in the proper time τ of the observed system using the observers' coordinates via

$$[K, z^j(\tau)] = i\hbar \frac{d}{d\tau} z^j(\tau) \quad , \quad [K, p_j(\tau)] = i\hbar \frac{d}{d\tau} p_j(\tau) \,. \tag{2.8}$$

These commutation relations are consistent with the proper time "Hamilton" equations given by

$$\frac{\partial K}{\partial p_j} = \frac{dz^j}{d\tau} \quad , \quad \frac{\partial K}{\partial z^j} = -\frac{dp_j}{d\tau} \,. \tag{2.9}$$

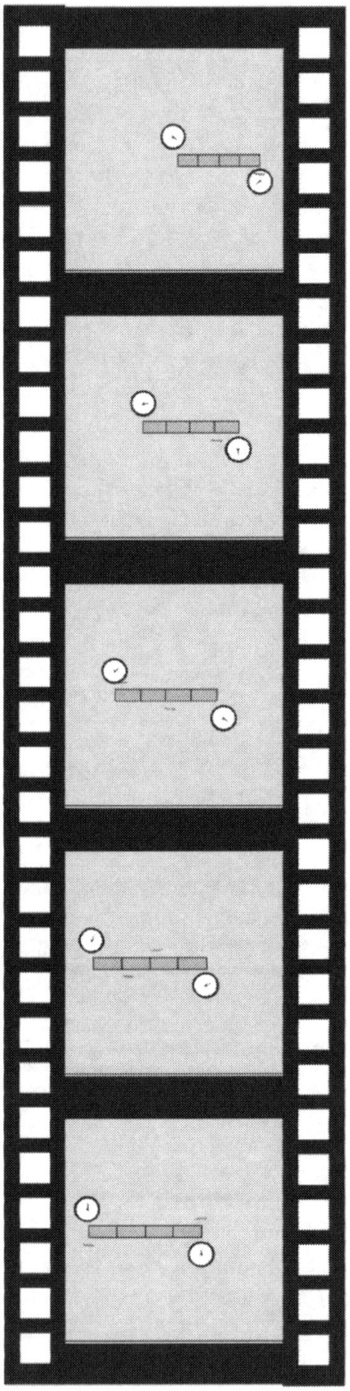

Figure 4. *The traversal of two photons between accelerating clocks.*

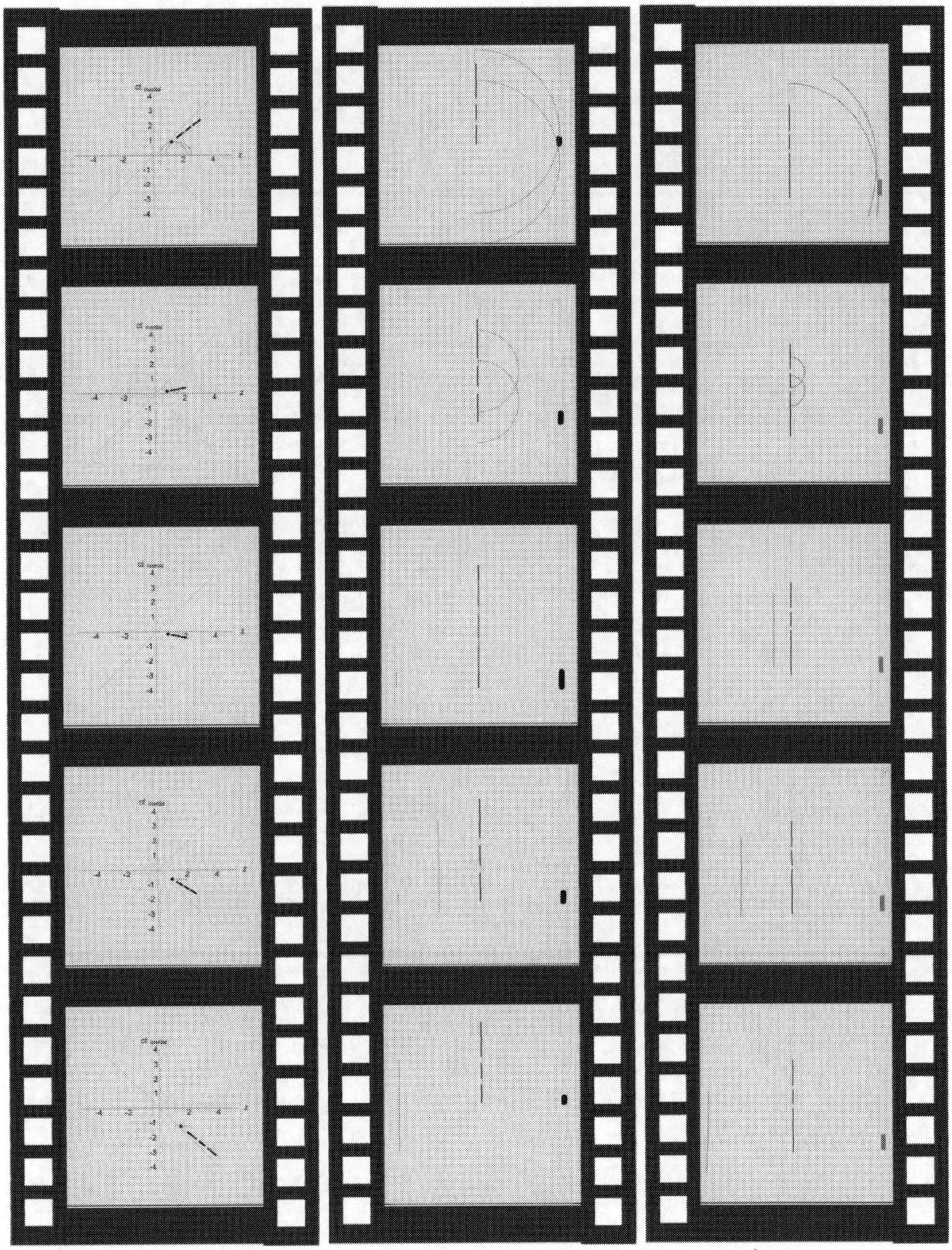

Figure 5. *Various depictions of an inertial wavefront incident upon an accelerating double slit barrier. The frames on the far left depict the wavefront and the accelerating barrier and detector on the Minkowski plane (ct, z) with the x-direction suppressed. In the middle frames, the inertial wavefront, accelerating barrier and accelerating detector are sequentially viewed by a stationary observer. In the frames on the right, the inertial wavefront is viewed relative to the barrier and detector.*

Journal of Physics: Conference Series **615** (2015) 012014 doi:10.1088/1742-6596/615/1/012014

If the relativistic momentum is to take the form $p^j = mu^j = m\frac{dz^j}{d\tau}$, and the relativistic force is of the conservative form $\frac{dp_j}{d\tau} = -\frac{\partial U(z)}{\partial z^j}$, this results in a proper time translation generator of the form

$$K = \frac{p^2}{2m} + U(z) + K_o. \tag{2.10}$$

Description of the dynamics using this canonical proper formulation has the following features:

- K generates translations in the proper time τ of the observed system, with an eigenvalue of the rest energy mc^2.
- The formulation looks very similar to non-relativistic physics, but is fully relativistic! This allows the use of techniques from non-relativistic quantum mechanics for solving relativistic systems.
- It should be noted that this is *not* a Lorentz transformation, rather it is a canonical transformation on the time variable. The temporal coordinate is the proper time of the system, while the momentum and position coordinates (p,x) are those assigned by the observer (which can be highly relativistic). The formulation generates equations of motion of the observers' coordinates for the system in terms of the proper time of that system, not the observers.

The question remains how one determines the constant K_o. This constant can be related to the standard energy E_o measured during production, detection, or a correspondence region of propagation of the quantum state, as will be illustrated in the following examples.

2.4.1. A Plane Wave One of the simplest of quantum systems is that describing a plane wave. A non-interacting quantum system with $U(x) = 0$ can generally have plane wave solutions

$$K = \frac{p^2}{2m} + K_o \quad , \quad K\Psi(x) = mc^2\Psi(x)\,. \tag{2.11}$$

The Hamiltonian formulation for this system has the usual square root form, which connects the constant K_o to the energy of the wave via

$$E_o = \sqrt{(pc)^2 + (mc^2)^2} \quad \Rightarrow \quad K_o = \frac{3(mc^2)^2 - E_o^2}{2mc^2}. \tag{2.12}$$

Thus, the canonical proper energy takes the form

$$\hat{K} = \frac{\hat{p}^2}{2m} + \frac{3(mc^2)^2 - E_o^2}{2mc^2}. \tag{2.13}$$

2.4.2. Canonical proper acceleration Next, consider the uniformly accelerating system examined in section 2.2. Since the solution is already known, one can instead determine the equations of motion.

From (2.5), one can make the identifications

$$\frac{dz}{d\tau} = c\sinh\frac{a\tau}{c} = \frac{\partial K}{\partial p_z}, \text{ and } \frac{dp_z}{d\tau} = \frac{d}{d\tau}\left(m\frac{dz}{d\tau}\right) = ma\cosh\frac{a\tau}{c} = -\frac{\partial K}{\partial z}. \tag{2.14}$$

These equations immediately imply that

$$\frac{\partial K}{\partial p_z} = \frac{p_z}{m}, \text{ and } \frac{\partial K}{\partial z} = -ma\left[\frac{a}{c^2}(z - z_{Po})\right], \tag{2.15}$$

which results in a generator for proper time translations of the form

$$K = \frac{p^2}{2m} - ma\left[\frac{a}{2c^2}z^2 + \left(1 - \frac{a z_{Po}}{c^2}\right)z\right] + K_o. \tag{2.16}$$

In this equation, z_{Po} is a coordinate of correspondence between the systems.

These steps can be analogously repeated to determine the Hamiltonian formulation for this system:

$$\frac{dz}{dt} = \frac{p_z c}{\sqrt{p_z^2 + (mc)^2}} = \frac{\partial H}{\partial p_z}, \text{ and } \frac{dp_z}{dt} = ma = -\frac{\partial H}{\partial z}. \tag{2.17}$$

This results in a generator for time translations of the form

$$H = \sqrt{(pc)^2 + (mc^2)^2} - maz. \tag{2.18}$$

The last term in (2.18) is akin to the usual mgz form for the gravitational potential energy in elementary physics descriptions of motion in a uniform gravitational field.

To evaluate the integration constant K_o, note that a correspondence can be found such that

$$E_o = \sqrt{(p_{\perp o}c)^2 + (mc^2)^2} - maz_{Po}, \tag{2.19}$$

which results in an integration constant K_o of the form

$$K_o = mc^2 - \frac{(E_o + m\,a\,z_{Po})^2 - (mc^2)^2}{2mc^2} + ma\left[\frac{a}{2c^2}z_{Po}^2 + \left(1 - \frac{a\,z_{Po}}{c^2}\right)z_{Po}\right], \tag{2.20}$$

where $z_{Po} = z_o$ is appropriately chosen when $p_z = 0$.

Writing $p^2 = p_x^2 + p_z^2$ in (2.16), one obtains solutions to the equations of motion

$$x(\tau) = x_o + \frac{p_{xo}\tau}{m} \text{ and } z(\tau) = z_{Po} + \frac{c^2}{a}\left(\cosh\frac{a\tau}{c} - 1\right). \tag{2.21}$$

The form of the space-time metric relates the proper time τ to the Minkowski time t via

$$(dc\tau)^2 = (dct)^2 - dx^2 - dz^2 \quad \Rightarrow \quad dct = dc\tau\sqrt{1 + \left(\frac{p_{xo}}{mc}\right)^2 + \sinh^2\frac{a\,\tau}{c}}. \tag{2.22}$$

Thus, the inertial and accelerating time coordinates can be related through the elliptic integral of the second kind:

$$ct = i\frac{c^2}{a}\sqrt{1 + \left(\frac{p_{xo}}{mc}\right)^2}\,E(i\frac{a\tau}{c}|\frac{1}{1 + (\frac{p_{xo}}{mc})^2}). \tag{2.23}$$

2.4.3. Semi-classical co-gravitation Consider next a set of freely co-orbiting quantum coherent (neutral) particles that are in composite spherically symmetric states. Experimentally, stationary gravitation itself does not break the coherence of any of the constituent particles. If one assumes that the local energy densities of the co-orbiting coherent quanta serve as gravitational sources generating the background field, then by Birkhoff's theorem (or Gauss' law for gravitational fields as demonstrated by Newton), only the interior (mass) energy density effects the local gravitation at a radial coordinate r. Such spherically symmetric shells can be incoherently combined to construct a co-gravitating system with quantum components.

A semi-classical depiction of this thought experiment analogous to Bohr's analysis of the hydrogen atom is shown in Figure 6. To the extent that any particles are measured interior to

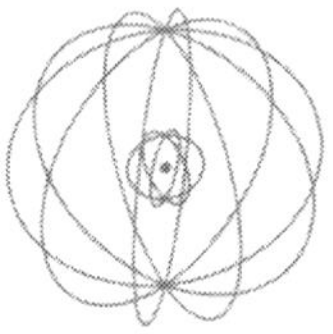

Figure 6. *A semi-classical incoherent sum of nested coherent gravitating particles acting as sources for exterior particles.*

a given co-gravitating particle, those measured particles serve as gravitational sources for any given exterior particle. Assume that the orbital angular momentum of a circularly orbiting mass m must be quantized using Bohr's criterion $L_n = m r_n v_n = n\hbar$. The gravitational source mass interior to r_n will be specified by M_n. Using arguments from elementary physics, the speed of the mass and radius parameterizing this orbital must satisfy

$$v_n = \frac{G_N m M_n}{n\hbar} \quad , \quad r_n = \frac{n^2\hbar^2}{G_N m^2 M_n} \quad . \tag{2.24}$$

As long as the dependence M_n assures that the radius increases with quantum number n, one can develop a model where each interior mass contributes as a gravitational source for the subsequent masses. A physical co-gravitating model is constructed by incoherently summing symmetric shells of mass m to give interior mass $M_n = n\, m$ for shell n. Equation (2.24) can then be re-written in terms of the reduced Compton wavelength $\lambda_m \equiv \frac{\hbar}{m\,c}$ of the mass m and the Planck length $L_P \equiv \frac{\hbar}{M_P c}$, giving

$$\begin{array}{ll} a_m = \left(\frac{\lambda_m}{L_P}\right)^2 \lambda_m \quad , & E_n = -\frac{1}{2}\left(\frac{L_P}{\lambda_m}\right)^4 m\, c^2 \quad , \\ v_n = \left(\frac{L_P}{\lambda_m}\right)^2 c \quad , & r_n = n\, a_m \quad , \end{array} \tag{2.25}$$

where a_m is a gravitational Bohr radius associated with the mass m, and E_n is the orbital energy of the mass in orbital n. In contrast to the behaviors of electrons in atomic physics, each successive shell of co-gravitating masses experiences increased attraction due to the interior masses. It is interesting to note that the semi-classical speeds of the masses, as well as their gravitational binding energies, are independent of the orbital in this model.

2.4.4. Proper time quantum co-gravitating particles Next, explore the co-gravitation of masses using the canonical proper time formulation. Replacing $U(r) = m\Phi(r)$ as the interaction potential for the mass m in (2.10), the canonical proper time generator K takes the form

$$K = \frac{p^2}{2m} + m\,\Phi(r) + K_o, \tag{2.26}$$

resulting in a canonical equation of motion of the form

$$\frac{d^2 x^j}{d\tau^2} = -\nabla_j \Phi(r). \tag{2.27}$$

If this equation of motion is to be consistent with geodesic motion on the geometry

$$\frac{d^2x^j}{d\tau^2} = -\Gamma^j_{\alpha\beta}\frac{dx^\alpha}{d\tau}\frac{dx^\beta}{d\tau}, \tag{2.28}$$

then for a stationary distribution with $\frac{dx^j}{d\tau}=0$ geodesic motion requires that

$$\frac{d^2x^j}{d\tau^2} = \frac{1}{2}g^{j\mu}\frac{\partial g_{00}}{\partial x^\mu}\left(\frac{dx^0}{d\tau}\right)^2. \tag{2.29}$$

As a quick calculation, consider a generalization of the exterior Schwarzschild geometry obtained by simply giving the mass a radial dependency:

$$ds^2 = -\left(1-\frac{2G_N M(r)}{c^2 r}\right)(dct)^2 + 2\sqrt{\frac{2G_N M(r)}{c^2 r}}\,dct\,dr + dr^2 + r^2\left(d\vartheta^2 + \sin^2\vartheta\,d\varphi^2\right). \tag{2.30}$$

If there is no radial dependency in the mass $M(r)$, then this form can be diagonalized into the Schwarzschild form using the Schwarzschild time $t_S \neq t$. However, (2.30) has no physical singularities except perhaps at the origin $r=0$ for a radially dynamic geometry. For this metric form, the potential that satisfies (2.29) and (2.27) has the form

$$\Phi(r) = -\frac{G_N M(r)}{r}, \tag{2.31}$$

which is the same as that for Newtonian gravitation, but includes space-time curvatures! Substitution of this metric into Einstein's equation gives

$$G^0{}_0 = \frac{1}{r^2}\frac{\partial}{\partial r}\frac{2G_N M(r)}{c^2} \quad \Rightarrow \quad M(r)c^2 = \int_0^r T^0{}_0(r')\,4\pi r'^2 dr', \tag{2.32}$$

which is precisely the interior mass corresponding to the energy density that curves the space-time.

A general quantum co-gravitating system is developed by considering a stationary mass m gravitating due both to a source mass M_ℓ which depends only on ℓ (the angular momentum quantum number associated with the stationary quantum state), as well as self gravitation. The proper energy form satisfies the stationary equation given by

$$\begin{gathered}\left(\frac{\hat{p}^2}{2m} - \frac{G_N m M_\ell(r)}{r} + K_\ell\right)\psi_{\ell\ell_z}(r,\theta,\phi) = mc^2\,\psi_{\ell\ell_z}(r,\theta,\phi)\quad, \\ \psi_{\ell\ell_z}(r,\theta,\phi) = R_\ell(r)Y_\ell^{\ell_z}(\theta,\phi).\end{gathered} \tag{2.33}$$

Re-writing the momentum squared operator as $\mathbf{p}\cdot\mathbf{p} = -\hbar^2\left[\frac{1}{r^2}\frac{\partial}{\partial r}\left(r^2\frac{\partial}{\partial r}\right) - \frac{\mathbf{L}\cdot\mathbf{L}}{\hbar^2 r^2}\right]$, one notes that spherically symmetric forms are constructed by summing over the $2\ell+1$ values of the index ℓ_z using the spherical harmonic addition theorem,

$$\sum_{\ell_z=-\ell}^{\ell}\left|Y_\ell^{\ell_z}(\theta,\phi)\right|^2 = \frac{2\ell+1}{4\pi}P_\ell(1).$$

The reduced radial wavefunctions defined by $u_\ell(r/a_m) \equiv rR_\ell(r)$ can be parameterized in terms of the dimensionless variable $\zeta \equiv r/a_m$, where as before $a_m \equiv \frac{\hbar^2}{G_N m^3}$. This allows the equation (2.33) to be re-written in a dimensionless form:

$$\frac{d^2u_\ell(\zeta)}{d\zeta^2} - \frac{\ell(\ell+1)}{\zeta^2}u_\ell(\zeta) + \left(\frac{2}{\zeta}\right)\left(\frac{M_\ell(\zeta)}{m}\right)u_\ell(\zeta) = \epsilon_\ell\,u_\ell(\zeta)\,, \tag{2.34}$$

where the dimensionless expression $\epsilon_\ell = -2\left(\frac{\lambda_m}{L_P}\right)^4 \frac{mc^2-K_\ell}{m\,c^2}$ parameterizes the gravitational binding energy of the mass m.

In analogy to the semi-classical case previously presented, the interior mass will be presumed to be composed of particles of mass m_ℓ:

$$M_\ell(\zeta) = \int_0^\zeta \rho_{mass}(\zeta')\,d\zeta' = \int_0^\zeta \sum_{\text{all }\ell'} n_{\ell'}\,m_{\ell'}\,u_{\ell'}^2(\zeta')\,d\zeta', \tag{2.35}$$

which represents an incoherent sum of the co-gravitating interior masses $m_{\ell'}$. This means that in (2.34), the particle masses appear only in ratios, an explicit demonstration of the elegance of the principle of equivalence.

For illustrative purposes, consider a single s-wave particle of mass m self-gravitating in a spherically symmetric manner. The gravitating distribution takes the form demonstrated in Figure 7. The metric component from (2.30) which has the form $g_{00} = -\left(1 - \frac{2G_N M(r)}{c^2 r}\right)$, can

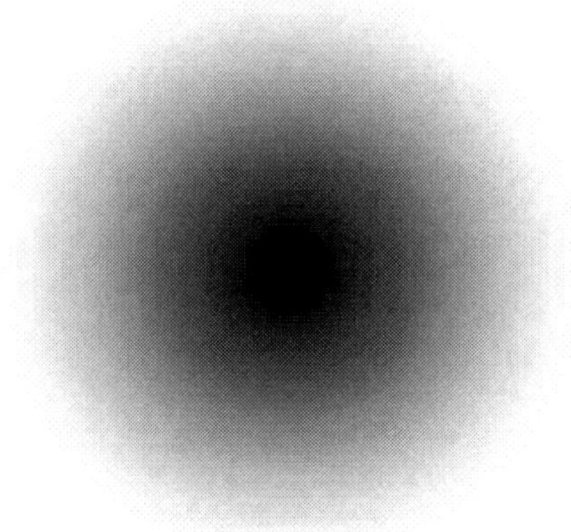

Figure 7. *Probability density of a single self gravitating proper coherent mass.*

be expressed in terms of the dimensionless functions

$$\frac{2G_N M(r)}{c^2 r} = 2\left(\frac{m}{M_P}\right)^4 \frac{\mathcal{P}_{interior}(\zeta)}{\zeta}, \tag{2.36}$$

where $\mathcal{P}_{interior}(\zeta)$ represents the integrated probability density interior to radial coordinate $\zeta = r/a_m$. Notice that in this expression, the particle mass appears only in terms of its ratio relative to the Planck mass M_P, with $\mathcal{P}_{interior}(\zeta)$ defining a *general* functional form in the coordinate ζ.

2.5. Global causal structure of gravitating mass

Quite often, diagrammatic techniques are useful for visualizing and interpreting important aspects of a physical model. Penrose diagrams are space-time diagrams that have the following properties:

Journal of Physics: Conference Series **615** (2015) 012014 doi:10.1088/1742-6596/615/1/012014

- Light rays travel along straight lines with slope ± 1
- The whole of space-time fits on a single page

These properties allow the diagrams to be used to determine the global causal structure of the given geometry.

Consider for example the Penrose diagram Figure 8 for flat Minkowski space-time. On the

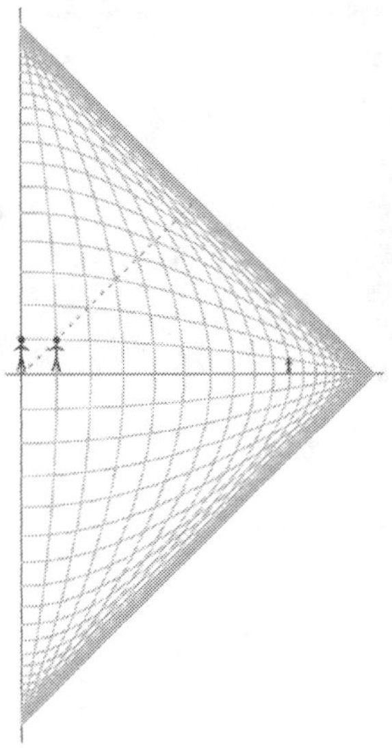

Figure 8. *Global causal structure of Minkowski space-time. The vertical axis represents ct (i.e., $r = 0$), and the horizontal axis represents r (i.e., $ct = 0$). The left boundary represents an observer at rest at $r = 0$. The lower right boundary (often referred to as* skri minus, *or past light-like infinity) represents an outgoing light ray emitted from $r = 0$ at $t = -\infty$, reaching $r = \infty$ at $t = 0$. The upper right boundary (often referred to as* skri plus, *or future light-like infinity) represents an ingoing light ray emitted from $r = \infty$ at $t = 0$, reaching $r = 0$ at $t = +\infty$. Each point on the diagram represents a spherical surface (i.e., all ϑ, φ) at a given time.*

diagram, observers of the same proper height are represented at $r = 0$, $r = 1$, and $r = 10$. The scale is seen to compress in distant locations and times in order to contain the whole of space-time on a finite diagram. An outgoing light-beam emitted from $r = 0$ at $t = 0$ traverses equal spatial and temporal displacements on the diagram as displayed as a dashed line of slope 1. No event above this light-like line can be a cause of an event below it. However, there is no spatial region in Minkowski space-time that is causally inaccessible from another over all times.

Energy density will curve the local space-time geometry, and modify the Penrose diagram, sometimes to the extent that it will change the global causal structure of the geometry (as with a black hole). For a dense star or self-gravitating mass such as in Figure 7, light rays are blue shifted when approaching the center $r = 0$, and red shifted when receding. Using the conformal space-time coordinates (i.e., coordinates that preserve the slope of light-like trajectories) needed to construct a Penrose diagram, a ingoing light beam from past light-like infinity that reflects off of (or passes through) $r = 0$ traverses a modified geometry, as illustrated in Figure 9. The curvature effect at the time-like center $r = 0$ is due to the high energy density, plotted as a proportionate density of shading on the diagram. For a high enough energy density, the curvature near the center is eventually large enough to form a black object (which has a trapping region within which outgoing light rays have decreasing radial coordinate) or a black hole (which

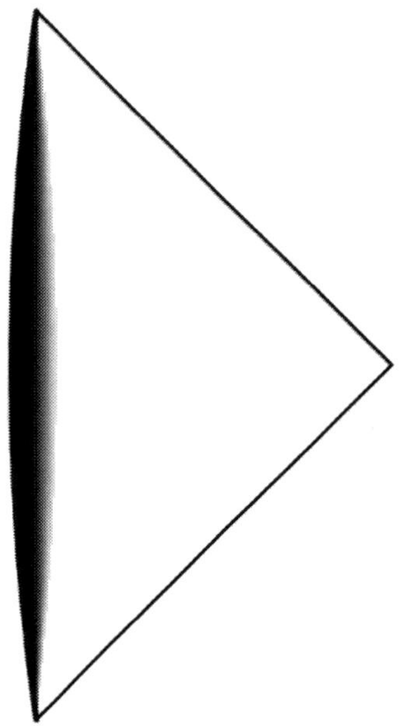

Figure 9. *Penrose diagram for a self gravitating high energy density. The global causal structure is the same as that of Minkowski space-time.*

manifests a *horizon*, within which there can be no cause that affects the exterior). In Figure 9, the Schwarzschild (vacuum) geometry is valid in the region exterior to any energy density. The past and future light-like infinities are familiar from the Minkowski space-time behavior of the geometry far away from its center.

3. A Non-Singular Black Hole

One notable feature of (2.36) is that the term calculated from the general equation of motion depends only upon the dimensionless radial variable $\zeta = r/a_m$, resulting in solutions for self gravitating masses that are always of the functional form demonstrated in Figure 7, regardless of the size of the mass m. This means that only the term involving the ratio $\left(\frac{m}{M_P}\right)^4$ determines when this distribution can result in a sign change in a metric component at some finite radial coordinate ζ. For the self-gravitating solution presented, no black hole is generated unless $g_{00} > 0$ which occurs only when $m \gtrsim 0.63 M_P$. Significantly for this calculation, *quantum non-locality prevents a singularity from ever forming regardless how large the self-gravitating mass!*

As a concrete example, consider a non-singular Planck mass black hole $m = M_P$. The metric component $-g_{00}(\zeta)$ is plotted in Figure 10. For large ζ far from the black hole, the metric takes the Minkowski form $g_{00} \to -1$. Similarly, very near the center $\zeta \to 0$ the metric likewise takes a Minkowski form. At the center, the metric and all curvatures are finite because of the quantum distribution consistent with non-locality and the uncertainty principle. However, in the region between $0.4 \lesssim \zeta \lesssim 2$, the component g_{00} is positive, defining a trapping region within which outgoing light rays will have decreasing radial coordinate ζ. Thus, the global geometry separates into three regions: an exterior region, a trapping region, and an interior region, as illustrated in the Penrose diagram for this system in Figure 11. On the diagram, the exterior region is labeled on the lower right hand part of the diagram, and the interior region is the

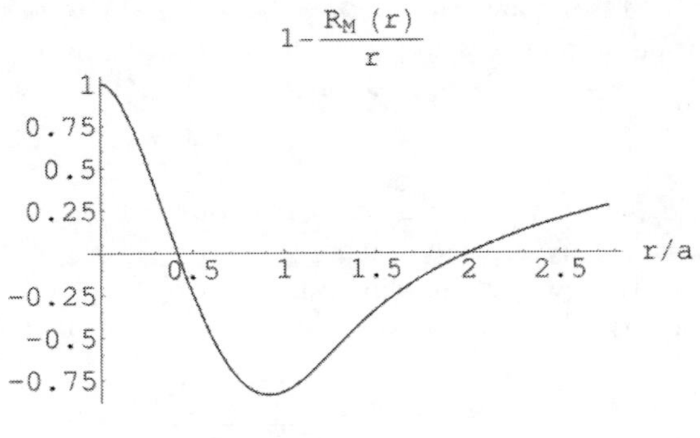

Figure 10. *Plot of the time-time component of the space-time metric for a self-gravitating Planck sized mass as a function of the dimensionless radial coordinate* $\zeta = \frac{r}{a_{M_P}} = \frac{r}{L_P}$.

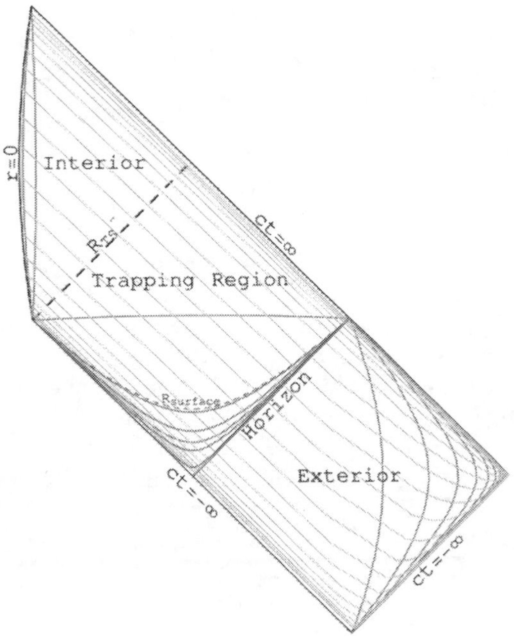

Figure 11. *Penrose diagram of a non-singular Planck mass black hole.*

upper left hand part of the diagram. The horizon of the black hole separates the exterior region from the trapping region, and the inner trapping surface R_{TS} separates the interior region from the trapping region. Curves of constant time are everywhere space-like volumes, all terminating on $r = 0$ (which is time-like and non-singular) and at the right apex of the diagram $r = \infty$ as shown. Curves of constant radial coordinate are vertical in the interior and exterior regions, but horizontal in the trapping region, as illustrated. Thus all light-like and time-like trajectories necessarily have decreasing radial coordinate in the trapping region. There are no exotic energy densities in the interior and exterior regions.

In the Penrose diagram 11, ingoing time-like or light-like trajectories can connect external causes to effects in the trapping region or interior. However, all time-like or light-like trajectories in the interior always terminate on a future boundary of the interior. This defines the object as a black hole.

3.1. General Stationary Spherically Symmetric Distributions

The Schwarzschild geometry only represents the vacuum solution of a stationary, spherically symmetric system exterior to any non-vanishing energy density. The most general, stationary, radially symmetric metric form that is also valid in the interior region with non-vanishing energy density can be written in the form[3]

$$ds^2 = -\left(1 + \frac{2[V(r) + V_o(r)]}{c^2}\right)(dct)^2 + 2\sqrt{\left(-\frac{2V(r)}{c^2}\right)\frac{\left(1+\frac{2[V(r)+V_o(r)]}{c^2}\right)}{\left(1+\frac{2V(r)}{c^2}\right)}}\, dct\, dr$$
$$+dr^2 + r^2\left(d\vartheta^2 + \sin^2\vartheta\, d\varphi^2\right), \tag{3.1}$$

which reduces to the form (2.30) in the exterior region $V_o \to 0$ and $V(r) \to -\frac{G_N M}{r}$.

The canonical proper time generator for the metric (3.1) consistent with previous assumptions takes the form

$$K = \frac{p^2}{2m} + m\,\Phi(r) + K_o \Rightarrow \frac{\partial}{\partial r}\Phi(r) = \frac{1+\frac{2V(r)}{c^2}}{1+\frac{2[V(r)+V_o(r)]}{c^2}}\frac{\partial}{\partial r}[V + V_o] \tag{3.2}$$

In the interior region, the solution for the canonical potential $\Phi(r)$ is not trivially related to general forms of the functions $V(r)$ and $V_o(r)$ as it was for the metric (2.30), but should be calculable in terms of an integral. In addition, the condition of isotropy in Einsteins equation, as well as any equation of state, relates V_o to V, which is again determined via the quantum distribution of a self-gravitating mass. This means that the calculation of a physically meaningful non-singular self-gravitating systems should be straightforward, though complicated. It is expected that even a black hole resulting from such a non-singular self-gravitating system should have no exotic energy density (i.e., no space-like energies, thereby violating energy conditions) anywhere on the geometry.

Substitution of the metric (3.1) into Einstein's equation directly relates the geometry to the energy-momentum densities consistent with that geometry. By direct calculation, the ${T^0}_0$ component of the energy-momentum tensor takes the form

$${T^0}_0 = \frac{1}{r^2}\frac{\partial}{\partial r}\left(\frac{rV(r)c^2}{4\pi G_N}\right) \tag{3.3}$$

This implies that if the function $V(r)$ is rewritten in the general form $V(r) = -\frac{G_N M(r)}{r}$, then

$$M(r)c^2 = -\int_0^r {T^0}_0\, 4\pi\tilde{r}^2 d\tilde{r}. \tag{3.4}$$

One can immediately interpret the function $M(r)$ as the mass interior to a spherical region of radius r.

4. Beyond the Canonical Proper Generator K: Scattering in Quantum Gravity

To end the discussion, a generalization of the canonical proper formulation beyond the eigenvalue problems discussed so far will be examined. Consider an invariant form for the generator K expressed as follows:

$$\hat{K}|\psi_m\rangle = i\hbar \frac{dx^\mu}{d\tau}\frac{\partial}{\partial x^\mu}|\psi_m\rangle = -u^\mu \hat{P}_\mu |\psi_m\rangle = mc^2|\psi_m\rangle. \tag{4.5}$$

Dynamics expressed in this form typically represent substantive quantum flows[12]. As previously discussed, although the canonical variables used to develop expressions for the generator K are generally those of observers in arbitrary motions, its eigenvalue is the invariant energy of the system.

An analogous expression is given by the Dirac equation[14] for spinor fields. Consider the expression on a locally flat space given by

$$\frac{1}{\Gamma}\mathbf{\Gamma}^\mu \hat{P}_\mu \mathbf{\Psi}^{(\Gamma)}_{(\gamma)}(m\vec{u}, J, s_z) = -\frac{(\gamma)}{\Gamma} mc\, \mathbf{\Psi}^{(\Gamma)}_{(\gamma)}(m\vec{u}, J, s_z), \tag{4.6}$$

where $\frac{(\gamma)}{\Gamma} = \pm 1$ for the $\Gamma = \frac{1}{2}$ representation. The substantive flow field for general coordinates can be expressed in terms of the curvilinear matrix-valued 4-velocity $\mathbf{U}^\beta$ defined using the principle of equivalence

$$\mathbf{U}^\beta(x) \equiv \frac{\partial x^\beta}{\partial \xi^\mu}\frac{\mathbf{\Gamma}^\mu}{\Gamma}. \tag{4.7}$$

In this expression, the ξ^μ are locally flat coordinates that can always be found. An algebra that consistently develops this dynamics is demonstrated in the next section.

4.0.1. Extended Lorentz Group Commutation Relations

The commutation relations of the Lorentz group algebra can be extended in a manner consistent with the Dirac matrices[15, 16] as follows:

$$\begin{aligned}
&[J_j\,,\,J_k] = i\hbar\,\epsilon_{jkm}\,J_m, \quad [J_j\,,\,K_k] = i\hbar\,\epsilon_{jkm}\,K_m, \quad [K_j\,,\,K_k] = -i\hbar\,\epsilon_{jkm}\,J_m,\\
&\left[\Gamma^0\,,\,\Gamma^k\right] = \tfrac{i}{\hbar}\,K_k, \qquad [\Gamma^0\,,\,J_k] = 0, \qquad [\Gamma^0\,,\,K_k] = -i\hbar\,\Gamma^k,\\
&\left[\Gamma^j\,,\,\Gamma^k\right] = -\tfrac{i}{\hbar}\,\epsilon_{jkm}\,J_m, \quad [\Gamma^j\,,\,J_k] = i\hbar\,\epsilon_{jkm}\,\Gamma^m, \quad [\Gamma^j\,,\,K_k] = -i\hbar\,\delta_{jk}\,\Gamma^0.
\end{aligned} \tag{4.8}$$

The operators Γ^β are chosen to be dimensionless, while the angular momentum J_k and boost K_j generators have dimensions of $\hbar$. Representations are labeled by the parameter Γ, which must take on integer or half-integer values. Microscopic causality[3] then requires that the associated spinor fields must satisfy boson statistics for integer values and fermi statistics for half-integer values of Γ.

4.0.2. A closed set of extended Poincare operators The introduction of operator valued group generators $\hat{\Gamma}^\beta$ that extend the Lorentz group algebra require the introduction of an additional scalar operator $\hat{\mathcal{M}}_T$ when space-time translations are included, in order to close the resulting extended group algebra[3]. Any closed group of operations must satisfy the algebraic Jacobi identities, $[A,[B,C]] + [C,[A,B]] + [B,[C,A]] = 0$. The non-vanishing commutators involving the operators $\hat{P}_\mu$ and $\hat{\mathcal{M}}_T$ that satisfy the Jacobi identities[17] and insure group closure are given by

$$\begin{aligned}
&[J_j\,,\,P_k] = i\hbar\,\epsilon_{jkm}\,P_m, \quad [K_j\,,\,P_0] = -i\hbar\,P_j, \quad [K_j\,,\,P_k] = -i\hbar\,\delta_{jk}\,P_0,\\
&[\Gamma^\mu\,,\,P_\nu] = i\,\delta^\mu_\nu\,\mathcal{M}_T c, \quad [\Gamma^\mu\,,\,\mathcal{M}_T] = \tfrac{i}{c}\,\eta^{\mu\nu}\,P_\nu.
\end{aligned} \tag{4.9}$$

An extended Poincare group Casimir operator is constructed using the Lorentz invariants

$$\mathcal{C}_m \equiv \mathcal{M}_T^2 c^2 - \eta^{\beta\nu} P_\beta P_\nu. \tag{4.10}$$

The label m in the Casimir $\mathcal{C}_m$ is a mass that parameterizes the eigenstates used to construct finite dimensional representations. Due to the form of the group Casimir, eigenvalues of the hermitian operator $\mathcal{M}_T$ will be referred to as the *transverse mass* of the state.

The transverse mass operator $\mathcal{M}_T$ introduced to close the algebra can be shown to have non-vanishing values only for massless particles $\eta^{\beta\nu} P_\beta P_\nu = 0$. For massless particles, $\mathcal{M}_T$ serves as the generator for affine translations along their light-like trajectories . Non-vanishing eigenvalues for $\mathcal{M}_T$ provide a mechanism for dynamic mixing of massless particles of differing transverse mass.

This formulation has direct applicability in relativistic cluster decomposable and unitary multi-particle scattering theory consistent with correspondence to classical, non-relativistic dynamics. However, the approach here utilized differs from that in reference [18], wherein space-time curvature is introduced via the generalization $\{\gamma^\mu, \gamma^\nu\} = -\eta^{\mu\nu}\mathbf{1} \rightarrow -g^{\mu\nu}(x)\,\mathbf{1}$ for the Dirac matrices γ^μ[19]. In general, $\{\mathbf{\Gamma}^\mu, \mathbf{\Gamma}^\nu\} \neq -\eta^{\mu\nu}\mathbf{1}$ (except for the $\Gamma = \frac{1}{2}$ representation which reproduces the Dirac equation). Rather, the approach presented here introduces space-time curvature directly through the principle of equivalence, along with the introduction of the Minkowski metric through the commutation relations between the operators $\hat{\Gamma}^\mu$, along with Lorentz invariance of the product $\hat{\Gamma}^\mu \hat{P}_\mu$.

5. Discussion and Conclusions

The physical universe is inherently quantum mechanical at its most fundamental level. Although quantum physics is non-deterministic, classical behaviors emerge when measurements cluster very near to quantum averaged values, providing validity to deterministic classical models.

Experiments demonstrate that the known gravitational effects cannot break the quantum coherence of gravitating systems in order to establish the needed space-time dependent phase relationships. In particular, gravitating quantum systems establish interference across regions of space-time described using classical ideas of geometric relationships. This means that the use of classical space-time relationships is sufficient to describe all known data, even for gravitating quantum coherent systems. Therefore, the approach taken here is to use expectation values of the energy-momentum tensor in Einstein's equation to generate the consistent curvatures of classical geometries upon which measurements take place. Covariance must connect such geometries to any other approach (e.g. using quantized coordinates, etc.) through appropriate correspondence, as long as Einstein's equation remains consistent with experiment.

Locally inertial coordinates can always be found for freely gravitating systems. Generally, one expects that the particle mass of a freely falling system is a parameter best described using the proper reference frame dynamics of that system. The proper time of a particle of defined mass provides a very convenient parameter for describing that dynamics, since it is the temporal variable of most relevance to the particle itself. This has been the primary motivation for the formulation of canonical proper gravitation as developed in this paper.

Since canonical proper gravitation embraces equivalence, it formulates quantum gravity in a way that is consistent with the formulation of classical gravitation by Einstein. It preserves quantum linearity, unitarity and classical correspondence. Expectation values of quantum energy densities generate space-time curvatures consistent with general relativity. Furthermore, it has been demonstrated that the formation of a singularity during gravitational collapse is not inevitable, due to quantum non-locality (i.e., the non-commutivity of position and momentum, resulting in the uncertainty principle). Since quantum non-locality is expected to continue to be a fundamental characteristic of known physical energies, exotic geometric objects (e.g. black holes, etc.) need not necessarily imply exotic or singular energy densities, in principle.

Acknowledgments
The author wishes to acknowledge the International Association for Relativistic Dynamics for its continuing support of novel approaches towards understanding outstanding problems in theoretical physics. In addition, the author appreciates the efforts of Tepper Gill in developing the canonical proper time formulation of relativistic systems.

References

[1] Einstein A, Podolsky B, and Rosen N 1935 *Phys. Rev.* **47**, 777
[2] Bell J S 1964 *Physics* **1**, 195
[3] Lindesay J 2013 *Foundations of Quantum Gravity* (Cambridge: Cambridge University Press)
[4] Colella R, Overhauser A W, and Werner S A 1975 Observation of Gravitationally Induced Quantum Interference *PRL* **34** 1472-4
[5] Muller H, Peters A, and Chu S 2010 A precision measurement of the gravitational redshift by the interference of matter waves *Nature* **463** 926-9
[6] Levi B G 2004 Ultracold gases of fermionic atoms offer another path to atom interferometry *Physics Today* August 25
[7] Roati G, de Mirandes E, Ferlaino F, Ott H, Modugno G, and Inguscio M 2004 *Phys. Rev. Lett.* **92**, 230402
[8] Will C M 2006 The Confrontation between General Relativity and Experiment *Preprint* arXiv:gr-qc/0510072v2
[9] Susskind L and Lindesay J 2005 *An Introduction to Black Holes, Information, and the String Theory Revolution: The Holographic Universe* (Singapore: World Scientific)
[10] Lindesay J and Gill T 2004 Canonical Proper Time Formulation for Physical Systems *Foundations of Physics* **34**(1), 169-82
[11] Gill T and Lindesay J 1993 Canonical Proper Time Formulation of Relativistic Particle Dynamics *International Journal of Theoretical Physics* **32**, 2087-98
[12] Lindesay J V and Morrison H L 1985 The Geometry of Quantum Flow *Mathematical Analysis of Physical Systems*, ed R E Mickens, (NewYork: Van Nostrand Reinhold) p 135
[13] Dirac P A M 1928 *Proc. Roy. Soc.* (London) **A117** 610
[14] Dirac P A M 1928 *Proc. Roy. Soc.* (London) **A118** 351
[15] Lindesay J 2003 Linear Spinor Field Equations for Arbitrary Spins *Preprint* arXiv:math-ph/0308003
[16] Lindesay J 2003 Group Structure of an Extended Lorentz Group *Preprint* arXiv:math-ph/0309060
[17] Lindesay J 2003 An Extended Poincare Algebra for Linear Spinor Field Equations *Preprint* arXiv:math-ph/0308015
[18] Brill D and Wheeler J 1957 Interaction of Neutrinos and Gravitational Fields *Rev. Mod. Phys.* **29** 465
[19] Bade W L and Jehle H 1953 An Introduction to Spinors *Rev. Mod. Phys.* **25**, 714

IARD 2014 IOP Publishing
Journal of Physics: Conference Series **615** (2015) 012015 doi:10.1088/1742-6596/615/1/012015

The manifestly covariant Aharonov-Bohm effect in terms of the 4D fields

T. Ivezić
Ruđer Bošković Institute, P.O.B. 180, 10002 Zagreb, Croatia
E-mail: ivezic@irb.hr

Abstract.
In this paper it is presented a manifestly covariant formulation of the Aharonov-Bohm (AB) phase difference for the magnetic AB effect . This covariant AB phase is written in terms of the Faraday 2-form F and using the decomposition of F in terms of the electric and magnetic fields as four-dimensional (4D) geometric quantities. It is shown that there is a static electric field outside a stationary solenoid with resistive conductor carrying steady current, which causes that the AB phase difference in the magnetic AB effect may be determined by the electric part of the covariant expression, i.e., by the local influence of the 4D electric field and not, as generally accepted, in terms of nonzero vector potential.

PACS numbers: 03.65.Vf, 03.30.+p, 03.50.De

1. Introduction

In a recent paper [1] the covariant generalizations of the Aharonov-Bohm (AB) effect [2] are considered. One of these generalizations, which will be investigated in this paper, is in terms of the space-time "area" integral of the electric and magnetic fields written in terms of the Faraday 2-form F, Eq. (6) in [1] or Eq. (1) here.

In this paper two important changes relative to [1] will be presented. The first change, which will be discussed in Sec. 2, refers to the mathematical formulation, whereas the second one refers to the physical interpretation of the AB phase shift and it will be discussed in Sec. 3, see also Sec. 10 in [3]. It is true that the expression for the AB phase difference, Eq. (6) in [1], is a covariant expression, but it is not the case with the decomposition of F in terms of the components of the 3-vectors $\mathbf{E}$ and $\mathbf{B}$, Eq. (7) in [1]. Instead of it a manifestly covariant decomposition of F, i.e., of $F_{\mu\nu}$, will be presented by Eq. (5). As can be seen from [4-9], in the four-dimensional (4D) spacetime, in contrast to the usual transformations (UT) of the 3-vectors $\mathbf{E}$ and $\mathbf{B}$, Eq. (2) here, or Eq. (11.148) in [10], according to which the transformed $\mathbf{E}'$ is expressed by the mixture of the 3-vectors $\mathbf{E}$ and $\mathbf{B}$, the mathematically correct Lorentz transformations (LT) always transform the 4D algebraic object representing the electric field only to the electric field; there is no mixing with the magnetic field, Eq. (4) here or Eqs. (42) and (43) in [3]. This is first shown by Minkowski in Sec. 11.6 in [11] and reinvented and generalized in terms of 4D geometric quantities in [4-9]. A brief discussion is given in [3]. Using such 4D electric and magnetic fields the manifestly covariant expression for the AB phase difference is given by Eq. (8), which replaces Eqs. (8) and (9) from [1].

In Sec. 3, we use the results from [3], particularly it refers to the discussion from Sec. 10 in [3]. There, it is mentioned that always there are external electric fields from stationary resistive conductors carrying constant currents, see, e.g., Sec. 4 in [12] and references therein. In Secs. 7-7.2 in [3] it is shown that in the 4D geometric approach to special relativity, the invariant special relativity (ISR), there is a static electric field outside a moving and a *stationary* solenoid with a steady current not only for resistive conductors but also for superconductors. Note that in the ISR an independent physical reality is attributed to the 4D geometric quantities and not, as usual, to the 3D quantities. Furthermore, in Sec. 8 in [3], it is discovered that there is such static 4D electric field not only outside a *moving* permanent magnet, as generally accepted in physical literature, but outside a *stationary* permanent magnet as well. As explained in [3] that result is based on the paper [13] in which the generalized Uhlenbeck-Goudsmit hypothesis is formulated, Eq. (9) in [13], i.e., Eq. (59) in [3]. The mentioned results for the existence of the 4D external electric fields may give the possibility to explain the experimentally observed fringe shift for the magnetic AB effect even in Tonomura's experiments [14], Sec. 10 in [3]: "in terms of forces, which so far have been overlooked." Here, in Sec. 3, these results from [3] are combined with the correct covariant formulation of the AB effect from Sec. 2, i.e., with Eqs. (10) and (11) for $\delta\alpha_E$, to explain the existence of the magnetic AB phase difference in terms of the overlooked 4D electric force and not, as usual, in terms of the vector potential.

The existence of the overlooked 4D external electric fields is one of the reasons why we do not consider the covariant AB phase in terms of the four-potentials, $\delta\alpha_{EB} = (e/\hbar)\oint A_\mu dx^\mu$, Eq. (5) in [1]. Another reason is that in [15] an axiomatic formulation of the electromagnetism is presented in which only the field equation for F is postulated, Eq. (4) in [15], i.e., Eq. (20) in [3]. It is shown in [15] that the electromagnetic field F can be taken as *the primary quantity* for the whole electromagnetism both in the theory and in experiments; F is a well-defined 4D *measurable* quantity. It yields the complete description of the electromagnetic field and there is no need to introduce either the potentials (thus dispensing with the need for the gauge conditions) or the field vectors. That formulation with the F field is a self-contained, complete and consistent formulation. The generalization of Eq. (4) in [15] to a moving medium is presented in [16]. There, [16], the field equations are written in terms of F and the generalized magnetization-polarization bivector $\mathcal{M}$ and not, as usual, in terms of F and the electromagnetic excitation tensor $\mathcal{H}$.

2. Covariant expression for the AB phase shift

The covariant expression for the AB phase difference in terms of the Faraday 2-form F is presented by Eq. (6) in [1], which is repeated here

$$\delta\alpha_{EB} = (-e/2\hbar)\int F_{\mu\nu}dx^\mu \wedge dx^\nu = (e/\hbar)\int F, \tag{1}$$

where $F = (-1/2)F_{\mu\nu}dx^\mu \wedge dx^\nu$. (The notation is the same as in [1]; dx^μ and dx^ν are differential four-vectors and throughout the paper we set $c = 1$.) In order to show that this covariant expression (1) reduces to the usual expressions with the 3-vectors, Eqs. (2) and (4) in [1], the Faraday 2-form F is decomposed using the components of the 3-vectors $\mathbf{E}$ and $\mathbf{B}$, Eq. (7) in [1]. It is worth mentioning that Eq. (7) in [1] is not mathematically correct; the expression $(-1/2)F_{\mu\nu}dx^\mu \wedge dx^\nu$ (it will be denoted as (F)) is covariant under the LT, but it is not the case with its decomposition $(E_x dx + E_y dy + E_z dz) \wedge dt + B_x dy \wedge dz + B_y dz \wedge dx + B_z dx \wedge dy$ (it will be denoted as (EB)). The expression (EB) is obtained from that one with $F_{\mu\nu}$, (F), using the usual identification of the components of $F_{\mu\nu}$ with the components of the 3-vectors **E** and **B**, e.g., Eq. (11.138) in [10], see also Eq. (3) and the comment on it in [3]. In all traditional approaches it is supposed that the same identification holds in a relatively moving inertial frame of reference, see Eq. (7) in [3]. This means that it is considered that the components of **E** and **B** transform under the LT as the components of $F_{\mu\nu}$ transform, i.e., that the LT of the components of **E** and **B** (for the boost in the x direction) are

$$\begin{aligned} E'_x &= E_x,\ E'_y = \gamma(E_y - \beta B_z),\ E'_z = \gamma(E_z + \beta B_y), \\ B'_x &= B_x,\ B'_y = \gamma(B_y + \beta E_z),\ B'_z = \gamma(B_z - \beta E_y), \end{aligned} \tag{2}$$

see, e.g., Sec. 11.10 and Eq. (11.148) in [10], or the discussion and equations (9) and (10) in [3]. The essential feature of the transformations (2) is that *the transformed components $E'_{x,y,z}$ are expressed by the mixture of the components of the 3-vectors* **E** *and* **B**, *and similarly for* **B**$'$. The electric field **E** in one inertial frame is "seen" as slightly changed electric field **E**$'$ and an *induced magnetic field* **B**$'$ in a relatively moving inertial frame. From the time of Einstein's fundamental paper [17], the transformations (2) are always considered to be the relativistically correct LT (boosts) of **E** and **B**, but we shall call them, as already said, the UT. As can be seen from Secs. 3.1 and 3.2 in [3], the above mentioned identification is synchronization dependent and it holds only if Einstein's synchronization [17] is used. There, it is also shown that the mentioned identifications are meaningless if only the Einstein synchronization is replaced by an asymmetric synchronization, the "radio" synchronization. That nonstandard synchronization is described in more detail in [18], see also [13]. This is also mentioned below, see Eq. (9) and the discussion with it. But, *different synchronizations are only different conventions and physics must not depend on conventions.*

Therefore, as first shown by Minkowski in Sec. 11.6 in [11] and independently reinvented and generalized in terms of the 4D geometric quantities in [4-9], $F_{\mu\nu}$ can be decomposed in a covariant manner

$$\begin{aligned} F_{\mu\nu} &= (v_\mu E_\nu - v_\nu E_\mu) + \varepsilon_{\mu\nu\alpha\beta} v^\alpha B^\beta, \\ E_\mu &= F_{\nu\mu} v^\nu, \quad B_\mu = (1/2)\varepsilon_{\mu\nu\alpha\beta} F^{\nu\alpha} v^\beta, \end{aligned} \tag{3}$$

where E_μ and B_μ are the components of the 4D electric and magnetic fields respectively, whereas v_μ are the components of the 4D velocity of a family of observers who measure electric and magnetic fields, see also Sec. 5 in [3]. Since $F_{\mu\nu}$ is antisymmetric it holds that $E_\mu v^\mu = B_\mu v^\mu = 0$, only three components of E_μ and B_μ are independent. In the 4D spacetime the mathematically correct decomposition of F into 4D electric and magnetic fields *and* the 4-velocity of the observer, Eq. (3), is already firmly theoretically founded and it is known to many physicists. The recent example is in [19]; it is only the electric part (the magnetic part is zero there). Similarly, in the component form as in (3), this decomposition is presented, e.g., in [20] and in the basis-free form with the abstract 4D quantities, e.g., in [21].

¿From the mathematical viewpoint it is trivially to see how, e.g., E_μ from (3) is transformed under the LT; in the mathematically correct LT the transformed components E'_μ are not determined only by $F'_{\mu\nu}$, as in all usual approaches, e.g., Eqs. (11.147) and (11.148) in [10], but also by v'^μ. This is first shown by Minkowski in Sec. 11.6 in [11]. Let v, E and B are 1×4 matrices and F is a 4×4 matrix; their components are implicitly determined in the standard basis. Minkowski first described how v and F separately transform under the LT A (the matrix of the LT is denoted as A). The LT of the 4-velocity v is $v' = vA$ and the LT of the field-strength tensor F is $F' = A^{-1}FA$, then, as shown by Minkowski, the mathematically correct LT of $E = vF$ is $E = vF \longrightarrow E' = (vA)(A^{-1}FA) = (vF)A = EA$. This means that under the LT *both* quantities, the field-strength tensor F (4×4 matrix) *and the 4-velocity* v (1×4 matrix) are transformed and their product transforms as any 1×4 matrix transforms. As already stated that mathematically correct procedure is reinvented and generalized using the 4D geometric quantities both in the tensor formalism and in the geometric algebra formalism in [4-9]. Particularly, the comparison with Minkowski's results, Sec. 11.6 in [11], is presented in [9]. The essential point is that *the 4D electric field* E *transforms by the LT again to the 4D electric field* E'*; there is no mixing with the 4D magnetic field* B, i.e., *the components* E_μ *transform by the LT again to the components* E'_μ *of the same 4D electric field and there is no mixing with* B_μ,

$$E'_0 = \gamma(E_0 + \beta E_1),\ E'_1 = \gamma(E_1 + \beta E_0),\ E'_{2,3} = E_{2,3}, \tag{4}$$

for a boost along the x^1 axis. It is easily seen that the UT, Eq. (11.148) in [10], i.e., Eq. (2) here, will be simply obtained in this 4D geometric approach if *only* the components $F_{\mu\nu}$ are transformed *but not* the components v^μ. Such procedure corresponds to the usual identifications of the components of $F_{\mu\nu}$ with the components of the 3-vectors **E** and **B** in both relatively moving inertial frames of reference. A short derivation of these results can be seen in [7]. In this case there is no need to write the transformations for the components B_μ since they transform as in (4). This means that it is proved in [4-9] that, contrary to the generally accepted opinion, *the UT of the 3-vectors* **E** *and* **B**, Eq. (2), *are not the LT*, but that the mathematically correct LT are given by Eq. (4). For a brief review see Sec. 5 in [3] or Sec. 3 in [22]. It is interesting that although Eq. (3) is known to many physicists, e.g., [20, 21], it is not noticed that the mathematically correct LT of, e.g., $E_\mu = F_{\nu\mu}v^\nu$, necessarily require that both $F_{\nu\mu}$ and v^ν have to be transformed and not only $F_{\nu\mu}$. In the 4D spacetime, from the mathematical viewpoint, the 4D electric and magnetic fields are correctly defined and they transform as any other 4-vector transforms, i.e., according to Eq. (4).

Hence, instead of Eq. (7) in [1] we have

$$\begin{aligned} F &= (-1/2)F_{\mu\nu}dx^\mu \wedge dx^\nu = \\ &\quad (-1/2)[(v_\mu E_\nu - v_\nu E_\mu) + \varepsilon_{\mu\nu\alpha\beta}v^\alpha B^\beta]dx^\mu \wedge dx^\nu. \end{aligned} \tag{5}$$

In Eq. (5) both expressions for F are manifestly covariant under the LT, which does not hold, as already stated, for Eq. (7) in [1]. In contrast to the usual treatment from [1], $\delta\alpha_{EB}$ that is given by Eq. (8) below *is the same for all relatively moving inertial observers and for all coordinate bases used by them; the principle of relativity is naturally satisfied.* This proves a mathematical and relativistic correctness of this manifestly covariant approach.

For the reader's convenience and for easier comparison with [1] we have written, e.g., Eq. (3), only with components, but as F is a 4D geometric quantity, a 2-form ($F = (-1/2)F_{\mu\nu}dx^\mu \wedge dx^\nu$), so is, e.g., the electric field E, a 4D geometric quantity, an 1-form ($E = E_\mu dx^\mu$). Both, F and E in these relations are written in a specific coordinate basis, the standard basis, with the Einstein synchronization of distant clocks and Cartesian space coordinates. In [1], as in all usual covariant approaches, the standard basis is exclusively used, but, as pointed out above, different systems of coordinates are allowed in an inertial frame and they are all equivalent in the description of physical phenomena. Thus, for example, one can use the above mentioned asymmetric synchronization, the "radio" synchronization. The important difference relative to the usual formulation with 3-vectors is that in the 4D spacetime a 4D geometric quantity is *the same 4D quantity* for all inertial observers and for all coordinate bases used by them, $E = E_\mu dx^\mu = E'_\mu dx'^\mu = E_{\mu,r}dx^{\mu,r} = \ldots$, where the primed quantities are the Lorentz transforms of the unprimed ones and the quantities with the index "r" are in the coordinate basis with the "radio" synchronization. Observe that in [18] and in the second and third papers in [23] the "radio" synchronization is used throughout the papers. Moreover, in Eq. (4) in [18] it is presented the transformation matrix that connects Einstein's system of coordinates with another system of coordinates in the same reference frame. Also, Eq. (1) in [18], it is derived such form of the LT, which is independent of the chosen system of coordinates, including different synchronizations. Since in the ISR every 4D geometric quantity is invariant under the LT the principle of relativity is automatically satisfied and there is no need to postulate it outside the mathematical formulation of the theory as in Einstein's formulation of SR, [17].

For simplicity and for easier comparison with [1] we shall introduce the inertial frame of "fiducial" observers ($v^{\mu} = (1, 0, 0, 0)$) with the standard basis (Einstein's synchronization) in it, which will be called the "f"-frame. In that frame it holds that $E_0 = B_0 = 0$ and only the spatial components of E_{μ} and B_{μ} remain. From (3) it follows that these components are

$$E_i = F_{0i}v^0 = F_{0i}, \quad B_i = (1/2)\varepsilon_{0ijk}F^{kj}; \tag{6}$$

the same components as in, e.g., Eq. (11.138) in [10]. Observe that the "f"-frame is not any kind of a preferred frame, because any inertial frame can be chosen to be that frame and it is usually taken that the laboratory frame is the "f"-frame. However, in any other relatively moving inertial frame, the S' frame, the "fiducial" observers are moving, and the components v^{μ} transform as in (4), $v'^{\mu} = (\gamma, -\beta\gamma, 0, 0)$. Hence, as already shown by Minkowski in Sec. 11.6 in [11], for the transformations from the "f"-frame, see [7], $(E_{\mu})' = [F_{\nu\mu}v^{\nu}]' = [F_{0i}v^0]' = F'_{\nu\mu}v'^{\nu} = E'_{\mu}$, and Eq. (4) is obtained; *the components* E_{μ} *transform by the LT again to the components* E'_{μ}. Let us take in (5) that $E_1 = E_x$, ... , $B_1 = B_x$, ... , $\varepsilon_{0123} = 1$, $dx^0 = dt$, ... , $dx^3 = dz$, then in the "f"-frame the second covariant expression in (5) corresponds to the expression (EB) that is used in [1]. In a relatively moving inertial frame S' the LT (4) will give that E'_0 and B'_0 will be different from zero and these terms cannot exist in the approach from [1], which deals with the expression (EB), i.e., with the fields as the 3-vectors.

In the usual formulation the physical meaning of 3-vectors $\mathbf{E}$ and $\mathbf{B}$ is determined by the the Lorentz force as a 3-vector $\mathbf{F} = q\mathbf{E} + q\mathbf{u} \times \mathbf{B}$ and by Newton's second law $\mathbf{F} = d\mathbf{p}/dt$, $\mathbf{p} = m\gamma_u\mathbf{u}$.

However, in the 4D spacetime, the Lorentz force K is not a 3-vector, but it is a 4D geometric quantity. K is the contraction of the electromagnetic 2-form F with particle's 4-velocity u (it is defined to be the tangent to its world line). The components of K in the standard basis are $K_{\mu} = qF_{\mu\nu}u^{\nu}$, where u^{μ} is the 4-velocity (components) of a charge q, or with E_{μ} and B_{μ}, using the decomposition of $F_{\mu\nu}$, (3), they become

$$K_{\mu} = q[(v_{\mu}E_{\nu} - v_{\nu}E_{\mu}) + \varepsilon_{\mu\nu\alpha\beta}v^{\alpha}B^{\beta}]u^{\nu}. \tag{7}$$

In the 4D spacetime, the physical meaning of E_{μ} *and* B_{μ} *is determined by the Lorentz force* K_{μ} *and by the 4D expression for Newton's second law* $K_{\mu} = dp_{\mu}/d\tau$, $p_{\mu} = mu_{\mu}$, where p_{μ} is the proper momentum (components) and τ is the proper time. All components E_{μ} and B_{μ}, thus E_0 and B_0 as well, are equally well physical and measurable quantities by means of the mentioned K_{μ} and the equation of motion, i.e., the 4D expression for Newton's second law. Obviously, regardless of the fact that majority of physicists believe that only the 3-vectors $\mathbf{E}$ and $\mathbf{B}$ are physical and measurable quantities, *in the 4D spacetime, the 4D geometric quantities are properly defined both theoretically and experimentally.* In view of this discussion it is obvious that the question what physically are E_0 and B_0 is equivalent to the question - what is the temporal component x_0 of the position 4-vector. This is particularly visible if the Einstein synchronization is replaced by the "radio" synchronization in which the space and time are not separated, see Eq. (9) below. Then, the usual 3-vector **r**, and similarly the 3-vectors **E** and **B**, are meaningless. This fundamental difference between the usual formulation with the 3D quantities and the formulation with the 4D geometric quantities is exposed in much more detail, e.g., in [24].

It is also shown in, e.g., [5, 6, 15] that *the LT of the 4D* E_μ *and* B_μ, (4), *are in a true agreement (independent of the chosen inertial reference frame and of the chosen system of coordinates in it) with all experiments in the electromagnetism, whereas it is not the case with the UT of the 3D* **E** *and* **B**, (2). Thus, for example, it is shown in [5] that the conventional theory with the 3D **E** and **B** and their UT yields different values for the motional emf ε for relatively moving inertial observers, $\varepsilon = UBl$ and $\varepsilon = \gamma UBl$, whereas the approach with 4D geometric quantities and their LT yields always the same value for ε, which is defined as a Lorentz scalar, $\varepsilon = \gamma UBl$. This result is very strong evidence that the usual approach is not relativistically correct, i.e., it is not in agreement with the principle of relativity. It is for the experimentalists to find the way to precisely measure the emf ε for the considered problem of a conductor moving in a static magnetic field and to see that in the laboratory frame $\varepsilon = \gamma UBl$ and not simply $\varepsilon = UBl$. That problem is of a considerable importance in practice. The similar discussion with the same conclusions was presented for the Faraday disk in [6]. In the already mentioned [15] and in [25] the Trouton-Noble paradox is considered. It is shown that in the geometric approach with 4D quantities *the 4D torques* will not appear for the moving capacitor if they do not exist for the stationary capacitor, which means that with 4D geometric quantities the principle of relativity is naturally satisfied and *there is not the Trouton-Noble paradox.* The same conclusion holds in the low-velocity approximation $\beta \ll 1$, or $\gamma \simeq 1$. It is also shown in the same geometric approach with 4D torques that there is no Jackson's paradox [24] and the "charge-magnet paradox" [22].

At this point it is worth noting that in the mathematically correct approach, in general, there is no room for the 3-vectors in the 4D spacetime. Let us better explain that statement. It is written in [1] after Eq. (7) that: "$F = B_x dy \wedge dz + B_y dz \wedge dx + B_z dx \wedge dy = \mathbf{B} \cdot d\mathbf{S}$ where the differential forms expression has been converted back to three-vector notation and $d\mathbf{S}$ is the differential area." However, such an equality is mathematically impossible and incorrect. Namely, in the mathematically correct formulation dx, dy, dz have to be understood as differential 4-vectors dx^1, dx^2, dx^3, respectively, the 4D geometric quantities that are properly defined on the 4D spacetime; the wedge product refers to such 4D quantities and not to the usual scalar differentials. On the other hand, **B** and $d\mathbf{S}$, as *geometric quantities in the 3D space*, are constructed from the components and *the unit 3-vectors* **i**, **j**, **k**, e.g., $\mathbf{B} = B_x\mathbf{i} + B_y\mathbf{j} + B_z\mathbf{k}$. The unit 3-vectors have nothing to do with the basis in the 4D spacetime. The LT are properly defined on the 4D spacetime and they cannot transform the 3-vectors. Hence, in the 4D spacetime it is not mathematically correct to state as in [1]: ".. the expression in (6) reduces to $\delta\alpha_{EB} = (e/\hbar)\int F = (e/\hbar)\int \mathbf{B} \cdot d\mathbf{S}$ which is equivalent to the 3-vector expression (2)." In the 4D spacetime the covariant expression $((e/\hbar)\int F)$ is the correct one, but it is not the case with the usual expression for the magnetic flux with the 3-vectors $((e/\hbar)\int \mathbf{B} \cdot d\mathbf{S})$; *they cannot be equal.* The same objection refers to all other relations with the 3-vectors in [1]. Hence, in this geometric approach, using (1) and (5), the manifestly covariant expression for the AB phase difference becomes

$$\delta\alpha_{EB} = (-e/2\hbar)\int [(v_\mu E_\nu(x) - v_\nu E_\mu(x)) + \varepsilon_{\mu\nu\alpha\beta} v^\alpha B^\beta(x)] dx^\mu \wedge dx^\nu. \tag{8}$$

In Sec. 3 in [1] it is investigated "the usual magnetic AB set-up of an infinite solenoid but with a time dependent magnetic field and vector potential, i.e., $\mathbf{B}(t)$ and $\mathbf{A}(t)$." As noted in [1] for that situation the scalar potential is still zero, $\phi = 0$. At first, it is worth mentioning that, as explained in Sec. 3 in [3], in a correct covariant formulation there is no static case. The 1-form A ($A = A_\mu dx^\mu$) and the Faraday 2-form F are both, always function of the position four-vector x; $A(x)$ and $F(x)$. If, for example, the usual 3-vector fields $\mathbf{A}(\mathbf{r})$, $\mathbf{B}(\mathbf{r})$ do not explicitly depend on the coordinate time t in one frame, then the LT will mix the time and space coordinates; they cannot transform the spatial coordinates from one frame only to spatial coordinates in a relatively moving inertial frame of reference. What is static case for one inertial observer is not more static case for relatively moving inertial observer, but a time dependent case. Furthermore, if an observer uses the "radio" synchronization and not Einstein's synchronization, then the space and time are not separated and the usual 3-vector $\mathbf{r}$ is meaningless. As can be seen from Eq. (13) in [3] the components of the position 4-vector x in the commonly used coordinate basis with Einstein's synchronization and that one with the "radio" synchronization are connected as

$$x_r^0 = x^0 - x^1 - x^2 - x^3, \quad x_r^i = x^i, \tag{9}$$

and the same relation holds, e.g., for (A_r^0, A_r^i), or (E_r^0, E_r^i).

This consideration suggests that the results from Sec. 3 in [1] for the time dependent, infinite solenoid, have to be reexamined using the correct covariant formulation (8). We shall only discuss the AB phase difference determined by Eqs. (8) and (9) in [1]. It is calculated using Eq. (7) from [1]. This will be compared with (8). As already mentioned above, in the 4D spacetime, Eq. (7) from [1] is not mathematically correct and the same holds for Eqs. (8) and (9) from [1], which deal with the 3-vectors. The part of the AB phase difference with B_μ from (8) is $\delta\alpha_B = (-e/2\hbar)\int \varepsilon_{\mu\nu\alpha\beta}v^\alpha B^\beta(x)dx^\mu \wedge dx^\nu$ and it replaces Eq. (8) from [1]. Only in the "f"-frame that part becomes $\delta\alpha_B = (-e/2\hbar)\int \varepsilon_{0ijk}v^0 B^k(x)dx^i \wedge dx^j$ and, as can be seen by the use of $B_1 = B_x$, etc. that mathematically correct expression corresponds to Eq. (8) from [1], i.e., to $\delta\alpha_B = (e/\hbar)\int \mathbf{B}\cdot d\mathbf{S}$. The essential difference is that all quantities in this covariantly defined $\delta\alpha_B$ are properly defined in the 4D spacetime and they correctly transform under the LT, like (4), which is not the case with the 3D quantities from Eq. (8) in [1].

The part of the AB phase difference with E_μ from (8) is

$$\delta\alpha_E = (-e/2\hbar)\int (v_\mu E_\nu(x) - v_\nu E_\mu(x))dx^\mu \wedge dx^\nu \tag{10}$$

and, the same as for $\delta\alpha_B$, it is the same quantity for all relatively moving inertial observers and for all bases used by them. Only in the "f"-frame $\delta\alpha_E$ from (10) becomes

$$\delta\alpha_E = (e/\hbar)\int v_0 E_i(x)dx^i \wedge dx^0 \tag{11}$$

and it can be compared with Eq. (9) from [1]. For that comparison $F_{\mu\nu}$ is written in terms of A_μ as $F_{\mu\nu} = \partial_\mu A_\nu - \partial_\nu A_\mu$. In that expression it is considered that A_μ are the primary quantities whereas $F_{\mu\nu}$ are derived from them. But, as clearly shown in [15], the F field is *the primary quantity* for the whole electromagnetism and not the four potential, which is gauge dependent. However, here, for the comparison with [1], we use the above relation with A_μ. Then, (3) is used to get E_μ in terms of A_μ, $E_\mu = F_{\alpha\mu}v^\alpha = (\partial_\alpha A_\mu - \partial_\mu A_\alpha)v^\alpha$. In the "f"-frame $E_0 = 0$ and $E_i = (\partial_0 A_i - \partial_i A_0)v^0$, what corresponds to the components of the usual three-vector $\mathbf{E}$, e.g., E_1 corresponds to $E_x = -\partial A_x/\partial t - \partial_x\phi$; remember that if A_μ is written in the usual notation it is $A_\mu = (\phi, -A_x, -A_y, -A_z)$ and in the "f"-frame $v^\mu = (1,0,0,0)$. Hence, in the "f"-frame, $\delta\alpha_E = (e/\hbar)\int(\partial_0 A_i - \partial_i A_0)dx^i \wedge dx^0$, which for $A_0 = 0$ becomes $= (e/\hbar)\int \partial_0 A_i dx^i \wedge dx^0$ and, by the procedure from [1], it corresponds to Eq. (9) in [1], i.e., to $\delta\alpha_E = (-e/\hbar)\int \mathbf{B}\cdot d\mathbf{S} = -\delta\alpha_B$. Thus, *only* in the "f"-frame and for $A_0 = 0$ "the two parts cancel exactly." Observe that the condition $A_0 = 0$ is not a Lorentz covariant condition; in a relatively moving inertial frame A_0 will be $\neq 0$. Furthermore, as seen from (9), in the basis with the "radio" synchronization the temporal and spatial components of A_μ cannot be separated, which means that in the 4D spacetime the condition $A_0 = 0$ has not a well-defined meaning. The similar objections hold for the whole discussion presented in [26].

3. The Aharonov-Bohm effect in terms of fields

It is really surprising that both in all numerous theoretical discussions, e.g., [1, 2, 27, 28, 26], in the experiments with microscopic solenoids [29] and also in the recent experiment with macroscopic solenoid [30], it is never noticed that in the rest frame of the solenoid there are *always* external *static* electric fields for *stationary, resistive* conductor carrying *constant current.* In an ohmic conductor there are quasistatic surface charges, which generate not only the electric field inside the wire driving the current, *but also a time independent electric field outside it.* That electric field is proportional to the current, see, e.g., Sec. 4 in [12] and references therein. As mentioned in [12] the existence of such quasistatic surface charges was first pointed out by Kirchhoff, Refs. [18-20] in [12]. There are no analytic solutions for these surface charges and the external electric fields for the case of finite solenoids; for an infinite solenoid see [31]. The distribution of the surface charges and the magnitude of the induced electric fields depend not only on the geometry of the circuit but even of its surroundings. *These external electric fields from steady currents are firmly experimentally confirmed,* see, e.g., [12], *and they are well-known in electrical engineering.* In [12], two other contributions to the external electric field are discussed, but, as explained in Sec. 10 in [3], they are of no concern here. It is worth mentioning that the expression from Sec. 4 in [12] is for a cylindrical wire of length l carrying a constant current I and that wire is a part of a square circuit. That expression is not appropriate for a finite solenoid with steady current. In [31] an infinite solenoid with steady current is considered and it is appropriate for the case considered in [1]. There, in [31], a uniform cylindrical resistive sheet of the radius a with a "line" battery with terminals at potentials $\pm V_0/2$ driving current *azimuthally* in it is considered. In Sec. IV, [31], it is presented (i) the magnitude of the electric field outside the solenoid, Eq. (11),

$$E = (V_0/\pi)(a/r\rho), \tag{12}$$

where r and ρ are the polar radii measured from the center (axis) and from the baterry (respectively), and (ii) the electric lines of force, Fig. 3. It is visible from Fig. 3. in [31] that the electric field has radial and poloidal components, where the latter ones follow the direction of the current just outside the solenoid in the same way as the magnetic vector potential.

In the recent experiment [30] the absence of electromagnetic forces outside the solenoid that are predicted by Boyer's force picture [32] has been experimentally investigated by means of a time-of-flight experiment for a macroscopic solenoid. It is looked for a time delay for electrons passing on opposite sides of the solenoid. As discussed above in the generally accepted theory the electron wave packets are influenced by nonzero vector potential, i.e., by the quantum action of the magnetic flux even when electrons pass through the field-free regions of space. On the other hand in Boyer's semiclassical theory [32] there is a back-action force of the solenoid on the electron, which gives rise to a time delay and to a phase shift that exactly matches the AB-phase shift. It is shown in [30] that there is no time delay and it is concluded that there are no fields predicted by Boyer's force picture [32]. In his comment on the results obtained in [30] Boyer [33] stated: "the Aharonov-Bohm phase shift has never been observed for such a macroscopic solenoid, ..." In [33], it is also argued that if the solenoid resistance is large, as in [30], then the back forces will be small and there is no time lag, but for the microscopic solenoids it is the opposite case. It has to be pointed out that neither the authors of [30] nor Boyer [32, 33] knew anything about the electric fields caused by the quasistatic surface charges that exist outside the resistive conductors carrying *constant* currents. This means that it is not true that the paper [30] shows *experimentally* that forces cannot be responsible for the magnetic AB phase shift. The electric forces caused by the mentioned quasistatic surface charges have nothing to do with Boyer's force picture, [32, 33]. Thus, the main result from [30] about the absence of the time delay does not imply that the electrons travel in a field-free region. Obviously, the electric fields from quasistatic surface charges have to be taken into account for the explanation of the AB phase difference in the magnetic AB effect as well, i.e., in the usual magnetic AB set-up of an infinite solenoid, which is considered in [1] and also in the case of finite macroscopic [30] and microscopic [29] solenoids. From the viewpoint presented here the AB phase difference in the magnetic AB effect *is not* due to the vector potential, i.e., according to Eq. (2) from [1] due to the quantum action of the magnetic flux, but *it is due to the mentioned external electric field from stationary solenoids with steady currents.* In that case, contrary to the generally accepted opinion, the electron does not travel in the field-free region, but the electron wave packets are *locally* influenced by the electric field. A similar expression as (11) is obtained in [34], Eq. (28), but their procedure is not relativistically correct and the 3D electric field that enters into their Eq. (28) is proportional to the square of the current.

In order to clarify the situation from the experimental viewpoint we consider that some new experiments are required: the measurement in a *single* experiment of the AB phase shift and the time delay, as suggested in [33], and the measurement of the mentioned external electric fields *separately* from AB-studies.

The above consideration implies that in the expression for $\delta\alpha_{EB}$ (8) there is no need to take into account the magnetic part $\delta\alpha_B$, i.e., the non-local effect of the magnetic field. In the 4D spacetime only the local effects are important and physically justified. This means that from the viewpoint of this approach with 4D geometric quantities *the AB phase difference is even for the magnetic AB effect exclusively determined by the covariant expression* $\delta\alpha_E$ from (10), i.e., by the local influence of the 4D electric field. If the rest frame of the solenoid, the laboratory frame, is taken to be the "f"-frame then $\delta\alpha_E$ is given by Eq. (11). In our opinion the magnetic part $\delta\alpha_B$ of $\delta\alpha_{EB}$ (8) could be taken into account only in the case that the solenoid's magnetic field is not entirely restricted to the coil's interior but exists in the coil's exterior as well, i.e., along the electron's trajectory. The same conclusion that only the local effect of the 4D electric field, i.e., $\delta\alpha_E$ (10) ((11)) is important and physically meaningful holds in the same measure for the time dependent set-up that is considered in Sec. 3 in [1]. Thus, in that case there is no cancellation of the non-local effect of the magnetic field, $\delta\alpha_B$, with the local effect of the electric field, $\delta\alpha_E$ (10) ((11)), because, as explained above, only the electric field from the solenoid with current exists in the region outside the solenoid and consequently it can locally influence the electron travelling through that region. It is interesting that, as can be seen from Sec. 4 in [35], if the current in the solenoid varies linearly with time then it creates a time independent external electric field, see Eq. (8) and Fig. 1 in [35]. Hence, for the solenoid with such a time-dependent current there will be no time-dependent AB phase shift although only $\delta\alpha_E$ (10) ((11)), the electric part of $\delta\alpha_{EB}$ (8), is considered to be physically correct and justified.

Note that in this approach with the 4D geometric quantities the 3D quantities from the usual approaches, e.g., from [1, 26, 12, 34, 35], etc. have to be interpreted in a different way. Thus, for example, the *components* of the electric field 3-vector in [1] have to be understood as the *spatial components* in the standard basis of the 4D electric field; the rest frame of the solenoid is taken to be the "f"-frame and therefore the temporal component $E_0 = 0$ (also $B_0 = 0$). Also, in this geometric approach the components K_μ of the Lorentz force are given by (7). As discussed above, in the case considered in [1] only the electric part of K from (7) is physically important.

4. Conclusions

As seen from the preceding discussion the correct covariant formulation of the AB phase shift (8) deals with the 4D geometric quantities that properly transform under the mathematically correct LT (4). In the 4D spacetime Eq. (8) replaces Eqs. (8) and (9) from [1], which deal with the 3D quantities that transform under the UT (2). Both, the 3D quantities and their UT (2) are ill-defined in the 4D spacetime. As *proved* in [4-9], contrary to the generally accepted opinion, the UT (2) are not the mathematically correct LT. The main result that is obtained in this paper is that *even for the magnetic AB effect* (a stationary solenoid with *resistive* conductor carrying either steady current or the current that varies linearly with time) *the AB phase difference is exclusively determined by the covariant expression* $\delta\alpha_E$ from (10), i.e., by the *local* influence of the 4D electric field. Thus, here, it is shown that in the 4D spacetime only the electric part of $\delta\alpha_{EB}$ (8), i.e., $\delta\alpha_E$ (10) ((11)) is physically correct and meaningful. The reason for it is that there are *static* electric fields outside a *stationary, resistive* conductor carrying *steady current*, which means that it is not true that, e.g., in experiments [29, 30], the electron travels in the field-free region of space. The existence of the mentioned electric fields is firmly experimentally confirmed; for some experiments see, e.g., [12] and references therein. All this together shows that the magnetic AB phase shift considered in [1] is not a topological phase shift.

In Sec. 7.1 in [3] it is shown that the external static electric fields, the "relativistic" second-order electric fields, would need to exist not only for resistive conductors with steady currents but even for superconducting solenoids with steady currents. In Sec. 7.2 in [3] different experiments for the detection of the second-order electric fields outside a stationary superconductor with steady current are discussed. Furthermore, what is very important for the explanation of the AB effect, in Sec. 8 in [3] such second-order electric fields are predicted to exist outside a *stationary* permanent magnet as well. As discussed in Sec. 10 in [3], these results could explain the experimentally observed fringe shift for the magnetic AB effect even in Tonomura's experiments [14] *in terms of previously overlooked electric forces and not, as generally accepted, in terms of nonzero vector potentials.*

Similarly, the qualitative theoretical explanations of the quantum phase shifts in terms of the classical forces as the 4D vectors in the Aharonov-Casher and the Röntgen effects are presented in [7, 36]. Furthermore, in [37], the dipole moments are quantized and it is shown that the expectation value for the quantum force 4D vector is not zero in the case of the Aharonov-Casher and the Röntgen effects and in the neutron interferometry. Hence, in these experiments too the phase shifts are not due to force-free interaction of the dipole, i.e., they also are not the topological phase shifts.

The covariant AB effect in terms of F and not in terms of a vector potential is also investigated in [38].

It is interesting to note that recently another local explanation of the AB effect is proposed in [39]. There, it is argued that if the solenoid in the AB effect is treated in the framework of quantum theory then the effect can be explained by the local action of the field of the electron on the solenoid. In some respects there is a similarity between Boyer's calculation [32, 33] and Vaidman's determination [39] of the AB phase shift. Boyer in [32, 33] calculates the force exerted by the electron on a solenoid (represented by a line of magnetic dipoles) and then relies on Newton's third law to obtain a back-action force of the solenoid on the electron. The same Boyer's force approach is investigated in [30] but a solenoid is considered as a stack of current loops. However, Newton's third law is violated for the electromagnetic interaction and to overcome this difficulty a hidden momentum is often introduced, particularly in the case with current loops, see, e.g. [30, 40] and references therein. But, as shown, e.g., in [3] and [22], if an independent physical reality is attributed to the 4D geometric quantities and not, as usual, to the 3D quantities, then there is no need for the introduction of some "hidden" 3D quantities and there are no electromagnetic paradoxes. Vaidman, in [39], see Fig. 4 in [39], considers that the electron produces change in the magnetic flux of the solenoid, which causes an electromotive force on charged solenoids (in his example). This leads to the change in their velocity and to the shift of the wave packet of the cylinders and finally to the correct expression for the AB phase, Eq. (5) in [39] (arXiv: 1301.6153). *Observe that this phase shift is for the source (solenoids) and not for the passing electron.* Then, Vaidman states: "Since in quantum mechanics the wave function is for all parts of the system together, the change of the wave function of the source leads to observable effect in the interference experiment with the electron." (See Eqs. (8) and (9) in the first paper in [39] for the change in the total wave function of the electron and the solenoid.) It is worth noting that in Boyer's picture [32, 33] it is impossible to detect the predicted force on the solenoid since it requires the detection of the force of a single electron on a macroscopic object. For the same reason, in Vaidman's picture [39], it is impossible to detect the mentioned electromotive force and the change in the angular velocity of the solenoids. Thus, both Boyer's force and Vaidman's electromotive force cannot be experimentally verified, which means that neither of these approaches have some physical, experimental, foundation.

On the other hand, the theory presented here is based on the existence of the static *electric fields* outside a *stationary, resistive* conductor carrying *steady current*, and these fields are already firmly experimentally verified.

Acknowledgments

It is a pleasure to acknowledge to Larry Horwitz and Martin Land for inviting me to the IARD conferences and for their continuos support of my work. I am also grateful to Zbigniew Oziewicz for numerous and very useful discussions during years and to him and to Alex Gersten for the continuos support of my work.

References

[1] Singleton D and Vagenas E C 2013 *Phys. Lett.* B **723** 241

[2] Aharonov Y and Bohm D 1959 *Phys. Rev.* **115** 485

[3] Ivezić T 2013 *J. Phys.: Conf. Ser.* **437** 012014

[4] Ivezić T 2003 *Found. Phys.* **33** 1339

[5] Ivezić T 2005 *Found. Phys. Lett.* **18** 301

[6] Ivezić T 2005 *Found. Phys.* **35** 1585

[7] Ivezić T 2007 *Phys. Rev. Lett.* **98** 108901

[8] Ivezić T 2008 *Preprint* physics.gen-ph/0809.5277

[9] Ivezić T 2010 *Phys. Scr.* **82** 055007

[10] Jackson J D 1998 *Classical Electrodynamics* 3rd ed (New York: Wiley)

[11] Minkowski H 1908 *Nachr. Ges. Wiss. Göttingen* 53;
Reprinted in: 1910 *Math. Ann.* **68** 472;
English translation in: M N Saha and S N Bose 1920 *The Principle of Relativity: Original Papers by A. Einstein and H. Minkowski* (Calcutta: Calcutta University Press).

[12] Assis A K T, Rodrigues Jr. W A and Mania A J 1999 *Found. Phys.* **29** 729

[13] Ivezić T 2010 *Phys. Scr.* **81** 025001

[14] Tonomura A N, Osakabe T, Matsuda T, Kawasaki T, Endo J, Yano S and Yamada H 1986 *Phys. Rev. Lett.* **56** 792

[15] Ivezić T 2005 *Found. Phys. Lett.* **18** 401

[16] Ivezić T 2011 *Preprint* physics. gen-ph/1101.3292

[17] Einstein A 1905 *Annalen der Physik* **17** 891; English translation in: W. Perrett and G. B. Jeffery 1952 *The Principle of Relativity* (New York: Dover)

[18] Ivezić T 2001 *Found. Phys.* **31** 1139

[19] Vanzella D A T 2013 *Phys. Rev. Lett.* **110** 089401

[20] Núñez Yépez H N, Salas Brito A L and Vargas C A 1988 *Revista Mexicana de Física* **34** 636; Esposito S 1998 *Found. Phys.* **28** 231;
Anandan J 2000 *Phys. Rev. Lett.* **85** 1354

[21] Wald R M 1984 *General Relativity* (Chicago: The University of Chicago Press); Ludvigsen M 1999 *General Relativity, A Geometric Approach* (Cambridge: Cambridge University Press);
Sonego S and Abramowicz M A 1998 *J. Math. Phys.* **39** 3158;
Vanzella D A T, Matsas G E A and Crater H W 1996 *Am. J. Phys.* **64** 1075; Oziewicz Z 2011 *J. Phys.: Conf. Ser.* **330** 012012;
Hehl F W and Obukhov Yu N 2003 *Foundations of Classical*

Electrodynamics: Charge, flux, and metric (Boston: Birkhäuser);

Blandford R D and Thorne K S 2002-2003 *Applications of classical physics* (California Institute of Technology) [http://www.pma.caltech.edu/Courses/ph136/yr2004/].

[22] Ivezić T 2012 *Preprint* physics.gen-ph/1212.4684

[23] Ivezić T 2002 *Found. Phys. Lett.* **15** 27;
Ivezić T 2001 *Preprint* physics/0103026;
Ivezić T 2001 *Preprint* physics/0101091

[24] Ivezić T 2006 *Found. Phys.* **36** 1511;
Ivezić T 2007 *Fizika A* **16** 207

[25] Ivezić T 2007 *Found. Phys.* **37** 747

[26] Macdougall J and Singleton D 2014 *J.Math.Phys.* **55** 042101

[27] Peshkin M and Tonomura A 1989 *The Aharonov-Bohm Effect* (New York: Springer)

[28] Aharonov Y and Rohrlich D 2005 *Quantum Paradoxes, Quantum Theory for the Perplexed* (Weinheim: Wiley-VCH).

[29] Möllenstedt G and Bayh W 1962 *Naturwissenschaften* **49** 81

[30] Caprez A, Barwick B and Batelaan H 2007 *Phys. Rev. Lett.* **99** 210401

[31] Heald M A 1984 *Am. J. Phys.* **52** 522

[32] Boyer T H 2002 *Found. Phys.* **32** 41

[33] Boyer T H 2008 *Found. Phys.* **38** 498

[34] Torres Silva H and Assis A K T 2001 *Revista Facultad de Ingenieria U. T. A. (Chile)* **9** 29

[35] Miller M A 1984 *Sov. Phys. Usp.* **27** 69

[36] Ivezić T 2007 *Phys. Rev. Lett.* **98** 158901

[37] Ivezić T 2007 *Preprint* hep-th/0705.0744

[38] Anandan J 1980 *Int. J. Theor. Phys.* **19** 537

[39] Vaidman L 2012 *Phys. Rev. A* **86** 040101 (R);

Vaidman L 2014 *Quantum Theory: A Two-Time Success Story - Yakir Aharonov Festschrift* ed D C Struppa and J M Tollaksen (Italia: Springer) pp 247-255; Vaidman L 2013 *Preprint* quant-ph/1301.6153

[40] McGregor S, Hotovy R, Caprez A and Batelaan H 2012 *New J. Phys.* **14** 093020

Variable mass theories in relativistic quantum mechanics as an explanation for anomalous low energy nuclear phenomena

Mark Davidson
Spectel Research Corporation, Palo Alto, CA USA
E-mail: `mdavid@spectelresearch.com`

Abstract. A recent theoretical explanation for observed anomalous low energy nuclear phenomena which have puzzled physicists for many years is expanded on. Based on covariant relativistic quantum mechanics and historical time wave equations, it explains a large number of observed anomalous effects by supposing that nuclear masses can vary in "nuclear active environments" in condensed matter settings. The modified quantum wave equation originally introduced by Fock and Stueckelberg in the 1930s with significant enhancements up to the present by Horwitz and others prove that variable masses are compatible with the principles of both quantum mechanics and relativity. They can explain all of these effects by modifying the kinematic constraints of the reaction, enhancing electron screening and quantum tunneling rates, and allowing for resonant tunneling. Some previous results are recounted, and experimental evidence based on variable radioactive decay rates and other evidence for variable masses is presented which adds some new potential support for this theory.
Keywords: hydrated palladium, deuterated palladium, LENR, fusion, radioactivity
PACS: 24.10.-i, 3.65.-w, 3.70.+k

1. Introduction

Experimental claims of anomalous nuclear reactions in condensed matter settings are increasing in number, and this poses both a problem and an opportunity for physics. These controversial observations today cover a wide spectrum of phenomena including unexplained heat production in deuterated Palladium with commensurate production of Helium-4 suggesting d-d fusion, excessive heat production in hydrated nickel and titanium as well, transmutation of elements in hydrated or deuterated metals, strong gamma ray and positron bursts from lightning storms, transmutation of elements possibly observed in certain bacteria, variable radioactive decay rates for radioactive isotopes implanted in metals in special circumstances, time varying nuclear decay rates of radioactive isotopes, anomalous radiations being emitted from LENR experiments, neutron bursts coming from deuterium gas in strong time-varying magnetic fields, anomalous Helium-3/Helium-4 ratios rising from the earth's interior, piezo-nuclear phenomenon, x-ray bursts emitted from peeling Scotch tape in a near vacuum, etc. Although some of these might be explainable with conventional physics, it is the author's opinion that either the experiments are mostly wrong or else they indicate new physics. The key ingredient that most of these experiments suggest is that in some special circumstances in condensed matter, where these

Journal of Physics: Conference Series **615** (2015) 012016 doi:10.1088/1742-6596/615/1/012016

reactions are taking place, the nuclear rest masses of the particles involved in the reaction are not constant, but can vary. These special regions have been called the "nuclear active environment" by Storms [1]. Although other explanations have been proposed, my opinion is that none of them provide as satisfactory and complete an explanation as this. Variable rest masses can make forbidden nuclear processes possible, because the kinematics depend on them. Moreover, mass variation can lead to greatly enhanced quantum tunneling, and can also lead to resonant tunneling [2]. Although it is common in solid state physics to talk about effective masses, especially for electrons and holes, these masses depend on context and are typically used in phenomenological transport equations and have little or nothing to do with nuclear decays which in conventional physics depend only on the standard isolated particle's rest mass. External electromagnetic fields will shift the 4-momentum of charged particles placed in that field by changing the 4-momentum stored in the total field around the particle, but this effect cancels out from the kinematic equations of nuclear reactions as explained in [2] due to charge conservation and therefore cannot change nuclear reaction rates either (of course external electromagnetic fields can change the kinetic energy of particles as well as the center of mass two-body collision energy, but there is no other influence than this on nuclear rates). The kind of rest mass variation we are talking about is completely ruled out by conventional nuclear and condensed matter physics, as was made clear in [2]. Nevertheless, there is a body of theoretical literature that proves that such variation is compatible with the fundamental principles of relativistic classical and quantum mechanics. The discrepancy between conventional physics and some experimental claims of d-d fusion amounts to 50 orders of magnitude for the deuterium fusion rate [3]. To most physicists this is absolute proof that these experiments are incorrect. It is my opinion too that conventional solid state physics and nuclear physics cannot explain these discrepancies. Variable masses, which would entail new physics, can make up for the discrepancy as was shown in [2]. The parent fields of both nuclear and condensed matter physics are particle physics and relativistic quantum theory, and we must look to these to find a way to incorporate variable masses. It must be made clear from the outset that what has been proposed in [2] is radical, and has never been directly observed, as was pointed out there. Nevertheless, the possibility that new physics may be required to describe such a huge discrepancy is simply too important for physics to ignore. There have been many attempts at explaining the observed LENR phenomena in terms of conventional theory. The subject is too vast to delve into here, but it is clear that none of these have gained any widespread acceptance by the physics community. I feel that the many approximate ad hoc attempts to explain this or that LENR effect with conventional nuclear physics models are hiding the true nature of the phenomenon, which requires radical new physics in the form of variable mass theories.

Most variable mass theories in modern physics are based on spontaneous symmetry breaking. These techniques are rooted in quantum field theory, and they do not seem to be appropriate for the origin of mass variation that is observed in LENR. This is because the condensed matter environment cannot be expected to modify in any significant way the Higgs mechanism that determines particle masses. Moreover, nuclear physics and condensed matter physics are described mainly with quantum wave mechanics as opposed to second quantized field theory, and it would thus be desirable to have a variable mass wave mechanical theory. It has been argued that relativistic quantum theory contains the option of having inherently variable mass behavior which is not caused by any sort of Higgs mechanism [4, 5, 6, 7]. Specific theories with variable masses were developed early on by Fock and Stueckelberg [8, 9, 10] which evolved into the modern historical time models described below in section 3.

We use units such that $\hbar = c = 1$, and a relativistic metric signature (-,+,+,+), unless otherwise noted.

Journal of Physics: Conference Series **615** (2015) 012016 doi:10.1088/1742-6596/615/1/012016

2. A list of nuclear anomalies that can likely be explained by variable nuclear rest masses in condensed matter settings

A listing of nuclear anomalies which have been observed in experiments are presented here. For the most part these phenomena are strongly rejected by the physics community. The reason is that they are believed to be obviously incompatible with existing theories of nuclear physics. I believe that all of these anomalies can be explained by variable masses, as we discuss below.

2.1. Unexplained excess heat in hydrated and deuterated metals.

Reviews of this topic, together with citations to the literature may be found in [1, 11, 12, 13, 14]. The most substantiated data is for the deuterium-Palladium system where deuterium loading is achieved through electrolysis with heavy water (D_2O). In highly deuterated palladium, with loading factors (ratio of the number densities of d and pd) greater than about 0.9, it is claimed that many electrolytic cells have produced heat in excess of what can be accounted for chemically by a number of independent laboratories and using different calorimetry and electrolysis equipment. They also claim to have observed Helium-4 production in the reactions in quantities which were approximately consistent with the excess heat if the reaction were nuclear fusion with two deuterons forming an alpha particle, and with the entire energy release going into heat. This is a surface phenomenon restricted to within a few thousand angstroms of the surface of the palladium.

Considering the possibility of d+d fusion in a d_2 molecule, Koonin and Nauenberg calculated the penetration factor through the Coulomb barrier at room temperature [3] which gave reaction rates 50 orders of magnitude smaller than the claimed experimental results of Fleischmann and Pons [15]. Leggett and Baym [16, 17] came to similar negative conclusions based on necessary but unobserved affinity enhancement for ^{4}He in deuterated palladium. They and others concluded that either the experiments must be wrong, or that the effects were not due to fusion. Recent experiments on deuteron beam scattering off of deuterated metals of various kinds have exhibited much larger fusion cross sections than these calculations [18, 19, 20, 21, 22, 23]. This enhancement has been attributed to unexpectedly large electron screening effects. The other main problem is to explain the extreme distortion of the branching ratios as compared with plasma fusion which has the well-measured ratios [24]

$$d + d \rightarrow t(1.01\,MeV) + p(3.02\,MeV),\ (Q = 4.04\,MeV),\ 51\% \qquad (1)$$

$$d + d \rightarrow^3 He(.82\,MeV) + n(2.45\,MeV),\ (Q = 3.27\,MeV),\ 49\% \qquad (2)$$

$$d + d \rightarrow^4 He(0.076\,MeV) + \gamma(23.77\,MeV),\ (Q = 23.85\,MeV),\ (3.8 * 10 - 3)\% \qquad (3)$$

where Q is the kinetic energy released by the reaction in the center of mass Lorentz frame. When excess heat is observed in LENR experiments, it is produced with practically no neutrons, tritium, or gamma rays.

As explained in [2], this effect can be explained by a combination of continuous electron mass increase along with continuous deuterium mass decrease in the nuclear active environment. The excess heat is produced by two tunneling effects acting together. The electron mass increase leads to enhanced electron screening and a commensurate increase in the d-d tunneling rate for neighboring deuterium nuclei. Then, if the deuteron masses drop to the point where the sum of the masses of these two equals the mass of Helium-4, resonant tunneling occurs, greatly enhancing the tunneling rate from the already enhanced rate due to electron mass increase. Because the mass of the deuterons is changing in this process, the usual branching ratios are greatly distorted in a time dependent way from the standard values above. If this process occurs in nature, then it is so radically different from ordinary on-mass-shell fusion that it deserves a different name. Perhaps the prosaic "variable mass fusion" is a reasonable term. If the electron masses increase with the deuteron masses staying on-shell, then one obtains ordinary "hot fusion" with approximately equal numbers of tritium and neutrons produced with the rates and kinetic energies given by (1) and (2).

It is possible that the enhanced electron screening is not due to the electron's mass changing as there might be other effects in a solid to enhance it , as has been proposed in the literature . The modified ratio of fusion channels can only be explained by a deuterium mass change. It's also possible that fusion is not occurring as in Widom-Larsen theory [25], but in that case the electron's mass must also increase in order for the reaction to proceed (see section 4.1.1).

2.2. Transmutation of elements in hydrated and deuterated metals.

There are extensive experimental claims of a variety of elemental transmutations taking place inside or on deuterated palladium and also in other metal-hydrogen alloys [1, 11, 26, 27, 28, 29, 30, 31]. With varying particle masses, transmutations can conceivably in theory occur in a number of ways in a palladium-deuterium lattice. These include electron capture, beta and alpha decay of nuclei whose mass has increased, resonant fusion of protons, deuterons, alpha particles with other nuclei including palladium, and even fission of palladium and/or other impurity elements after a mass increase. Also, there is the possibility of neutron creation and subsequent capture as in the Widom-Larsen theory[25], leading to many possible reactions with no Coulomb barrier to be overcome. In short a world of possibilities exist. The fusion possibilities mentioned in this list have a much higher Coulomb barrier than the deuterium molecule. Nevertheless, if resonant tunneling occurs, there might be small numbers of such events occurring, as has been reported in [28, 31].

2.3. Neutron and tritium production.

In the following subsections only experiments that showed either neutrons or tritium significantly above background levels are included. The neutrons detected were all energetic and non-thermal, but the kinetic energy of the tritium was typically not measured. There were experiments which showed results consistent with background levels. The number of positive experiments is large enough that they represent a critical mass establishing a high probability of an effect. If nuclear rest mass variation is the cause that enables the anomalous production of neutrons and tritium, then the amounts will depend on how the electron and deuteron or hydrogen masses vary, and this may vary from experiment to experiment yielding different rates and branching ratios.

2.3.1. Neutron production at low levels in hydrated and deuterated metals. These papers have reported low levels of neutrons produced in various metals which have been loaded with hydrogen or deuterium or both [32, 33, 34, 35, 36, 37, 38, 39, 40, 41, 42, 43, 44, 45, 46, 47, 48, 49]. In order to produce neutrons in our model, the electron mass must increase while the deuterium mass must remain close to the mass shell. The increased electron mass can account for the 20 order of magnitude enhancement required by the data as was shown in [3]. Notice that if the deuterium mass moves quickly off the mass shell to a lower value where neutrons and tritium can no longer be produced, the experiment will show nothing - no excess heat, no neutrons, and no tritium. Perhaps a few gamma rays. Nevertheless, the system can still be considered as nuclear active. Only when the adjacent deuteron masses combined are equal to the mass of helium-4 will resonant tunneling lead to excess heat and no radiation [2].

2.3.2. Neutron bursts produced in deuterium gas at liquid Nitrogen temperature in time-varying magnetic fields This is an intriguingly simple experiment performed by Tadahiko Mizuno [43]. He observed neutron emissions from pure deuterium gas after it was cooled in liquid nitrogen and placed in a magnetic field. Neutron emissions were observed in ten out of ten test cases. It is not clear whether the neutrons are produced in the bulk D_2 gas or at the interface between the pyrex container and the D_2. If the reaction occurs in the bulk, then it is extremely simple. Just a time-varying magnetic field and the deuterium gas cooled to liquid nitrogen temperature. Assuming that this is evidence for d-d fusion, then it would be explained by an increase in the electron mass in the gas. This experiment is tantalizingly simple. Mizuno has speculated that perhaps cosmic muons or neutrinos may be playing a role here because the neutrons come in bursts. I would like to see this experiment replicated, and redesigned to determine whether the neutrons are coming from the surface or bulk of the gas. This simple experiment eliminates the material preparation uncertainties of the metal in traditional LENR experiments, and so it should be easier to understand what is happening. The neutrons are produced sporadically, and this non-Poisson temporal distribution is quite surprising for a bulk reaction in the D_2 gas. In muon catalyzed fusion, a single muon can catalyze a number of fusions in a deuterium gas, limited by the alpha-sticking problem and the lifetime of the muon[50]. Not more than about 100 $d - d$ fusions can be catalyzed by a single muon. The mean energy of muons from cosmic rays at sea level is $\sim 4GeV$, and so these muons would be moving too fast to be captured into a deuterium molecule that could undergo muon catalyzed fusion. The average muon flux on a horizontal detector at sea level is about $1/cm^2min$. It seems quite unlikely that muons could be playing a role in these neutron bursts through muon catalyzed fusion. There is no known physical mechanism by which neutrinos could catalyze fusion or produce neutrons in the quantity that has been reported in this experiment either. This suggests that the reaction is probably not occurring in the bulk D_2 gas, but rather at the pyrex-gas interface where sporadic effects related to the glass chemistry and structure can be envisioned that might lead to neutron bursts. But for this to happen, the electron mass would have to increase to reduce the Coulomb barrier.

2.3.3. Tritium production at low levels in hydrated and deuterated metals. Tritium is generally produced in larger amounts than neutrons, and has been observed in many experiments [51, 52, 34, 53, 54, 55, 56, 57, 58]. There are two mechanisms for tritium production: $d+d \to t+p$ and $d + p + e^- \to t + \nu_e$. The production of tritium in $d + d \to t + p$ was discussed in [2].

2.4. Anomalous nuclear radiation emissions from hydrated and deuterated metals

A large number of uncatergorized radiation has been observed in numerous experiments. See Storms [1] for a review.

2.5. Much larger than expected d-d fusion rates observed in deuterated Palladium in electrolysis experiments and low energy deuterium beams incident on Palladium surface.

There were early observations of neutrons, with energies characteristic of anomalous d-d fusion, being produced in electrolytic cells with heavy water and Palladium cathodes [32]. The magnitude of the discrepancy between theory and experiment was estimated to be about 20 orders of magnitude in [3], whereas the discrepancy between theory and the excess heat reports were 50 orders of magnitude [3]. There are now several experiments that show anomalous d-d fusion and neutron production in scattering experiments with low energy deuterons incident on a Palladium target [18, 19, 20, 21, 22, 23], and these show much higher d-d fusion rates than were expected as well, adding credence to the earlier results in [32].

2.6. Helium-4 produced in electrolysis of Palladium electrodes in heavy water where the amount is consistent with the excess heat produced if the reaction is deuterium fusion.

A number of experiments have looked for evidence of production of ^4He in deuterated palladium systems [1, 59, 60, 61]. The excess heat in these experiments agreed approximately with the number of ^4He produced if the reaction were $d + d \rightarrow^4 He$+23.85MeV of heat.

2.7. Reduction of tritium decay rate when adsorbed onto nanocrystal Titanium particles and heated as observed by Otto Reifenschweiler at Philips laboratories, Eindhoven.

These experiments were performed in the late 1950s at the Philips Research Laboratories, Eindhoven, The Netherlands. The results were not published until 1994 [62, 63, 64, 65, 66]. Quoting from [62]:

"By heating a $TiT_{0.0035}$ preparation [0.0035 is the number ratio of tritium atoms to Titanium atoms] consisting of extremely small monocrystalline particles ($\phi \approx 15$ nm) a decrease of the radioactivity by 40% was observed [on heating the sample from 115° C to 275° C]. In further experiments the concentration of tritium in such preparations was varied (TiT_x experiments) showing that the radioactivity of the tritium increased less than proportionally to its concentration [which also indicates a reduction of decay rate]. Careful analysis of the experiments seems to rule out the possibility of trivial errors."

"In our experiments the titanium preparation was made by evaporation of the metal in argon at a suitable pressure, e.g. 0.5 to 2 cm Hg.... It was deposited as a kind of soot, consisting of monocrystalline particles about 15 nm in diameter, arranged in chains, on the inner wall of the measuring vessel. ... After the argon was pumped out the tritium was added and was completely absorbed within a few seconds. This was confirmed by hundreds of experiments. In one part of the experiments the electron current (β-particles and secondaries) was measured by a vibrating reed electrometer via a cylindrical electrode placed inside the vessel. In several other experiments there was a thin steel window to enable the x-radiation accompanying the β-decay (internal and external bremsstrahlung and characteristic X-rays...) to be measured by a

Journal of Physics: Conference Series **615** (2015) 012016 doi:10.1088/1742-6596/615/1/012016

GM- tube.... Various experiments established for both detection systems the linear relationship between the read-out and the activity."

These experiments have no explanation in conventional nuclear physics. The conclusions were criticized [67], but this criticism was addressed by Reifenschweiler [63]. The reduction in the decay rate has a very simple interpretation in terms of the tritium mass variation. If the tritium mass is decreasing during the heating phase, then the phase space of the tritium decay will be reduced, and consequently the decay rate will be reduced. Unfortunately there have been no attempts to replicate these experiments as far as I know.

Tritium decays 100% by beta decay in the reaction.

$$^3H \rightarrow ^3 He + e^- + \overline{\nu}_e,\ Q = 18.59\, keV \tag{4}$$

The Fermi form for the electron energy spectrum of tritium β-decay is [68]

$$\frac{d\Gamma_F}{dE_e} = \frac{G_F^2}{2\pi^3} \left|\mathcal{M}\right|^2 F(Z, R_e, E_e) p_e E_e (E_e^{max} - E_e)^2 \tag{5}$$

In this formula the neutrino mass has been approximated by zero. where G_F is the Fermi constant p_e, E_e, and E_e^{max} are the momentum, energy, and maximum endpoint energy, respectively, of the electron, and $|\mathcal{M}|^2$ is the absolute square of the nuclear matrix element. The Fermi function, $F(Z, R_e, E_e)$, provides a correction to the decay rate which is due to the Coulomb interaction of the decay electron with the charge 2e of the daughter nucleus ^3He. If the tritium mass can vary, then the maximum electron energy in the decay can vary by the formula

$$E_e^{max} = (mass\, of\, tritium)\ -\ (mass\, of\, ^3He) = Q + M_e \tag{6}$$

$$Q = E_e^{max} - M_e \tag{7}$$

The kinematic factor

$$R = p_e E_e (E_e^{max} - E_e)^2 \tag{8}$$

controls the decay rate as $\frac{d\Gamma_F}{dE_e} \propto R$. But since $Q \ll M_e$, we have the approximate result

$$\frac{d\Gamma_F}{dE_e} \propto p_e (E_e^{max} - E_e)^2 \tag{9}$$

the total decay rate can be obtained by integration

$$\Gamma_F = \int_{M_e}^{E_e^{max}} p_e (E_e^{max} - E_e)^2 dE_e = \int_{M_e}^{E_e^{max}} \sqrt{2M_e\,(E_e - M_e)}(E_e^{max} - E_e)^2 dE_e \tag{10}$$

Journal of Physics: Conference Series **615** (2015) 012016 doi:10.1088/1742-6596/615/1/012016

Clearly, if the mass of the tritium drops then the decay rate will also. It the tritium mass reaches the mass of $M_{^3He} + M_e$ or lower, then the decay rate will stop altogether. According to Reifenschweiler, the tritium decay rate fits a statistical model in which there are two states for the tritium in his experiments, either decaying with the usual expected rate or not decaying at all [69]. The tritium mass only has to drop 18.59 keV to cause the tritium decay rate to be zero. This is quite small compared to the approximate 12MeV drop that is required for resonant tunneling of deuterium. So if the deuterium mass can drop 12 MeV in a deuterated Palladium undergoing vigorous electrolysis, then it seems quite reasonable that tritium, another isotope of Hydrogen after all, can experience a mass drop of just 18.59 keV. Thus tritium might be useful as a tracer element that can be placed into hydrated metals to test if its mass is varying in a candidate nuclear active location.

2.8. Radiogenic and nucleogenic isotope geology

If LENR effects occur in a laboratory, then undoubtedly they will occur naturally in planetary matter, and this means that they could have an impact on isotope geology.

2.8.1. Helium-3 to Helium-4 ratio The atmospheric ratio of the natural abundances of ^3He over ^4He is $R_a = 1.37 \times 10^{-6}$. In different locations in the earth this ratio can vary and differ considerably from R_a. The conventional explanation for this is complicated [70]. Whereas ^4He is readily formed in alpha decay of many radioactive isotopes, ^3He is harder to form and therefore rarer.

(i) ^3He can be trapped in the earth at formation or added by meteoric dust.
(ii) It can result from cosmic ray scattering, and this is called cosmogenic.
(iii) Lithium spallation is the process by which a high-energy neutron (typically from Uranium or Thorium decay in the earth) bombards a lithium atom, creating a ^3He and a ^4He ion.
(iv) The ^3He that reaches the atmosphere is continually lost into the vacuum of space at a relatively high rate due to its low mass.
(v) It has been observed that the ratio of ^3He over ^4He in thermally active locations on the earth often tends to be larger than R_a. Such sites include volcanic emissions, geothermal hot springs, oceanic ridges, and deep lakes. These sources of high ratio are assumed to indicate that the ^3He comes from the mantle and is largely promordial in nature.

It has been suggested that part of the reason for higher ratios of ^3He over ^4He coming from volcanoes and deep lakes are LENR reactions occurring in the earth's mantle or crust [71, 72, 73]. This is absolutely not accepted by the geophysical community, but we cannot ignore it here because if masses can vary then the prospect that LENR is occurring in the earth is quite possible.

2.8.2. Deuterium to hydrogen ratio in the oceans and the origins of water on earth It is thought today that most water in earth's hydrosphere probably comes from Chondritic meteorites [74, 75, 76]. Comets tend to have too high a value of D/H (by about a factor of two) compared with the earth's oceans, and analysis of the ratio $^{15}N/^{14}N$ is the earth's atmosphere and in comets limits the contribution of comets to the earth's water to a small fraction [75]. The ratio of deuterium to hydrogen can vary from one comet to the next, and the average is not known. The total amount of water in the earth's hydrosphere has been estimated to be about $1.4 \times 10^9 km^3$ as estimated by Igor Shiklomanov for the United Nations [77]. The annual hydrologic cycle of evaporated water amounts to about $5.05 \times 10^5 km^3/yr$. Some of this evaporation ends up flowing through the earth, and there it will come in contact with various metals (like Palladium, Nickel, and Titanium) which might allow for some LENR to occur. If deuterium fusion is possible, as many LENR experiments have suggested and which could be possible with variable masses as we are discussing here, then some deuterium could be removed permanently in each cycle, and over time the reduction in deuterium might be measurable. The energy released in these reactions could also add to the energy balance of the earth. If the D/H ratio is declining in time on the earth, then this factor may enter into the debate over origins of water on the earth.

The ratio of deuterium to hydrogen on the moon might shed light on this issue. The age of the moon is about 4.5 billion years. Since both would receive the same flux of similar incoming meteorites and comets, they should have about the same ratio of deuterium to hydrogen incrementally added. But the moon would not experience LENR because it doesn't have a hydrological cycle like earth. Recent evidence suggests that the amount of deuterium vs hydrogen on the moon is higher than on the earth [78]. But this could be due to preferential loss of light hydrogen to the vacuum of space rather than to the fact that comets were the source of water on the moon.

2.9. Unexpected gamma ray and positron bursts from lightning storms.

Scientists using NASA's Fermi Gamma-ray Space Telescope have detected beams of antimatter produced above thunderstorms on Earth, a phenomenon never seen before [79]. These are produced in addition to gamma ray bursts [80, 81, 82]. These events are not occurring in metals, and they probably are not of LENR origin, but they are as yet unexplained and so we include them in this list of anomalies.

2.10. Variation in beta and alpha decay radionuclides when implanted into metals and other materials

The decay rates of various radioactive isotopes embedded into various materials have been observed to change from their natural standard values [83, 84]. The standard explanation for this effect is that electron screening lengths and electron densities of materials vary, and these will affect nuclear decays which proceed through tunneling through a Coloumb barrier or by electron capture. Our LENR theory suggests that in a nuclear active environment the electrons may gain mass. This will have an effect similar to enhanced electron screening. However, the electron mass gain effect, if it exists at all, can be expected to occur only in small localized volumes of the material, and even then sporadically. In addition to the electron mass change, our theory could even support the possibility that the radioactive isotope's mass can change. This would affect the decay rate by changing the available phase space for the decay. An extreme example of this was the Reifenschweiler experiment with tritium decay mentioned above. There have been experiments in which the change in the decay rate was very large, but which have not been repeatable. For example, ^{22}Na decay was measured in palladium [85] at room temperature and at 12K, and the half-life was 1.2% shorter at the lower temperature. A similar but smaller result was observed in [86] where the half-life at 15K was 0.46% shorter than at room temperature. A third experiment showed no change with temperature at all [87], but the embedding metal was gold instead of palladium. Perhaps this variability is due to an electron mass variation in nuclear active environments in the Palladium which does not occur in gold, and as in LENR, the

Journal of Physics: Conference Series **615** (2015) 012016 doi:10.1088/1742-6596/615/1/012016

properties of the material are important and different conditions can produce different results. Moreover, the expected nuclear active zone is expected to occur on the surface of the palladium only, and so the depth of embedding of the radionuclide would be a critical factor. The biggest effect should occur near the surface.

It would be extremely interesting to measure radioactive decay rates in highly deuterated palladium where any mass change effects, if present, could make a large impact on the decay rates. Perhaps an interesting experiment would be to perform the ^{22}Na decay experiment in highly deuterated palladium which is undergoing gaseous diffusion. It is expected that this decay rate will depend on the diffusion rate of deuterium through the palladium, and also on the loading factor. The temperature might also be varied. The half-life should get shorter as the mass of the electrons increase which would be expected to happen with increased loading. Many experiments along these lines suggest themselves. One might vary the depth of the radionuclide into the surface of the palladium, or try other metals, try alpha decays as well as beta decays, study the effect of temperature, pressure, magnetic fields, etc.

2.11. Unaccounted Temporal variation of radionuclide decay rates and LENR without any apparent variation in the environmental conditions.

There are a number of experiments that show radioactive decay rates changing cyclically with time for no apparent cause [88, 89, 90, 91, 92, 93, 94]. This has been attributed to an extraterrestrial source, possibly solar neutrinos [93, 89, 88]. Periods of 1 year, 27 days, and 1 day have been reported. Other experiments do not show such variations, and so there is again experimental uncertainty. But small oscillations in the rest masses of electrons or the radioactive isotopes themselves can clearly account for such varying decay rates. Temporal variation of LENR experiments have also been reported [95, 96], and the experimenters have felt they were hard to explain as due to the variation in temperature and pressure in the environment. Another feature of LENR experiments is their sporadic nature. Periods of nuclear activity tend to appear suddenly and unpredictably, like earthquakes, and then last an unpredictable amount of time with varying intensity, and then nothing again for a while, and then perhaps again activity appears unpredictably again.

2.12. Sonoluminescence and sonofusion

Sonoluminescence, the production of light by collapsing bubbles in a fluid, is a completely reproducible and accepted experimental phenomenon [97, 98, 99]. Sonofusion, on the other hand, the claimed $d-d$ fusion of deuterium inside collapsing bubbles undergoing forced cavitation is quite controversial [100]. The temperature rise inside the collapsing bubble is quite high $\sim 15,000K$, but this is not high enough to produce significant hot fusion in deuterium. If electron masses could increase inside the bubble, then the tunneling rate could increase, and it might be possible to produce neutrons and tritium in measurable quantities, and if the deuterium mass could change, then it could changed the branching ratios of the neutron and tritium channels.

2.13. X ray production from peeling scotch tape - triboelectric X ray source

Researchers at UCLA found X rays being produced by peeling transparent tape in a vacuum of about 10^{-3} Torr of air [101, 102, 103, 104]. The X ray photons range in energy from zero to 100keV with a mean about 20 keV. The photons are produced in nanosecond bursts with many photons per burst suggesting an avalanche discharge effect. The usual explanation for triboluminescence, which normally produces visible wavelengths, is that charges build up between two different materials that are being rubbed together, and then electron currents can arc though air and produce radiation through bremsstrahlung. Such theories look promising to describe this phenomenon, provided the necessary large charging of the peeling tape and roll can be explained. The angular distribution is more sharply peaked than simple bremsstrahlung however [104], and so we tentatively list it as an anomaly. At the very least, these experiments definitely show that unexpectedly high energy radiation can be produced by very mild conditions at room temperature in everyday objects. These experiments are also 100% reproducible and they work on demand. Products have already been developed which use this technology in commercial x-ray imaging systems. They illustrate beyond any doubt that nuclear scale energies are possible in seemingly innocuous condensed matter settings at room temperature. It turns out that the output of x-rays depend critically on the tape speed, and on the pressure in the atmosphere surrounding the tape. Outside of a range in both of these parameters there are no x-rays produced. This was not apparent at the start of the experiment, and so some experimenters could miss the effect altogether if they were not persistent. LENR is similar in that with knowledge and experience more things that can affect the performance are being found.

3. Theoretical basis for variable mass particle theories.

Fixed particle rest masses are not a requirement of special relativity, general relativity, quantum mechanics, or quantum field theory, but experiments show that all elementary particles of a given type,in isolation, always have the same mass. Quantum field theory tells us that particles in continuous interaction must have variable masses though. Historical time models, introduced by Fock and Steuckelberg [8, 9, 10] are one prominent and simple example of a variable mass theory. They were introduced in order to solve some of the problems encountered early on in developing a relativistic wave mechanics. Modern variants of this theory offer the most well-developed variable mass theories. Other possibilities exist though. An analysis of the possibility of variable mass particle theories, in the context of general relativity and the principal of equivalence, was carried out by Greenberger [4, 5, 6, 7]. He treated the proper time of a particle as a quantum operator with coordinate time serving the role of describing the causal evolution. This is the reverse of the historical time approach where the coordinate time is an operator, but the Lorentz-invariant historical time (which is formally similar to proper time) is simply a real-valued numerical parameter which parametrizes the causal evolution of the system. His conclusion is that variable mass theories are compatible with the requirements of general relativity. For the case of particles with spin, another approach for variable mass particles was developed by Corben [105, 106]. The standard model of particle physics incorporates spontaneous symmetry breaking and the Higgs mechanism to produce masses for the bosons of the weak interaction, and these masses are dependent on the Higgs field, and therefore somewhat variable as well.

Consider the classical relativistic particle equation of motion for particle orbit in Minkowski space.

$$\frac{dP^{\mu}(\tau)}{d\tau} = F^{\mu}(\tau) \tag{11}$$

where τ is proper time and $P = MU(\tau)$ with $u(\tau)$ being the proper 4-velocity $U(\tau) = dx(\tau)/d\tau$. In general, the mass of the particle will change by the formula

$$\frac{dP^\mu P_\mu}{d\tau} = 2P^\mu F_\mu \tag{12}$$

There is no requirement that the mass be any particular fixed value in general. If it so happens that $P^\mu F_\mu = 0$, then the mass will be fixed on an orbit, but not constrained to any particular value initially. The electromagnetic Lorentz force $\frac{dP^\mu(\tau)}{d\tau} = qU_\nu F^{\mu\nu}(x)$ satisfies this because the Faraday tensor $F^{\mu\nu}$ is antisymmetric in its indices, and thus conserves the mass. So relativity alone does not require fixed mass particles. In conventional relativistic quantum mechanics for a single particle, time is taken to be a real number which parametrizes the evolution of the state as it is in non-relativistic quantum mechanics, but the position operators and their conjugate momenta are operators in a Hilbert space that do not commute. Therefore, ab initio, time and space are fundamentally different in quantum wave mechanics, and this contravenes relativity which asserts that they must reside in a 4D manifold and can mix under Lorentz transformations. If we mix a real-numbered time with an operator valued position to get the time in a new frame, then it will become a non-commuting operator. Thus time is required to play two roles simultaneously, first as a c-number parametrization of causally related events unfolding in space-time, and second as an operator in a Hilbert space that does not commute with the energy operator. Thus, combining relativity and quantum mechanics is fundamentally difficult. Field theory is different, because in a field the position 4-vector is a c-number vector for both time and space components, and the dynamical variables are the fields and not the particle coordinates. Thus manifest covariance is easier to achieve. As a consequence, the modern approach in physics has been to give up on trying to develop a relativistically covariant multi-particle wave function approach and appeal to second-quantized field theory as the only acceptable relativistic particle theory. Attempts at relativistic wave mechanics, such as the Dirac equation or the Klein-Gordon equation, have unresolved problems with negative energy states or the impossibility of defining a positive definite probability density. However, in historical time models the rest masses of particles can vary and manifest covariance can be achieved, and this is why they were and are still studied.

3.1. Historical time models

In quantum field theory, the standard method of introducing variable masses is via spontaneous symmetry breaking and the Higgs mechanism, but in nuclear physics or condensed matter physics, the standard theoretical analysis is done in terms of multi particle wave functions in the Schrdinger representation. To introduce variable masses in this context, one naturally turns to the historical time models which are the most well-known and developed versions of wave-function-based variable mass theories.

The application of historical time models by Feynman to the relativistic path integral problem is illustrative [107](see Appendix A). In order to put space and time on a manifestly covariant footing, he chose to introduce a second time variable, call it τ. The usual Klein-Gordon equation is

$$(i\partial - A)^\mu (i\partial - A)_\mu \Psi = -M^2\Psi \tag{13}$$

where M is the particle's rest mass, and where A is the vector potential for an external electromagnetic field. There are well known problems with the probabilistic interpretation of this equation because the conserved current is $j_\mu = i\left[\phi^*\partial_\mu\phi - \phi\partial_\mu\phi^*\right]$ and j_0 takes on both positive and negative values. Moreover, a localizable Schrdinger type position operator cannot be defined for this equation [108, 109]. Feynman wished to apply his path integration method of quantization to this. In his words "... *we try to represent the amplitude for a particle to get from one point to another as a sum over all trajectories of an amplitude exp(iS) where S is the classical action for a given trajectory. To maintain the relativistic invariance in evidence the idea suggests itself of describing a trajectory in space-time by giving the four variables* $x^\mu(\mu)$ *as functions of some fifth parameter* μ *(rather than expressing* x_1, x_2, x_3*in terms of* x_4*)*". So Feynman replaces 13 with (we use $x^0 = t$, and τ instead of μ and we use spacelike metric, whereas Feynman used timelike)

$$i\frac{\partial\varphi(x,\tau)}{\partial\tau} = \frac{1}{2}\left(i\partial - A\right)_\mu \left(i\partial - A\right)^\mu \varphi(x,\tau), \text{ where } x = \{x^0, x^1, x^2, x^3\} \tag{14}$$

It is very similar to the time dependent Schrdinger equation, but with the "historical time" τ replacing the usual time variable t, and with the four coordinates of space-time x^μ replacing the usual three coordinates of space. Because of this, it is quite easy to formulate the path integration method for this system by following the same steps as used for the Schrdinger equation. Feynman points out that if A^μ does not depend on τ, then separable solutions exist so that $\varphi(x,\tau) = \Psi(x)exp(i\frac{1}{2}M^2\tau)$ and ψ is the solution to the usual Klein-Gordon equation. Equations of this type were first studied by Fock [8] and Stueckelberg [9, 10, 110]. The path integral solution includes all paths in space-time connecting two space-time points and parametrized by τ, with no restrictions on the path, so that off-mass-shell paths are included in the path integral solutions. This reflects that fact that Feynman diagrams contain virtual particles which are not on the mass shell. The reformulation of quantum mechanics along these lines was initially called the "proper time formalism", and was also studied by Nambu [111], [112], Cooke [113], and others. A good theoretical justification for studying this type of theory as a replacement for fixed-mass relativistic wave mechanics was given by Pearle [114] who also addressed the two-body scattering problem. A general multi-body theory was developed by Horwitz and Piron [115]. A 5D generalization of electrodynamics and quantum electrodynamics called "Pre Maxwell theory" was introduced by Saad, Horwitz, Arshansky, and Land [116, 117, 118]. The literature on historical time models is new quite extensive. A good source for much of the subject is Fanchi [119, 120].

3.2. Horwitz-Piron theory

Horwitz and Piron [115, 121] supposed that a single historical time τ simultaneously parametrizes multiple particles, with their motion determined by a Hamiltonian K, conjugate to τ, that had only simultaneous (in τ) interactions. This simplifies the calculation of many-particle systems considerably. They postulated a Hamiltonian K, conjugate to the historical time τ. They chose K as follows

$$K = \sum_{i=1}^{n}\left[\frac{1}{2M_i}\left(p_i - eA(x_i)\right)^\mu \left(p_i - eA(x_i)\right)_\mu\right] + \sum_{i<j} V_{ij}(|x_i - x_j|) \tag{15}$$

where A^μ is an external electromagnetic four vector potential, and V is an inter-particle potential energy. The "dot" notation denotes τ derivatives . Hamilton's equations of motion become

$$\dot{x}_i^\mu = \frac{\partial K}{\partial p_{i\mu}} = (p_i - eA(x_i))/M_i, \quad \dot{p}_i^\mu = -\frac{\partial K}{\partial x_{i\mu}} \tag{16}$$

$$\frac{dK}{d\tau} = 0 \tag{17}$$

This leads to the Lorentz force equation

$$M_i \ddot{x}_i^\mu = eF^{\mu\nu}\dot{x}_{i\nu} - \sum_{j \neq i} \frac{\partial V_{ij}(|x_i - x_j|)}{\partial x_{i\mu}}, \quad where\ M_i\dot{x}_i^\mu = p_i^\mu - eA^\mu \tag{18}$$

Because of the similarity to Newtonian mechanics, the solution of these equations (the Cauchy problem) is no more difficult, except for the extra dimension. The mass of a particle is given by

$$m_i^2 = -M_i^2 \dot{x}_{i\mu}\dot{x}_i^\mu = -(p_i - eA(x_i))^\mu (p_i - eA(x_i))_\mu \tag{19}$$

The particle's mass m_i is not necessarily equal to M_i, and is not even necessarily a constant of the motion if V is not zero. Particles can scatter and end up off the mass shell (ie. so that $m_i \neq M_i$), and then move apart into isolated regions where there are no fields, and they can remain off the mass shell there. There is no obvious restorative force that returns particles to the mass shell in this theory, however, since the Hamiltonian K is independent of historical time τ, whenever the particles are far apart and well separated so that the V_{ij} can be ignored, and if the fields vanish there too, we have

$$K = -\sum_{i=1}^{n} \frac{m_i^2}{2M_i} \tag{20}$$

Therefore, if some particles increase their mass, then others will have to decrease their mass in order to keep K constant. In the absence of an external force, the total 4-momentum $P = \sum_i p_i$ is conserved as well.

3.3. Pre-Maxwell 5D theory

Off-shell electrodynamics was generalized to a five dimensional gauge invariant theory by Saad, Horwitz, Arshansky, and Land [116, 117, 118]. Here we follow the notation in [118]. In this theory, τ-dependent gauge invariance was assumed so that the wave equation is invariant under local gauge transformations of the form

$$\psi(x,\tau) \longrightarrow e^{ie_0\Lambda(x,\tau)}\psi(x,\tau) \tag{21}$$

$$\left(i\frac{\partial}{\partial\tau} + e_0 a_5(x,\tau)\right)\psi(x,\tau) = \frac{1}{2M}(p - e_0 a(x,\tau))^\mu (p - e_0 a(x,\tau))_\mu \psi(x,\tau) \tag{22}$$

This leads to a non-trivial generalization of electromagnetism. The parameter e_0 has units of length and is proportional to the electric charge e

$$e = \frac{e_0}{\lambda} \tag{23}$$

where the constant parameter λ is new. The fields transform as

$$a^{\mu}(x,\tau) \to a^{\mu}(x,\tau) + \partial^{\mu}\Lambda(x,\tau) \qquad a_5(x,\tau) \to a_5(x,\tau) + \partial_{\tau}\Lambda(x,\tau) \tag{24}$$

Non-vanishing values of a_5 leads to mass variation. The Schrdinger equation 22 leads to a five dimensional conserved current

$$\partial_{\mu}j^{\mu} + \partial_{\tau}j^5 = 0 \tag{25}$$

$$j^5 = \rho = |\psi(x,\tau)|^2 \qquad j^{\mu} = \frac{-i}{2M}\left\{\psi^*\left(\partial^{\mu} - ie_0a^{\mu}\right)\psi - \psi\left(\partial^{\mu} + ie_0a^{\mu}\right)\psi^*\right\} \tag{26}$$

The usual Maxwell theory is recovered by integrating over τ, a process termed "concatenation".

$$J^{\mu}(x) = \int_{-\infty}^{+\infty} j^{\mu}(x,\tau)d\tau \qquad A^{\mu}(x) = \int_{-\infty}^{+\infty} a^{\mu}(x,\tau)d\tau \tag{27}$$

where A^{μ} and J^{μ} are the usual Maxwell potential fields and 4-current respectively. Saad et al [116] suggested an action which had higher 5-dimensional spacetime symmetry so that the Lorentz group O(3,1) is a subgroup. This requires either O(4,1) or O(3,2) symmetry. Both are considered in the literature, and to handle this, a 5D metric tensor is introduced $g^{\alpha\beta} = diag(-1,1,1,1,\sigma)$, where $\sigma = +1$ for O(4,1) and $\sigma = -1$ for O(3,2).

This theory has been studied extensively by Horwitz in particular along with a number of co-authors [116, 122, 123, 124, 125, 118, 126, 117, 127, 128]. Path integral quantization has been analyzed [129] as has more canonical second quantization approaches to off-shell quantum electrodynamics for spinless charged particles [OSQED][118]. A theory of relativistic wave functions generalizable to arbitrary spin and based on Wigner's induced representations of the homogeneous Lorentz group has been described by Horwitz in [130] and references therein. This theory produces relativistic wave fuctions which satisfy all the desired requirements by introducing an arbitrary timelike vector n^{μ} for massive particles, and which is then used to define the induced representation and Wigner's little group for the particle.

Land and Horwitz [118] have applied perturbative quantum field theory methods to off-shell pre-Maxwell electromagnetism in interaction with spinless charged particles. They have developed Feynman rules and applied them to various scattering processes. This is a 5D theory, and there are two possibilities for the 5D symmetry group that contains the Lorentz group as a subset. The Lorentz group O(3,1) is a subgroup of either O(4,1) or O(3,2), and these two possibilities lead to two different theories with different physical properties. The results for particle-particle scattering show that in general the masses will change for particles in a scattering process. They present detailed cross-section calculations for both Mller and Compton scattering, and both calculations show mass changes after scattering. This is an extremely interesting paper as it provides ample off-shell behavior. The mass of a particle in this theory is a function of the past history of the particle.

The self interaction of a classical charged particle in pre-Maxwell theory has also been studied. The solutions are more complicated than the Lorentz-Dirac equation [128, 131, 132]. The runaway solutions are replaced by chaotic nonlinear equations which include variation of the classical particle's mass with time.

There are no rigorous bounds on how far off-shell the massive particles can wander in the various historical time theories for realistic condensed matter systems, especially in non-equilibrium situations.

3.4. How to perhaps reconcile these ideas with the standard model

The established mechanism for mass fixing in the standard model of particle physics is spontaneous symmetry breaking and the Higgs mechanism. It is probably unrealistic to imagine that the vacuum Higgs field could be modified in a condensed matter setting. But, it is conceivable that these fixed mass values are the on-shell parameters that appear in the Horwitz-Piron or pre-Maxwell theories which might be effective Lagrangian approximations to the standard model in some circumstances.

In relativistic field theory all of the loop momenta in a Feynman diagram are off the mass shell. Since all matter is interacting with other matter all the time through long range forces, we can deduce that all particles in our universe are very slightly virtual all of the time, and are never exactly on the mass shell. The question is how far off the mass shell can they wander. The LENR experiments suggest that there exist situations in condensed matter settings where this mass variation can be much larger than expected.

The nuclear active environment for LENR in metals is known to occur near the surface of palladium or other metals, and in areas where the lattice has been deformed. The surface morphology and chemistry are also known to be critical factors - rough surfaces on the nano-scale are much more active, and of course high loading of deuterium or hydrogen is also a requirement. In general, they also require statistical non-equilibrium induced by outside influences like continuous electrolytic loading, pulsing electrical signals injected into the material, optical stimulation of the surface, thermal cycling, magnetic cycling, acoustic waves, incoming particle beams, etc.

Effective Lagrangian approximations are routinely employed in nuclear physics. Many of these methods predate the standard model. They often exploit the low values for mass of the u and d quarks which when taken to zero lead to chiral-isospin symmetry which can then justify utilizing group theoretical methods to derive general forms for interacting potentials [133, 134, 135]. Most of theoretical nuclear physics calculations are based on the Schrdinger equation with effective nuclear and electromagnetic potential functions which do not allow any off-mass-shell behavior, even though the relativistic field theory perturbation does. Up until LENR effects were discovered, there was no need for such behavior in nuclear physics. It is my opinion that there is such a need now since conventional approaches have failed by more than 50 orders of magnitude [3] to explain what is being seen in LENR experiments, and it seems that off-mass-shell behaviour of one type or another can explain all of the experimental anomalies.

In multi-dimensional tunneling for many-body systems, many tunneling paths compete for the "path of least barrier" to reach the true ground state from a metastable one [136]. Condensed matter systems are always in fact quasi-stable systems. The true ground state - after allowing for all nuclear reactions - is at a much lower energy level, but large Coulomb barriers prevent their decay. Of course, it is usually observed that tunneling to the true ground state through nuclear reactions happens at such a slow rate that it can be ignored, and this agrees with calculations of tunneling rates using quantum mechanics. LENR phenomenon suggest that in some special unique situations a nuclear active environment can arise in which the tunneling rate is much larger. Mass deviation, chaos [137], resonances [138, 139], and driving forces [140] are all mechanisms for enhanced tunneling rates. Given all these possibilities, there is certainly no rigorous theorem based on the standard model that says the experimental claims of LENR cannot occur. There may exist effective off-mass-shell Lagrangians for such systems which facilitate calculations of multi-dimensional tunneling, which give the correct S-matrix when all particles are separated spatially in the initial and final state, and which allow for LENR behavior.

So in this paper, as in [2], it is proposed that in a condensed matter setting in special regions or domains (nuclear active domains), the electromagnetic interaction of charged particles can be described by an effective off-mass-shell Lagrangian. The best choice seems to be the historical time models, but a completed derivation of this from the standard model in the future may reveal extra terms in the effective Lagrangian that have not yet been studied or anticipated. In the absence of such a complete derivation, experiments can help to guide the way to implementing this approach.

3.5. A Phenomenological working model

The historical time approach to relativistic quantum mechanics provides an excellent starting point for attempting to develop an acceptable theory of variable masses in LENR. Extra terms in the Lagrangian or some other modification or understanding will need to be found to clarify how particles get back onto the mass shell after they have departed form it, and also to describe both the nuclear active environment and ordinary non-nuclear-active environments with the same theory. Obviously this will involve one or several parameters which control the degree of nuclear activity in a given condensed matter setting. I think that experiments should lead the way in this search. To have a starting point we have proposed a phenomenological approach. In order to describe the experiments in LENR, the following assumptions were made about mass variation in [2]

(i) Mass variation is rare and it happens at best at very localized micro-regions of a material which have been termed "nuclear active environments" in the LENR literature. The precise requirements for such regions to exist are not known. The assumption here is that the only thing that distinguishes a nuclear active environment is that it is a region where masses can change. The Pre-Maxwell theory is an electromagnetic phenomenon, and it seems quite reasonable to assume that all charged particles can experience mass variation, and in general the mass can increase or decrease.

(ii) The mass distribution of a particular particle type can be a time varying probability distribution. The mean value and the variance of this distribution can change with time. For fusion to occur, it was assumed in [2] that as the two neighboring deuterons approach resonance in the Helium-4 channel from above, their combined mass is sharply peaked and changes slowly with time. Such adiabatic behavior ensures that the combined mass will pass very close to the alpha particle's mass, which is a requirement for resonant tunneling to occur.

(iii) Immediately after a nuclear reaction involving strong or weak processes, the resulting particles are assumed to be exactly on the mass shell. This assumption was deemed necessary to make a sensible theory. The Horwitz-Piron theory and the Pre-Maxwell theories, which describe electromagnetic interactions only, do not have this property. They allow off-mass-shell particles in the final scattering state as well as the initial state of the scattering process. A general problem with these theories is that if a particle has gotten off the mass shell by a succession of scattering processes, it is not clear how that particle will get back onto the mass shell. Nuclear decays involving strong and weak forces are assumed to co-exist with a Horwitz-Piron or pre-Maxwell type of model for the electromagnetic interaction. But, for electromagnetic decays, this assumption still is a bit problematical. What is needed is a mechanism to allow particles to be forced back onto the mass shell if they are off of it, and this effect must probably be mediated by some parameter that interpolates continuously between a pure historical time model and conventional on-mass-shell quantum mechanics. Such a parameter was suggested in [2].

Ultimately I hope that the historical-time models can be modified to exhibit these assumptions rigorously, to clarify the return to the mass-shell after excursions, and to characterize the unique distinguishing parameters that determine the degree of nuclear activity in a region of condensed matter.

4. Possible mechanisms for anomalous nuclear behavior

Here we enumerate a variety of effects that can lead to nuclear reactions in nuclear active environments.

4.1. Electron mass change

4.1.1. Widom-Larsen effect enabling the reaction $p + e \Rightarrow n + \nu_e$ Low energy neutrons can be produced if the electron mass changes which can cause neutron activated changes in the isotopic mix [25]. This reaction was apparently first proposed and discussed by Swartz [141] to explain experimental results observed in nickel-hydrogen electrolysis experiments. In conventional nuclear physics, the velocity of electrons needs to be 92% of the speed of light in order to be able to form a neutron in collision with a proton. The presence of electromagnetic fields inside of a solid can change the speed of the electrons, but not their nuclear mass [2]. If electrons of this velocity were inside a solid, they would produce bremsstrahlung x-rays and gamma-rays. These are not typically observed in LENR experiments in sufficient quantities to explain excess heat production. In a historical time model, however, the rest mass of a particle depends on its past history, and the possibility of off-mass-shell electrons or protons cannot be ruled out a priori. In these theories the electron's nuclear rest mass might be large (the threshold mass for neutron production is about $2.5 \times$ *electron mass*), and yet its velocity small so that there could be no x-ray bremsstrahlung. Thus I accept the Widom-Larsen process as possible. As it is mediated by the weak interaction, it must compete with other processes which may have higher reaction rates in selected instances. The historical time model must be such that inside the bulk metal, there is no significant off-mass-shell behavior. Only in a rare surface state - the nuclear active environment - can mass variation be occurring. Ongoing experiments are gaining knowledge about the requirements for a nuclear active environment or domain.

4.1.2. Enhanced electron screening leading to greatly enhanced tunneling rates for fusion reactions In the Born-Oppenheimer approximation to molecules and solids, the size scale is determined by the inverse of the electron mass [2]. Therefore, an electron mass increase will reduce the inter-nuclear distance, and greatly enhance quantum tunneling and fusion rates [3, 2]. This is a form of electron screening. The enhancement of deuterium fusion in a d_2 molecule was rigorously quantified in [3], where it was shown that a mass of ten electron masses was necessary to explain the reaction rates measured by Fleischmann and Pons [15], and a mass of five electron masses was needed to explain Jones' results [32]. With simply a large electron mass, one cannot explain the discrepancy in the branching ratio, but with a simultaneous reduction in the deuterium mass, one can explain that as well [2], and in this case the electron mass increase factor can be much smaller than 10 since the resonant tunneling effect greatly enhances the fusion rate. Electron mass increase factor of 5 are sufficient to explain the results of Jones [32][3], and this could be adequate, when combined with resonant tunneling, to explain the excess heat experiments that have been observed [2].

4.1.3. Enhanced electron screening leading to modified radioactive decay rates Alpha decay rates of radioisotopes are increased (half lives reduced) by enhanced electron screening because the potential barrier that the alpha particle must overcome to get out of the nucleus is reduced by the more effective electron screening cloud surrounding the nucleus[142]. Beta^{+} decay rates are enhanced by electron screening, whereas beta^{-} decay rates are reduced [143].

4.1.4. Electron capture by nuclei not normally allowed. The modified electron mass could combine with a nucleus (whose mass might also be off shell) to form a different nuclear isotope or isomer which would be forbidden by energy conservation requirements for on-shell electrons.

4.2. Deuterium mass change

4.2.1. Decrease in the mass leads to a change of branching ratios in d+d fusion reactions as described in [2]. If the deuterium nucleus experiences a decreased mass, then the first result in a d+d fusion reaction is that the available energy released decreases for each of the reaction channels. The various reaction products have reduced kinetic energies. If the deuterium mass decreases enough, there will no longer be enough center of mass energy to produce neutrons, and the main reaction becomes the tritium channel. If the mass decreases still more, eventually there is not enough energy to produce tritium and the only channel that is open is the gamma-ray channel. As the deuterium mass decreases still further, eventually it reaches the resonant point where the center of mass energy is exactly equal to ^{4}He, and resonant tunneling occurs.

4.3. tritium mass decrease can lead to a reduction of the tritium decay rate as has been observed by Reifenschweiler

As described in 2.7, if the tritium mass decreases, the result is a reduction on the available phase space for the three particle beta decay.

4.4. Alpha particle mass change could lead to resonant fusion involving alpha particles and other nuclei

The same tunneling effects that can allow d-d fusion [2] might apply to fusion of nuclei, and the easiest such reactions would occur with lighter elements for which the Coulomb barriers are lower. The combination of increases electron mass - leading to enhanced electron screening - with modified alpha particle or nucleus mass resulting in resonant tunneling can in theory at least allow for fusion reactions to occur which would normally be almost unimaginably unlikely because of the extremely large Coulomb barriers that must be overcome. These types of reactions can also occur through neutron capture as well, as in the Widom-Larsen model.

4.5. Radioactive isotope mass variation causing decay rates to change since available phase space depends on the mass of the parent radio-isotope.

The decay rate of any particle into a final state of decay product particles is given by the general formula

$$d\Gamma = \frac{S\,|\mathcal{M}|^2}{2M} d\Phi_n(P, p_1, p_2, ..., p_n) \tag{28}$$

where $M = \sqrt{|P_\mu P^\mu|}$ is the mass of the decaying particle, n is the number of decay particles, S a combinatoric factor accounting for identical particles, $\mathcal{M}$ is the quantum transition matrix element, and $d\Phi_n$is the phase space of the decay given by

$$d\Phi_n(P, p_1, p_2, ..., p_n) = (2\pi)^4 \delta^4(P - \sum_{i=1}^{n} p_i) \prod_{i=1}^{n} \frac{d^3\overrightarrow{p}_i}{2\,(2\pi)^3\,E_i} \tag{29}$$

where P is the 4-momenta of the decaying particle, p_i the 4-momenta of the decay products, and where E_i is the total energy of particle i. Examples of phase space calculations can be found in [144]. The phase space increases monotonically as a function of the decaying particles mass M. Thus, any increase or decrease in this mass will directly increase or decrease the decay rate, however the decay matrix element $\mathcal{M}$ may also depend on this mass, and so the phase space may not give the whole picture, but I would expect the phase space to be the dominant factor. So, observations of varying radioactive decay rates in condensed matter settings may indicate mass variation if no other explanations seem plausible.

5. Conclusion

We have listed here a number of anomalous nuclear reactions which have not been explained by conventional nuclear physics. Many of these are probably impossible to explain without something new. But, if nuclear rest masses can vary in special "nuclear active environments", then all of these reactions can be allowed with the right mass variation. After surveying the literature on mass variation in relativistic physics, the historical time models pioneered by Fock and Stueckelberg are prominent. The modern refinements of this theory especially by Horwitz and co-workers stand out as the most likely candidate theories around which to construct a model for LENR. The task before us is to try and quantify what physical effects cause a nuclear active environment to come into existence. The theories for mass variation are not yet complete. In particular, we don't have a theory which interpolates between theories in which the rest mass is quite freely variable, like for instance pre-Maxwell theory, and ones that have fixed mass, like conventional QED or the standard model (at least as they are normally understood). I have in mind a new phenomenological parameter that would depend on position inside a material and control how easy or hard it is for particles to drift off the mass shell. In ordinary matter or in vacuum, this parameter would make deviation from standard rest masses exceedingly small, but in the nuclear active environments it would allow significant deviation of mass from the expected value. It is my opinion that this is the most likely way to reach an understanding of the LENR experiments in total.

Once again, as in [2], I remind the reader that there is no direct evidence that masses can change as has been suggested here. This is a radical proposal which deserves to be treated with great skepticism. The only reason I am proposing it is because it is the only way I can see to explain all these anomalous experiments, and I feel that they must be explained. In the words of Sherlock Holmes "When you have eliminated the impossible, whatever remains, however improbable, must be the truth" [145] (in chapter 6).

Many of the objections to LENR experimental claims in the past were based on the fact that neutrons and tritium were not observed in experiments involving deuterated palladium in the proper amounts, energies, and ratios as deuterium fusion would require [146, 147]. But as we have seen, if masses of electrons and deuterons can vary, then this not a valid justification for rejecting these results out of hand. So, the rejection of LENR has been based largely on a theoretical bias. Since all particles in condensed matter are slightly virtual (in the relativistic Feynman diagram sense) all the time, they are slightly off the mass shell all the time. Therefore we know that assuming the particles in condensed matter to be exactly on the mass shell cannot be exactly correct. What we don't know is how far off-shell they may wander. It is routinely found (or at least assumed) in most situations that they don't wander off shell enough to make a difference for nuclear kinematics, or for chemistry and solid state physics for that matter, but relativistic quantum mechanics permits the possible existence of anomalous regions in condensed matter where masses can vary significantly. So, if one is to rule out a theory like the present one, the change in the ratios of tritium to neutrons or their energies cannot be used to dismiss experiments. Rather, they might be viewed as experimental evidence in support of variable masses. Experiments like the x-ray producing scotch tape [101, 102, 103, 104] prove with certainty that unexpected energetic excursions can occur in ordinary condensed matter. Perhaps mass excursions can occur as well to explain other anomalies.

Much of modern physics has separated from close laboratory experimental feedback. Fields like quantum gravity, string theory, interpretations of quantum mechanics, emergent theories of gravity, etc. are not dealing with specific discrepancies with laboratory experiments. They have become to some extent excursions in pure mathematics, beautiful and interesting in their own right, but lacking the compass of frequent experimental guidance. Here we have a large discrepancy which may be resolved by a fundamental modification of our theories for nuclear physics and condensed matter, and we also have lots of laboratory feedback to aid us. It seems that physicists ought to re-examine this subject in the light of these new developments, and determine with certainty whether or not the nuclear anomalies are real.

Acknowledgements

I thank Larry Horwitz for extensive valuable correspondence and Martin Land, Philip Mannheim, Peter Hagelstein, Michael McKubre, Michael Melich, David Nagel, Mitchell Swartz, Fran Tanzella, Vladimir Kresin, Bob Perlmutter, Peter Schattner, and Andrew Davidson for informative discussions and correspondence.

References

[1] Storms E 2007 *Science of Low Energy Nuclear Reaction: A Comprehensive Compilation of Evidence and Explanations about Cold Fusion* (World Scientific Publishing Company) ISBN 9812706208
[2] Davidson M 2014 *Found. Phys.* **44** 144–174
[3] Koonin S E and Nauenberg M 1989 *Nature* **339** 690–691
[4] Greenberger D M 1970 *Journal of Mathematical Physics* **11** 2329–2340 ISSN 00222488
[5] Greenberger D M 1970 *Journal of Mathematical Physics* **11** 2341–2347 ISSN 00222488
[6] Greenberger D M 1974 *Journal of Mathematical Physics* **15** 395 ISSN 00222488

[7] Greenberger D M 1974 *Journal of Mathematical Physics* **15** 406 ISSN 00222488
[8] Fock V 1937 *Phys. Z. Sowjetunion* **12** 404–425
[9] Stueckelberg E 1941 *Helv. Phys. Acta* **14** 588–594
[10] Stueckelberg E 1941 *Helv. Phys. Acta* **14** 322–323
[11] Srinivasan M, Miley G and Storms E 2011 Low-energy nuclear reactions: Transmutations *Nuclear Energy Encyclopedia* ed Krivit S B, Lehr J H and Kingery T B (John Wiley & Sons, Inc.) pp 503–539 ISBN 9781118043493
[12] Hagelstein P L, McKubre M C H, Nagel D J, Chubb T A and Hekman R J 2006 *Condensed matter nuclear science: Proceedings of the 11th international conference on cold fusion*
[13] Nagel D J 1998 *Radiat. Phys. Chem.* **51** 653–668
[14] Hubler G K 2007 *Surface and Coatings Technology* **201** 8568–8573 ISSN 0257-8972
[15] Fleischmann M, Pons S and Hawkins M 1989 *Journal of Electroanalytical Chemistry* **261** 301–308
[16] Leggett A J and Baym G 1989 *Phys. Rev. Lett.* **63** 191–194
[17] Leggett A J and Baym G 1989 *Nature* **340** 45–46 ISSN ${footerJournalISSN}
[18] Czerski K, Huke A, Biller A, Heide P, Hoeft M and Ruprecht G 2001 *Europhysics Letters (EPL)* **54** 449–455 ISSN 0295-5075, 1286-4854
[19] Czerski K, Huke A, Heide P and Ruprecht G 2004 *EPL (Europhysics Letters)* **68** 363–369
[20] Czerski K, Huke A, Heide P and Ruprecht G 2006 *Eur. Phys. J. A* **27** 83–88
[21] Czerski K, Huke A, Martin L, Targosz N, Blauth D, Górska A, Heide P and Winter H 2008 *Journal of Physics G: Nuclear and Particle Physics* **35** 014012 ISSN 0954-3899, 1361-6471
[22] Czerski K 2009 Enhanced electron screening and nuclear mechanism of cold fusion *ICCF-15* vol 15 (Rome, Italy: ENEA) pp 197–202
[23] Tsyganov E N 2012 *Physics of Atomic Nuclei* **75** 153–159 ISSN 1063-7788, 1562-692X
[24] Atzeni S and Meyer-ter Vehn J 2004 *The Physics of Inertial Fusion:BeamPlasma Interaction, Hydrodynamics, Hot Dense Matter: BeamPlasma Interaction, Hydrodynamics, Hot Dense Matter* (Oxford University Press) ISBN 9780198562641
[25] Widom A and Larsen L 2006 *The European Physical Journal C - Particles and Fields* **46** 107–111 ISSN 1434-6044, 1434-6052
[26] Mizuno T 1997 *Nuclear Transmutation: The Reality of Cold Fusion* (Infinite Energy Press) ISBN 1-892925-00-1
[27] Iwamura Y, Itoh T, Sakano M, Yamazaki N, Kuribayashi S, Terada Y, Ishikawa T and Kasagi J 2006 Observation of nuclear transmutation reactions induced by d2 gas permeation through pd complexes *ICCF-11* vol 11 (Marseilles, France) pp 339–350
[28] Iwamura Y, Itoh T, Terada Y and Ishikawa T 2012 Transmutation reactions induced by deuterium permeation through nano-structured pd multilayer thin film *Transactions of the American Nuclear Society, Vol. 107* (San Diego, CA USA) pp 422–425
[29] Takahashi A, Ohta M and Mizuno T 2001 *Japanese Journal of Applied Physics* **40** 7031–7046
[30] Zhang W S and Dash J 2007 Excess heat reproducibility and evidence of anomalous elements after electrolysis in pd/d_2+h_2SO_4 electrolytic cells *Proceedings, ICCF-13* (Sochi, Russia) p 202
[31] Iwamura Y, Itoh T and Tsuruga S 2014 *Journal of Condensed Matter Nuclear Science* **13** 242
[32] Jones S E, Palmer E P, Czirr J B, Decker D L, Jensen G L, Thorne J M, Taylor S F and Rafelski J 1989 *Nature* **338** 737–740
[33] Jones S E e a 2003 Neutron emissions from metal deuterides *in Tenth International Conference on Cold Fusion* (Cambridge, MA: LENR-CANR.org)
[34] Claytor T N, Tuggle D G and Menlove H 1991 Tritium generation and neutron measurements in pd-si under high deuterium gas pressure *in Second Annual Conference on Cold Fusion* (Como, Italy: Societa Italiana di Fisica, Bologna, Italy)
[35] Jorne J 1991 *Fusion Technol.* **19** 371–374
[36] Jorne J 1994 *Fusion Technol.* **26** 244–247
[37] Jorne J 1996 *Fusion Technol.* **29** 83–89
[38] Cirillo D, Germano R, Tontodonato V, Widom A, Srivastava Y, Del Giudice E and Vitiello G 2011 *Key Engineering Materials* **495** 104–107 ISSN 1662-9795
[39] Cirillo D, Germano R, Tontodonato V, Widom A, Srivastava Y N, Giudice E D and Vitiello G 2012 *Key Eng. Mater.* **495** 104–107
[40] Domenico C 2012 *Transactions of the American Nuclear Society* **107** 418–421
[41] Lipson A G, Lyakhov B F, Roussetski A S, Akimoto T, Mizuno T, Asami N, Shimada R, Miyashita S and Takahashi A 2000 *Fusion Technol.* **38** 238–252
[42] Mizuno T, Akimoto T and Sato N 1989 *Denki Kagaku* **57** 742–743
[43] Mizuno T, Akimoto T, Takahashi A and Celani F 2004 Neutron emission from d</sub>2</sub> gas in

magnetic fields under low temperature *Eleventh International Conference on Condensed Matter Nuclear Science* (Marseille, France.)

[44] Mosier-Boss P A, Gordon S S n F E and Forsley L P G 2008 Detection of energetic particles and neutrons emitted during pd/d co-deposition *Low-Energy Nuclear Reactions Sourcebook (Vol. 1)* vol 1 ed Marwan J (Washington, USA: Oxford University Press) pp 311–334 ISBN 9780841269668, 0841269661

[45] Shyam A e a 1995 Observation of high multiplicity bursts of neutrons during electrolysis of heavy water with palladium cathode using the dead-time filtering technique. in 5th international conference on cold fusion *in 5th International Conference on Cold Fusion.* (Monte-Carlo, Monaco: IMRA Europe, Sophia Antipolis Cedex, France)

[46] Oya Y e a 1998 Material conditions to replicate the generation of excess energy and the emission of excess neutrons *in The Seventh International Conference on Cold Fusion* (Vancouver, Canada: ENECO, Inc., Salt Lake City, UT.)

[47] Srinivasan M 2009 *J. Sci. Explor.* **23** 477–482

[48] Iyengar P K, Srinivasan M, Sikka S K, Shyam A, Chitra V, Kulkarni L V, Rout R K, Krishnan M S, Malhotra S K, Gaonkar D G, Sadhukhan H K, Nagvenkar V B, Nayar M G, Mitra S K, Raghunathan P, Degwekar S B, Radhakrishnan T P, Sundaresan R, Arunachalam J, Raju V S, Kalyanaraman R, Gangadharan S, Venkateswaran G, Moorthy P N, Venkateswarlu K S, Yuvaraju B, Kishore K, Guha S N, Panajkar M S, Rao K A, Raj P, Suryanarayana P, Sathyamoorthy A, Datta T, Bose H, Prabhu L H, Sankaranarayanan S, Shetiya R S, Veeraraghavan N, Murthy T S, Sen B K, Joshi P V, Sharma K G B, Joseph T B, Iyengar T S, Shrikhande V K, Mittal K C, Misra S C, Lal M and Rao P S 1990 *Fusion Technol.* **18** 32–94

[49] Prelas M and Lukosi E 2014 *J. Condensed Matter Nucl. Sci.* **13** 455

[50] Jackson J 1957 *Physical Review* **106** 330–339 ISSN 0031-899X, 1536-6065

[51] Storms E and Talcott C 1990 *Fusion Technol.* **17** 680–695

[52] Srinivasan M 2010 Wide-ranging studies on the emission of neutrons and tritium by LENR configurations: an historical review of the early BARC results *Low-Energy Nuclear Reactions Sourcebook (Vol. 2)* vol 2 ed Marwan J and Krivit S (Washington, USA: Oxford University Press) pp 35–57 ISBN 9780841224544

[53] Claytor T N, Schwab M J, Thoma D J, Teter D F and Tuggle D G 1998 TRITIUM PRODUCTION FROM PALLADIUM ALLOYS (Vancouver, Canada: ENECO, Inc., Salt Lake City, UT) p 88

[54] Claytor T N, Jackson D D and Tuggle D G 1996 *Journal of new energy* **1** 111–118

[55] Szpak S, Mosier-Boss P A, Boss R D and Smith J J 1998 *Fusion Technol.* **33** 38–51

[56] Packham N J C, Wolf K L, Wass J C, Kainthla R C and Bockris J O M 1989 *J. Electroanal. Chem.* **270** 451–458

[57] SANKARANARAYANAN T K, SRINIVASAN M, BAJPAI M B and GUPTA D S 1995 Evidence for tritium generation in self-heated nickel wires subjected to hydrogen gas absroption/desorption cycles *5th International Conference on Cold Fusion. 1995.* (Monte-Carlo, Monaco: IMRA Europe, Sophia Antipolis Cedex, France.)

[58] Sankaranarayanan T K, Srinivasan M, Bajpai M B and Gupta D S 1996 *Fusion Technol.* **30** 349–354

[59] Chien C C, Hodko D, Minevski Z and Bockris J O M 1992 *J. Electroanal. Chem.* **338** 189–212

[60] Bush B F, Lagowski J J, Miles M H and Ostrom G S 1991 *J. Electroanal. Chem.* **304** 271–278

[61] Miles M H, Hollins R A, Bush B F, Lagowski J J and Miles R E 1993 *J. Electroanal. Chem.* **346** 99–117

[62] Reifenschweiler O 1994 *Lett. A* 149–153

[63] Reifenschweiler O 1995 *Z. Naturforsch* **50a** 973–974

[64] Reifenschweiler O J A 1996 *Fusion technology* **30** 261–272 ISSN 0748-1896

[65] Reifenschweiler O J A 1997 *Fusion technology* **31** 291–299 ISSN 0748-1896

[66] Reifenschweiler O 1998 *Radiation Physics Chemistry* **51** 327–334

[67] Wicke E 1994 *Z. Naturforsch* **49a** 1259–1261

[68] Robertson R G H and Knapp D A 1988 *Annual Review of Nuclear and Particle Science* **38** 185–215

[69] Reifenschweiler O 2004 *Infinite Energy* 14–22

[70] Dickin A P 2005 *Radiogenic Isotope Geology* 2nd ed (Cambridge, UK ; New York: Cambridge University Press) ISBN 9780521530170

[71] Jiang S, Liu J and He M 2010 *Naturwissenschaften* **97** 655–662 ISSN 0028-1042, 1432-1904

[72] Jiang S, He M, Yue W and Liu J 2008 *Journal of Fusion Energy* **27** 346–354 ISSN 0164-0313, 1572-9591

[73] Jiang S and He M 2008 *Chinese Science Bulletin* **53** 540–547 ISSN 1001-6538, 1861-9541

[74] Alexander C M O, Bowden R, Fogel M L, Howard K T, Herd C D K and Nittler L R 2012 *Science* **337** 721–723 ISSN 0036-8075, 1095-9203

[75] Hutsemékers D, Manfroid J, Jehin E and Arpigny C 2009 *Icarus* **204** 346–348 ISSN 0019-1035

[76] Cleeves L I, Bergin E A, Alexander C M O, Du F, Graninger D, Öberg K I and Harries T J 2014 *Science* **345** 1590–1593 ISSN 0036-8075, 1095-9203

[77] Villiers M D 2003 *Water: The Fate of Our Most Precious Resource* (Toronto: M&S) ISBN 9780771026416

[78] Greenwood J P, Itoh S, Sakamoto N, Warren P, Taylor L and Yurimoto H 2011 *Nature Geoscience* **4** 79–82 ISSN 1752-0894
[79] Briggs M S, Connaughton V, Wilson-Hodge C, Preece R D, Fishman G J, Kippen R M, Bhat P N, Paciesas W S, Chaplin V L, Meegan C A, von Kienlin A, Greiner J, Dwyer J R and Smith D M 2011 *Geophysical Research Letters* **38** L02808 ISSN 1944-8007
[80] Fishman G J, Bhat P N, Mallozzi R, Horack J M, Koshut T, Kouveliotou C, Pendleton G N, Meegan C A, Wilson R B, Paciesas W S, Goodman S J and Christian H J 1994 *Science (New York, N.Y.)* **264** 1313–1316 ISSN 0036-8075
[81] Köhn C and Ebert U 2014 *Atmospheric Research* **135–136** 432–465 ISSN 0169-8095
[82] Smith D M, Lopez L I, Lin R P and Barrington-Leigh C P 2005 *Science* **307** 1085–1088 ISSN 0036-8075, 1095-9203
[83] Emery G T 1972 *Annual Review of Nuclear Science* **22** 165–202
[84] Shu-Hua Z 2011 *Chinese Physics C* **35** 449 ISSN 1674-1137
[85] Limata B, Raiola F, Wang B, Yan S, Becker H W, D'Onofrio A, Gialanella L, Roca V, Rolfs C, Romano M, Schürmann D, Strieder F and Terrasi F 2006 *The European Physical Journal A - Hadrons and Nuclei* **28** 251–252 ISSN 1434-6001, 1434-601X
[86] Gang L, Jun S, Bao-Xiang W, Chao J, Xi-Xiang B, Sheng Z, Yong-Nan Z, Sheng-Yun Z, Li-Hua Z, Wei-Ping L, Zhi-Hong L, You-Bao W, Bing G, Yun-Ju L and Xing Q 2008 *Chinese Physics Letters* **25** 70–72 ISSN 0256-307X, 1741-3540
[87] Ruprecht G, Vockenhuber C, Buchmann L, Woods R, Ruiz C, Lapi S and Bemmerer D 2008 *Phys.Rev.* **C77** 065502
[88] Jenkins J H and Fischbach E 2009 *Astroparticle Physics* **31** 407–411 ISSN 09276505
[89] Jenkins J H, Fischbach E, Buncher J B, Gruenwald J T, Krause D E and Mattes J J 2009 *Astroparticle Physics* **32** 42–46 ISSN 0927-6505
[90] Jenkins J H, Fischbach E and Mundy D W 2009 *Nucl.Instrum.Meth.* **A620** 332–342
[91] Jenkins J H, Herminghuysen K R, Blue T E, Fischbach E, Javorsek II D, Kauffman A C, Mundy D W, Sturrock P A and Talnagi J W 2012 *Astroparticle Physics* **37** 81–88 ISSN 0927-6505
[92] Jenkins J H, Fischbach E, Lee R H, Javorsek D and Sturrock P A 2012
[93] Fischbach E, Buncher J B, Gruenwald J T, Jenkins J H, Krause D E, Mattes J J and Newport J R 2009 *Space Science Reviews* **145** 285–335 ISSN 0038-6308, 1572-9672
[94] Veprev D P and Muromtsev V I 2012 *Astroparticle Physics* **36** 26–30 ISSN 0927-6505
[95] Nagel D, Mizuno T and Letts D 2010 Diurnal variations in LENR experiments, in 15th int. conf. of low energy nuclear reactions, roma, italy, 2010. *in 15th Int. Conf. of Low Energy Nuclear Reactions, Roma, Italy, 2010.* (Rome, Italy)
[96] Scholkmann F, Mizuno T and Nagel D 2012 *J. Condensed Matter Nucl. Sci.* **8** 37–48
[97] Barber B P, Hiller R A, Löfstedt R, Putterman S J and Weninger K R 1997 *Physics Reports* **281** 65–143 ISSN 0370-1573
[98] Brenner M P, Hilgenfeldt S and Lohse D 2002 *Reviews of Modern Physics* **74** 425–484
[99] Putterman S J and Weninger K R 2000 *Annual Review of Fluid Mechanics* **32** 445–476
[100] Camara C G, Hopkins S D, Suslick K S and Putterman S J 2007 *Physical Review Letters* **98** 064301
[101] Camara C G, Escobar J V, Hird J R and Putterman S J 2008 *Nature* **455** 1089–1092 ISSN 0028-0836
[102] Hird J R, Camara C G and Putterman S J 2011 *Applied Physics Letters* **98** 133501 ISSN 0003-6951, 1077-3118
[103] Camara C G, Escobar J V, Hird J R and Putterman S J 2010 *Applied Physics B* **99** 613–617 ISSN 0946-2171, 1432-0649
[104] Constable E, Horvat J and Lewis R A 2010 *Applied Physics Letters* **97** 131502 ISSN 0003-6951, 1077-3118
[105] Corben H C 1962 *Proceedings of the National Academy of Sciences of the United States of America* **48** 1559–1566 ISSN 0027-8424
[106] Corben H C 1962 *Proceedings of the National Academy of Sciences of the United States of America* **48** 1746–1752 ISSN 0027-8424
[107] Feynman R P 1950 *Physical Review* **80** 440–457
[108] Newton T D and Wigner E P 1949 *Reviews of Modern Physics* **21** 400–406
[109] Wightman A S 1962 *Reviews of Modern Physics* **34** 845–872
[110] Lacki J, Ruegg H and Wanders G 2008 *E.C.G. Stueckelberg, An Unconventional Figure of Twentieth Century Physics: Selected Scientific Papers with Commentaries* (Springer) ISBN 9783764388775
[111] Nambu Y 1950 *Progress of Theoretical Physics* **5** 82–94
[112] Schwinger J 1951 *Physical Review* **82** 664–679
[113] Cooke J H 1968 *Physical Review* **166** 1293–1298
[114] Pearle P M 1968 *Physical Review* **168** 1429–1444

[115] Horwitz L P and Piron C 1973 *Helv. Phys. Acta* **46** 316–326
[116] Saad D, Horwitz L P and Arshansky R I 1989 *Foundations of Physics* **19** 1125–1149
[117] Land M C 1998 *Foundations of Physics* **28** 1479–1487 ISSN 0015-9018, 1572-9516
[118] Land M C and Horwitz L P 1996 *arXiv:hep-th/9601021*
[119] Fanchi J R 1993 *Parametrized Relativistic Quantum Theory* (Springer-Verlag GmbH) ISBN 9780792323761
[120] Fanchi J R 1993 *Foundations of Physics* **23** 487–548 ISSN 0015-9018, 1572-9516
[121] Reuse F 1979 *Foundations of Physics* **9** 865–882 ISSN 0015-9018, 1572-9516
[122] Horwitz L P, Arshansky R I and Elitzur A C 1988 *Foundations of Physics* **18** 1159–1193 ISSN 0015-9018, 1572-9516
[123] Horwitz L and Shnerb N 1998 *Foundations of Physics* **28** 1509–1519 ISSN 0015-9018
[124] Land M C and Horwitz L P 1991 *Foundations of Physics* **21** 299–310 ISSN 0015-9018, 1572-9516
[125] Land M C and Horwitz L P 1991 *Foundations of Physics Letters* **4** 61–71 ISSN 0894-9875, 1572-9524
[126] Land M C 1997 *Foundations of Physics* **27** 19–41 ISSN 0015-9018, 1572-9516
[127] Land M 2011 *Journal of Physics: Conference Series* **330** 012015 ISSN 1742-6596
[128] Aharonovich I and Horwitz L P 2012 *Journal of Mathematical Physics* **53** 032902–032902–29 ISSN 00222488
[129] Seidewitz E 2009 *Annals of Physics* **324** 309–331 ISSN 0003-4916
[130] Horwitz L 2013 *Journal of Physics A: Mathematical and Theoretical* **46** 035305 ISSN 1751-8121
[131] Horwitz L P, Katz N and Oron O 2004 *Discrete Dynamics in Nature and Society* **2004** 179–204 ISSN 1026-0226, 1607-887X
[132] Roitgrund A and Horwitz L 2010 *Discrete Dynamics in Nature and Society* **2010** 1–36 ISSN 1026-0226, 1607-887X
[133] Weinberg S 1990 *Physics Letters B* **251** 288–292 ISSN 03702693
[134] Weinberg S 1979 *Physica A: Statistical Mechanics and its Applications* **96** 327–340 ISSN 0378-4371
[135] Epelbaum E 2010 *arXiv:1001.3229*
[136] Coleman S 1977 *Physical Review D* **15** 2929–2936
[137] Tomsovic S and Ullmo D 1994 *Physical Review E* **50** 145–162
[138] Hagelstein P L 2005 Resonant tunneling and resonant excitation transfer *Proceedings, ICCF-12* (Cambridge, MA, USA) pp 871–886
[139] Li X Z 1999 *Czech. J. Phys.* **49** 985–992
[140] Grifoni M and Hänggi P 1998 *Physics Reports* **304** 229–354 ISSN 0370-1573
[141] Swartz M 1996 *Journal of New Energy* **1** 68–79
[142] Czerski K, Heide P, Huke A, Kilic A I, Kulesza I and Targosz-Sleczka N 2009 *ACTA PHYSICA POLONICA B* **40** 903–907
[143] Czerski K, Heide P, Huke A, Martin L and Ruprecht G 2006 Enhanced electron screening in nuclear reactions and radioactive decays *International Symposium on Nuclear Astrophysics – Nuclei in the Cosmos – IX* (CERN, Geneva, Switzerland)
[144] Jackson J D 2000 *The European Physical Journal C - Particles and Fields* **15** 211–214 ISSN 1434-6044, 1434-6052
[145] Doyle A C 1890 *Lippincotts Monthly Magazine*
[146] 1989 Cold fusion research, november 1989 - a report of the energy research adivsory board to the united states department of energy DOE Report DOE/S-0071 (August, 1989) and DOE/S-0073, DE90, 005611 Department of Energy
[147] Huizenga J R 1994 *Cold Fusion: The Scientific Fiasco of the Century* (Oxford; New York: Oxford University Press) ISBN 9780198558170

IARD 2014 IOP Publishing
Journal of Physics: Conference Series **615** (2015) 012017 doi:10.1088/1742-6596/615/1/012017

The center of mass and center of charge of the electron

Martin Rivas
Theoretical Physics Department, University of the Basque Country,
Apdo. 644, 48080 Bilbao, Spain

E-mail: martin.rivas@ehu.es

Abstract. The goal of this contribution is to show that the hypothesis that the center of mass(CM) and the center of charge(CC) of a classical electron are two different points is only compatible with a relativistic description. The existence of two separated points is analyzed by the different dynamical behaviour of the angular momenta with respect to both points. It shows, from the classical point of view, that the angular momentum with respect to the CC of the electron satisfies the same dynamical equation as Dirac spin operator in the quantum case. In the free motion, the CC follows a helix at the speed of light, and the CM the axis of the helix. The particle, if its electromagnetic structure is reduced to a total charge e located at CC, has a magnetic moment and also an electric dipole moment with respect to the CM, like in Dirac's theory. The analysis of Dirac spin operator in the quantum case, shows that the electron is a particle where the CM and CC are necessarily different points.

1. Introduction

Real elementary particles like electrons have, among other properties, mechanical properties and also electromagnetic properties. Associated to the mechanical properties there is, from the classical point of view, a geometric point, called the center of mass $\boldsymbol{q}$, and similarly, associated to the electromagnetic properties another point $\boldsymbol{r}$, called the center of charge. Only two things can happen, that both points are exactly the same or that they are different. Although it is not explicitely assumed, the spinless point particle description considers that both points are the same. In this paper we are going to explore what physical consequences we shall obtain by assuming that perhaps the electron has some extensive, but unknown structure, such that both points are different. To our knowledge no similar analysis has been published before. What we are going to show is that, when compared with Dirac theory, the electron structure appears to be more likely associated with a classical object with two different centers than with the usual point particle assumption. The kinematical theory developed by the author leads to spinning particle models with two separate centers, as summarized in section 5.

1.1. The CM and the CC

If a real elementary particle is exactly a geometrical point then all their mechanical and electromagnetic properties would be associated to that point. If elementary particles are not exactly geometrical points, classical mechanics defines for any material system a geometrical point $\boldsymbol{q}$, called the center of mass, such that the linear momentum of the particle takes, in a nonrelativistic framework, the form $\boldsymbol{p} = m\boldsymbol{v}$, where m is the mass and $\boldsymbol{v} = d\boldsymbol{q}/dt$, or $\boldsymbol{p} = \gamma(v)m\boldsymbol{v}$, in a relativistic one, where $\gamma(v) = (1 - v^2/c^2)^{-1/2}$.

From the electromagnetic point of view, the electromagnetic structure of any charge and current distribution can always be reduced to a single point $\boldsymbol{r}$ where we locate the total charge of the system and the electric and magnetic multipoles defined with respect to that point. This point is in general arbitrary, but if the charge distribution of an elementary particle has a spherical symmetry around a point, the electric structure around that point is reduced to the value of the total charge and no other electric multipoles. If the current densities are also symmetrically distributed around that point, the magnetic multipoles also vanish. We can call this geometrical point, where the electromagnetic structure takes the simplest form of a total charge e and no multipoles, the center of charge. If we consider another point $\boldsymbol{k}$ in the particle, different from $\boldsymbol{r}$, the electromagnetic structure of the particle with respect to this point will be reduced to the same total charge e located at $\boldsymbol{k}$ and also an electric dipole $\boldsymbol{d} = e(\boldsymbol{r} - \boldsymbol{k})$ and a magnetic dipole $\boldsymbol{\mu} = e(\boldsymbol{r} - \boldsymbol{k}) \times \boldsymbol{w}/2$, provided $\boldsymbol{w}$ is the relative velocity between the point $\boldsymbol{r}$ with respect to the point $\boldsymbol{k}$.

We do not know what is the exact electromagnetic structure of the electron, how the charge and its possible internal currents are distributed. What we know is that the usual coupling of quantum electrodynamics $j^\mu A_\mu$, between the particle current field j^μ and the external potentials A_μ, is obtained in quantum field theory by the local gauge invariance hypothesis. We see that no multipole interactions with the external fields appear in this coupling. For a strict point particle of charge e, that coupling is reduced in the classical description to $-e\phi + e\boldsymbol{u} \cdot \boldsymbol{A}$, where $\boldsymbol{u}$ is the velocity of the charge, and ϕ and $\boldsymbol{A}$ the external scalar and vector potentials, respectively, defined at that point. Conversely, if we describe the classical interaction of the electron with an external field in this form, we are implicitely assuming that the interacting properties of the electron are reduced to a total charge e located at a point $\boldsymbol{r}$, where the potentials are evaluated. This would imply for an electron of unknown structure that the external fields have a smooth variation along the very small region where the charge is distributed such that the total external force can be replaced by the evaluation of the fields at the point CC where we locate the total charge, as we shall assume here.

2. Lagrangian description of a particle under an EM field

Let us consider the following Lagrangian of a particle under some external electromagnetic field

$$L = L_0 + L_I, \quad L_I = -e\phi(t, \boldsymbol{r}) + e\boldsymbol{u} \cdot \boldsymbol{A}(t, \boldsymbol{r}),$$

where e is the charge of the particle, $\phi(t, \boldsymbol{r})$ and $\boldsymbol{A}(t, \boldsymbol{r})$ are the external scalar and vector potentials, respectively, defined at the point $\boldsymbol{r}$, $\boldsymbol{u}$ the velocity of this point, and L_0 the relativistic or nonrelativistic free Lagrangian of a particle of mass m.

The Lagrangian L_0 describes the mechanical properties of the particle and L_I its interaction. L_0 is expressed as usual in terms of the CM velocity $\boldsymbol{v}$, but if there is some relationship between $\boldsymbol{q}$ and $\boldsymbol{r}$ it will be finally written in terms of $\boldsymbol{r}$ and its derivatives. The last part L_I suggests that the particle, from the electromagnetic point of view, can be considered as an object with a charge e located at point $\boldsymbol{r}$ where the external potentials are defined, and no further multipoles.

From L_0, the different mechanical properties are defined. For instance the mechanical energy H and the mechanical linear momentum $\boldsymbol{p}$, which, according to the special relativity are expressed as $H = \gamma(v)mc^2$ and $\boldsymbol{p} = \gamma(v)m\boldsymbol{v}$, respectively. The vector $\boldsymbol{v} = d\boldsymbol{q}/dt$ represents the velocity of the center of mass $\boldsymbol{q}$ and $\gamma(v) = (1 - v^2/c^2)^{-1/2}$. In the nonrelativistic limit $\boldsymbol{p} = m\boldsymbol{v}$ and $H = mv^2/2$, respectively. If the particle is exactly a geometrical point then $\boldsymbol{q} = \boldsymbol{r}$, but if it is not exactly a point these two points could be different, as we shall assume here. The interacting part suggests a spherical symmetry for its electromagnetic structure around the CC $\boldsymbol{r}$, but the mechanical part says nothing about the mass distribution. Only the existence of a point $\boldsymbol{q}$ which represents the location of the CM such that the mechanical energy and linear momentum can be expressed in terms of its velocity.

2.1. The free motion of the center of mass and center of charge

From the L_0 part we get the free dynamical equation $d\boldsymbol{p}/dt = 0$, and from the L_I part the external Lorentz force, such that the complete dynamical equation will be

$$\frac{d\boldsymbol{p}}{dt} = e(\boldsymbol{E}(t,\boldsymbol{r}) + \boldsymbol{u} \times \boldsymbol{B}(t,\boldsymbol{r})), \qquad (1)$$

where $\boldsymbol{E} = -\nabla\phi - \partial\boldsymbol{A}/\partial t$, $\boldsymbol{B} = \nabla \times \boldsymbol{A}$ and $\boldsymbol{u} = d\boldsymbol{r}/dt$. The fields are being evaluated at the CC $\boldsymbol{r}$, while the left hand side is the time derivative of $\boldsymbol{p} = \gamma(v)m\boldsymbol{v}$, or $\boldsymbol{p} = m\boldsymbol{v}$, in the nonrelativistic approach. To integrate (1) to obtain the trajectory of the CM $\boldsymbol{q}(t)$, we need to know the trajectory of the CC $\boldsymbol{r}(t)$. We have to evaluate the external fields at $\boldsymbol{r}$ and it is the velocity of point $\boldsymbol{r}$, $\boldsymbol{u}$ which is included in the magnetic force. But the CC position $\boldsymbol{r}$ will be in the neighborhood of $\boldsymbol{q}$, closely related to it, so we must make some *ansatz* about their relationship to conveniently express the linear momentum in terms of the variable $\boldsymbol{r}$ and its derivatives. Let us see first how they are related in the free case.

Let us make the analysis in an inertial reference frame, either relativistic or non-relativistic. If the particle is free $\boldsymbol{p}$ is conserved so that the CM position $\boldsymbol{q}$, moves along a straight line with a constant velocity $\boldsymbol{v}$. But, what about the trajectory of the point $\boldsymbol{r}$?

From the geometrical point of view in three-dimensional space, the trajectory of a point which follows a continuous and differentiable path, can be described as the evolution of its Frenet-Serret triad. We can parameterize the trajectory in terms of the arc length s parameter, $\boldsymbol{r}(s)$, or in terms of the time of the inertial observer $\boldsymbol{r}(t)$, if we know the kinematics of the trajectory $s(t)$. Let us consider the location of its Frenet-Serret triad at some particular time t of the inertial observer. This triad is displaced an arc length $ds = u(t)dt$ in time dt along the unit tangent vector $\boldsymbol{t}$, and rotates an angle $\kappa(t)ds$ around the binormal $\boldsymbol{b}$, and also an angle $\tau(t)ds$ around the tangent $\boldsymbol{t}$ in the same time, where $u(t)$ is the absolute value of the instantaneous velocity of the point and $\kappa(t)$ and $\tau(t)$ are the instantaneous curvature and torsion of the trajectory at time t, respectively. But if the motion is free it means that we cannot obtain a different kinematical behaviour of the CC motion at two different times. Otherwise, a different kinematical behaviour will mean that something different is happening at different times, which is contradictory with the assumption that the particle is free. The above infinitesimal displacement and angles must be independent of the time so that the CC follows a trajectory of constant curvature and torsion at a velocity of constant absolute value $u = ds/dt$, in this inertial reference frame.

The CC follows a three-dimensional helix at a constant speed. The CM seems to be the central point of the helix, i.e., the CM motion describes the axis of the helix, such that the projection of the velocity $\boldsymbol{u}$ of the CC along the axis will be the CM velocity $\boldsymbol{v}$ in this inertial reference frame.

There is a possibility that both constants κ and τ will be zero, and thus the two points $\boldsymbol{q}$ and $\boldsymbol{r}$ follow parallel straight lines with the same velocity in this frame. Different velocities will mean that both points separate from each other and this makes no sense for an elementary particle description. The CC is just displaced a constant distance from the CM and when analyzed in the reference frame where $\boldsymbol{q}$ is at rest, also $\boldsymbol{r}$ is at rest and the particle does not rotate and there is no angular momentum in this frame. Elementary matter has angular momentum and thus, what we are going to analyze is the more general situation in which the particle could possibly rotate, and therefore the constants κ and τ should be different from zero.

2.2. Nonrelativistic analysis
Let us consider the class of Galilei inertial observers, i.e., those observers related to the previous one by means of a transformation of the Galilei group. What we assume is that the above description of a CM moving along a straight line at a constant velocity and a CC describing a helix of constant curvature and torsion at a constant speed, has to be valid in every inertial reference frame, although the velocities, curvature and torsion will take different constant values in the different inertial frames. For the inertial observer of the previous analysis the curly motion of the CC, with a nonvanishing curvature and torsion, is an accelerated motion at a constant speed and therefore it is also an accelerated motion at a constant speed for the remaining inertial observers.

Since the relative velocity among inertial observers is unrestricted, let us consider some particular inertial observer which at a certain instant is at rest with respect to the CC. For this observer the velocity of the CC is $u = 0$ at this instant, and the above requirement implies that u will be zero for ever. This is impossible because the motion of the CC in this inertial frame is accelerated.

The requirement that the motion of the CC will be at a velocity of constant absolute value in every inertial frame is not verified, because this constant velocity has to be always different from zero in every frame. This means that for this requirement to be valid the velocity of the CC has to be an unreachable velocity for all inertial observers, which is not the case if the transformations among observers are those of the Galilei group.

The nonrelativistic analysis is not compatible with the existence of an elementary particle with two different CM and CC centers. For the nonrelativistic elementary particle necessarily the CM and the CC are the same point.

2.3. Relativistic analysis
Let us consider now the class of Poincaré inertial observers, i.e., those observers related to the previous one by means of a transformation of the Poincaré group.

The nonrelativistic analysis above is suggesting that the possibility of a curly motion of the CC at a constant speed in every inertial frame is allowed if the velocity of the CC is an unreachable velocity for any inertial oberver. The special theory of relativity suggests that the speed of light is a good candidate for this unreachable and constant velocity limit. In special relativity, if a point $\boldsymbol{r}$ is moving at the speed c for some particular inertial observer, then it moves with the same speed c, for the remaining ones. Only a relativistic description is compatible with the assumption of two different points CC and CM, but necessarily the CC has to be moving at the speed of light c. The free motion of a particle with two centers, implies that the CC follows a helix at the speed of light. In this case the CM velocity $\boldsymbol{v}$ will be the projection of the velocity $\boldsymbol{u}$ of the CC, along the axis of the helix, and therefore $v < c$.

2.4. Summary and discussion
This description of an elementary particle with two centers is incompatible with a nonrelativistic framework, but not in special relativity. The hypothesis of two separate points requires the existence of velocities of physical points, constant and unreachable for all inertial observers. Although the CM does not move at the speed of light, the CC does. It is accelerated and therefore classical electrodynamics implies that the particle has to radiate. From the CM point of view the particle will behave though it has a magnetic moment because there always exists a relative velocity between $\boldsymbol{r}$ and $\boldsymbol{q}$ and also an oscillating electric dipole with respect to the CM. For every observer the particle has an internal frequency ω.

The particle rotates and therefore it will have angular momentum. If the description is done in the CM frame, the motion of the CC will be a circle of constant radius R, and this implies a constant and unique angular velocity $\omega = c/R$ for all identical particles at rest, like the

Journal of Physics: Conference Series **615** (2015) 012017 doi:10.1088/1742-6596/615/1/012017

electrons, and also a constant and unique angular momentum with respect to the CM, which will be conserved in the free case. Otherwise, if the particle radiates, the radius will change and if the CC velocity remains constant the angular momentum of the particle must also change. For the model to be consistent, this object with a unique angular momentum has to produce no radiation whenever the CM is at rest or moving with a constant velocity. Radiation has to be produced when the particle gets some energy from the external world, i.e., when the CM is accelerated.

Classical electrodynamics is an incomplete theory because does not allow radiationless motions of finite bound systems of charged particles. We have to complement classical electrodynamics with an extra statement of no radiation for free spinning particles with a unique and unmodified angular momentum, although the CC is accelerated, provided the CM is not. Something similar to the no radiation hypothesis of the stationary orbits of quantized orbital angular momentum in Bohr's atomic formalism. We have a theory of radiation of point particles. The corresponding theory of radiation of spinning particles with two separate centers is not yet done.

We have postulated the possibility of two different centers and this leads to a unique internal relative motion between them. Are there any possibility of analysis to show if this hypothesis is right or wrong? Are there any specific observable which is able to distinguish between both points? The answer is positive. This observable is the angular momentum. Its dynamical behaviour will shed light about this conjecture.

3. The angular momentum of the particle

The angular momentum of any mechanical system is a magnitude which is defined with respect to some prescribed point. If the particle has two different characteristic points, the center of charge $\boldsymbol{r}$ and the center of mass $\boldsymbol{q}$, the angular momentum of the particle can be defined with respect to both points. Let us call $\boldsymbol{S}$ the angular momentum with respect to the CC and $\boldsymbol{S}_{CM}$ the angular momentum with respect to the CM, for some inertial observer. Even more, let us assume that $\boldsymbol{k}$ is another geometrical point of the particle, different from $\boldsymbol{q}$ and $\boldsymbol{r}$, and let us call $\boldsymbol{S}_k$ the angular momentum with respect to this point (see figure 1).

Let $\boldsymbol{J}$ be the total angular momentum of the particle with respect to the origin in this inertial reference frame. It can be written in the following alternative forms in terms of the previous angular momenta, either

$$\boldsymbol{J} = \boldsymbol{r}\times\boldsymbol{p} + \boldsymbol{S}, \quad \text{or} \quad \boldsymbol{J} = \boldsymbol{q}\times\boldsymbol{p} + \boldsymbol{S}_{CM}, \quad \text{or} \quad \boldsymbol{J} = \boldsymbol{k}\times\boldsymbol{p} + \boldsymbol{S}_k,$$

where $\boldsymbol{p}$ is the linear momentum of the particle in this inertial reference frame. The center of mass observer is defined as that inertial frame where $\boldsymbol{p} = 0$. Then, the angular momentum with respect to any point in that frame takes the same value, so that $\boldsymbol{S} = \boldsymbol{S}_{CM} = \boldsymbol{S}_k = \boldsymbol{J}$, for the center of mass observer.

If the particle is under some external electromagnetic force $\boldsymbol{F}$ evaluated and applied at the charge position $\boldsymbol{r}$, $\boldsymbol{J}$ and $\boldsymbol{p}$ are not conserved but satisfy

$$\frac{d\boldsymbol{p}}{dt} = \boldsymbol{F}, \quad \frac{d\boldsymbol{J}}{dt} = \boldsymbol{r}\times\boldsymbol{F}.$$

Taking the time derivatives of the above three expressions of the total angular momentum $\boldsymbol{J}$, and taking into account these dynamical equations, we arrive at

$$\frac{d\boldsymbol{S}}{dt} = \boldsymbol{p}\times\frac{d\boldsymbol{r}}{dt}, \quad \frac{d\boldsymbol{S}_{CM}}{dt} = (\boldsymbol{r}-\boldsymbol{q})\times\boldsymbol{F}, \quad \frac{d\boldsymbol{S}_k}{dt} = \boldsymbol{p}\times\frac{d\boldsymbol{k}}{dt} + (\boldsymbol{r}-\boldsymbol{k})\times\boldsymbol{F},$$

because the linear momentum is pointing along the direction of the velocity of point $\boldsymbol{q}$ and not in general along the direction of the velocities of the other two points $\boldsymbol{r}$ and $\boldsymbol{k}$.

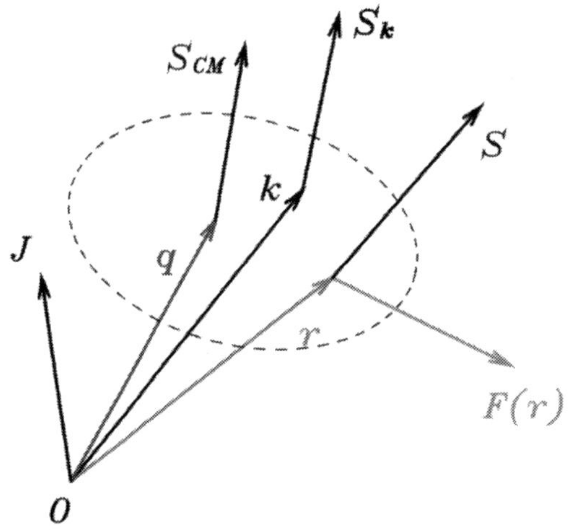

Figure 1. Different angular momenta $\boldsymbol{S}$, $\boldsymbol{S}_{CM}$, $\boldsymbol{S}_k$ and $\boldsymbol{J}$ of the electron defined with respect to different points in an inertial reference frame with origin at point O. The external force $\boldsymbol{F}$ defined at CC is also depicted. The dotted line tries to suggest some possible, but unknown, extensive structure of the electron.

These three angular momenta satisfy three different dynamical equations. In the free case $\boldsymbol{F} = 0$, and $\boldsymbol{S}_{CM}$ is conserved while $\boldsymbol{S}$ and $\boldsymbol{S}_k$ evolve in a direction orthogonal to $\boldsymbol{p}$, so that only their projections on $\boldsymbol{p}$, $\boldsymbol{S} \cdot \boldsymbol{p}$ and $\boldsymbol{S}_k \cdot \boldsymbol{p}$, respectively, are conserved.

The dynamical equation of the CC angular momentum $\boldsymbol{S}$ is independent of whether the particle is free or not. Its evolution is always orthogonal to $\boldsymbol{p}$. This means that if we compute the angular momentum of the particle with respect to some point, the dynamical behavior of this angular momentum will also give us information about whether this point is the center of charge, the center of mass or a point different from them.

If the particle is not free but both points are the same $\boldsymbol{r} = \boldsymbol{q}$, and also $\boldsymbol{S} = \boldsymbol{S}_{CM}$ and thus these angular momenta (but not $\boldsymbol{S}_k$) are necessarily conserved. But, conversely, if the angular momentum with respect to the CC or with respect to the CM is not conserved it means that this point $\boldsymbol{r}$ is necessarily a different point than the center of mass $\boldsymbol{q}$.

The CC angular momentum $\boldsymbol{S}$ does not satisfy a torque dynamical equation like $\boldsymbol{S}_{CM}$. It is not conserved for a free particle. It seems to precess around the direction of the linear momentum and its dynamical equation is independent of the external force.

The analysis of a particle with two centers produces a unique description of a motion of the CC at the speed of light. The angular momentum with respect to the CC satisfies the same dynamical equation than Dirac's spin operator in the quantum case. This is going to be compared with Dirac's analysis of the electron.

4. Dirac Analysis

In his original papers in 1928, Dirac [1],[2] analyzes an electron interacting with an external electromagnetic field, through a minimal coupling and arrives to the Hamiltonian

$$H = (\boldsymbol{p} - e\boldsymbol{A}(t, \boldsymbol{r})) \cdot c\boldsymbol{\alpha} + \beta mc^2 + e\phi(t, \boldsymbol{r}),$$

where $\boldsymbol{\alpha} = \gamma^0\boldsymbol{\gamma}$ and $\beta = \gamma^0$, are Dirac's hermitian matrices. In the original papers Dirac uses a different notation for the above matrices but we have kept today's more accepted one. The usual minimal coupling interaction, where the potentials, ϕ and $\boldsymbol{A}$, are defined at point $\boldsymbol{r}$, suggests that point $\boldsymbol{r}$ represents the location of the CC. But, is it also the location of the CM?

When computing the velocity of point $\boldsymbol{r}$, Dirac obtains

$$\boldsymbol{u} = \frac{d\boldsymbol{r}}{dt} = \frac{i}{\hbar}[H, \boldsymbol{r}] = c\boldsymbol{\alpha},$$

irrespective of whether the particle is free or not. The eigenvalues of the hermitian matrices $\boldsymbol{\alpha}$ are ± 1. It writes on page 262 of his book [3]: "*... a measurement of a component of the velocity of a free electron is certain to lead to the result $\pm c$. This conclusion is easily seen to hold when there is a field present*".

This point $\boldsymbol{r}$ is moving at the speed of light. In Dirac's Nobel lecture [4] he says: "*It is found that an electron which seems to us to be moving slowly, must actually have a very high frequency oscillatory motion of small amplitude superposed on the regular motion which appears to us. As a result of this oscillatory motion, the velocity of the electron at any time equals the velocity of light. ... one must believe in this consequence of the theory, since other consequences of the theory ..., are confirmed by experiment.*"

To see if point $\boldsymbol{r}$ represents also the location of the CM, let us analyze the dynamical behaviour of the angular momentum of the electron with respect to this point. If it represents the location of the CM, it will be conserved in the free or interacting motion. Otherwise, the CM will be a different point than the point $\boldsymbol{r}$. The total angular momentum of the electron with respect to the origin of the observer frame is

$$\boldsymbol{J} = \boldsymbol{r} \times \boldsymbol{p} + \boldsymbol{S}, \quad \boldsymbol{S} = \frac{\hbar}{2}\begin{pmatrix} \boldsymbol{\sigma} & 0 \\ 0 & \boldsymbol{\sigma} \end{pmatrix},$$

where Dirac spin operator $\boldsymbol{S}$, written in terms of Pauli matrices, represents the angular momentum of the electron with respect to the point $\boldsymbol{r}$. Both parts $\boldsymbol{r} \times \boldsymbol{p}$ and $\boldsymbol{S}$ are not separately conserved for the free electron. In the introduction of [1], Dirac writes: "*The most important failure of the model seems to be that the magnitude of the resultant orbital angular momentum of an electron moving in an orbit in a central field of force is not a constant, as the model leads one to expect.*"

The spin part $\boldsymbol{S}$ satisfies

$$\frac{d\boldsymbol{S}}{dt} = \frac{i}{\hbar}[H, \boldsymbol{S}] = \boldsymbol{p} \times c\boldsymbol{\alpha} = \boldsymbol{p} \times \boldsymbol{u}$$

even under the external interaction. This no torque dynamical equation for an angular momentum suggests that the point $\boldsymbol{r}$ where the external fields are defined is a different point than the CM of the electron.

The linear momentum is not along this velocity $\boldsymbol{u}$ but is related to some average value: "*... the x_1 component of the velocity $c\alpha_1$, consists of two parts, a constant part $c^2p_1H^{-1}$, connected with the momentum by the classical relativistic formula, and an oscillatory part, whose frequency is at least $2mc^2/h$,...*". Point $\boldsymbol{r}$ is not the position of the CM. This frequency predicted by Dirac for the motion of the point $\boldsymbol{r}$ is just twice de Broglie's postulated frequency.

When expanding the Hamiltonian in powers of $\boldsymbol{p}$, i.e., which could be intrepreted as the expression of the energy in terms of the CM motion, he finds, in addition to the kinetic energy terms and the interacting term $e\phi$, two new interaction terms:

$$\frac{e\hbar}{2mc}\boldsymbol{\Sigma} \cdot \boldsymbol{B} + \frac{ie\hbar}{2mc}\boldsymbol{\alpha} \cdot \boldsymbol{E} = \boldsymbol{\mu} \cdot \boldsymbol{B} + \boldsymbol{d} \cdot \boldsymbol{E}, \quad \boldsymbol{\Sigma} = \begin{pmatrix} \boldsymbol{\sigma} & 0 \\ 0 & \boldsymbol{\sigma} \end{pmatrix}$$

He says: "*The electron will therefore behave as though it has a magnetic moment* $(e\hbar/2mc)\boldsymbol{\Sigma}$, *and an electric moment* $(ie\hbar/2mc)\boldsymbol{\alpha}$. *The magnetic moment is the just assumed in the spinning electron model. The electric moment, being a pure imaginary, we should not expect to appear in the model. It is doubtful whether the electric moment has any physical meaning ...*".

In his book [3] gives the same expression but he never mentioned, even in subsequent works like in the Nobel conference [4], the existence of this electric dipole. He analyzes the magnetic interaction but devotes no single line to the electric dipole interaction, which has appeared on the same footing as the magnetic one. He disliked that the electron would have an electric dipole structure. The absolute value of this term is the charge e times a distance $\hbar/2mc$. The operator $(i\hbar/2mc)\boldsymbol{\alpha}$, represents the relative position quantum operator of the CC with respect to the CM, as has been shown in [5]. It simply means that the magnetic dipole and the electric dipole represent the electromagnetic electron structure with respect to the CM, provided its structure with respect to the CC is that of a total charge e, without further multipoles.

The electron has two different centers separated by half Compton's wavelength. The motion of the CC around the CM, in the center of mass frame, is at a frequency twice the frequency postulated by de Broglie and at the speed of light. In the quantum case, the free electron does not radiate.

5. Kinematical theory

In the kinematical formalism for describing spinning particles developed by the author [5], this assumption about the existence of two separated centers is not postulated. What is postulated is that an elementary particle cannot have excited states and its internal structure cannot be modified by any interaction. The mathematical requirement of this postulate is that the boundary variables of its classical Lagrangian description span necessarily a homogeneous space of the Poincaré group. The description of a classical elementary spinning particle in this formalism, is in terms of a single point $\boldsymbol{r}$ where the external potentials are defined. What is found is that the Lagrangian which describes an elementary spinning particle, in the most general case, has to be expressed in terms of the velocity and acceleration of this point, and thus dynamical equations are fourth order differential equations for the variable $\boldsymbol{r}$. The linear momentum is found to be not along the velocity of this point. The angular momentum with respect to this point $\boldsymbol{S}$, satisfies the same dynamical equation as Dirac's spin operator. General spinning particles have a CM which is different point than the CC although for its dynamical description it is sufficient to give the evolution of a single point $\boldsymbol{r}$, the CC.

The only spinning model in this kinematical formalism which satisfies Dirac's equation when quantized corresponds to the model such that the velocity of point $\boldsymbol{r}$ is the speed of light [6]. Quantization of all other spinning models does not produce Dirac's equation. The classical description of the electron at rest is given in figure 2. It is a mechanical system of six degrees of freedom. Three $\boldsymbol{r}$, represent the location of the CC, and another three $\boldsymbol{\alpha}$, represent the orientation of a local cartesian frame located at the CC position and which is not depicted in the figure. What is depicted is the angular velocity $\boldsymbol{\omega}$ of this local frame. The radius of this motion is $R = |S_z|/mc = \hbar/2mc$, half Compton's wavelength, and the angular velocity in this frame $\omega = 2mc^2/\hbar$ is twice De Broglie's postulated frequency. In this frame, the total angular momentum $\boldsymbol{S} = \boldsymbol{S}_{CM}$ is conserved and the dynamical equation satisfied by the CC of the particle is

$$\boldsymbol{r} = \frac{1}{mc^2}\boldsymbol{S} \times \boldsymbol{u}, \quad \boldsymbol{u} = d\boldsymbol{r}/dt, \quad u = c.$$

For the antiparticle the motion of the CC is the reversed one, with the same orientation of the spin. The total angular momentum is the addition of two terms, $\boldsymbol{S} = \boldsymbol{Z} + \boldsymbol{W}$, one $\boldsymbol{Z}$ related to the relative motion of the CC around the CM, known as the *Zitterbewegung*, and another $\boldsymbol{W}$ related to the change of orientation and thus along the angular velocity $\boldsymbol{\omega}$.

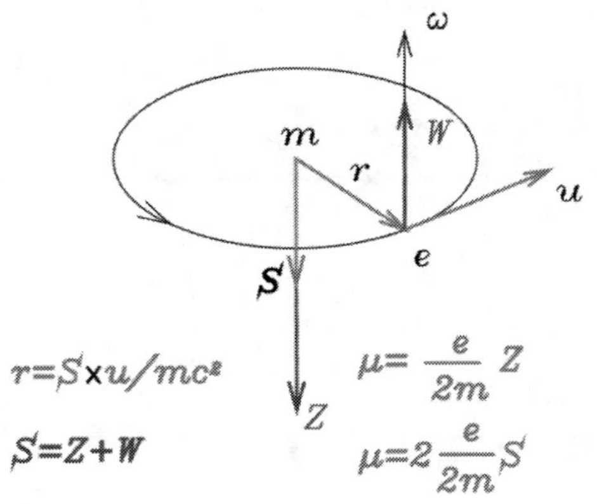

Figure 2.

In this formalism, the separation between the CM and CC is responsible for a classical interpretation of the formation of bound pairs of spinning electrons [7] (see figure 3) and the clarification of the mechanism of tunneling [8]. Quarks are considered Dirac particles. Since Dirac spin operator represents the angular momentum of the quark with respect to the CC, it is impossible that the addition of the three Dirac spin operators of the three quarks gives rise to the spin of the proton. This apparent failure, (see figure 4) known in the literature as the proton spin crisis, has been recently pointed out [9].

6. Conclusions

In this paper we have shown that the hypothesis of two separate centers for elementary spinning particles is consistent only with special relativity and that these objects have a unique angular momentum that cannot be modified. In the free motion, the CC moves along a helix at the speed of light and the CM describes the axis of the helix. The quantization of this model satisfies Dirac's equation, such that Dirac spin operator represents the angular momentum with respect to the CC.

Acknowledgments

This work has been partially supported by Universidad del País Vasco/Euskal Herriko Unibertsitatea grant 9/UPV00172.310-14456/2002.

References

[1] Dirac PAM 1928 The Quantum Theory of the Electron, *Proc. Roy. Soc. Lon.* **A117**, 610
[2] Dirac 1928 The Quantum Theory of the Electron. Part II, *Proc. Roy. Soc. Lon.*, **A118**, 351
[3] Dirac PAM 1967 *The Principles of Quantum mechanics* (Oxford: Oxford Univ. Press.) 4th ed
[4] Dirac PAM 1933 Nobel Lecture: Theory of Electrons and Positrons, `http://www.nobelprize.org/nobel_prizes/physics/laureates/1933/dirac-lecture.html`
[5] Rivas M 2001 *Kinematical theory of spinning particles* (Dordrecht: Kluwer)
[6] Rivas M 1994 Quantization of generalized spinning particles. New derivation of Dirac's equation, *J. Math. Phys.*, **35**, 3380

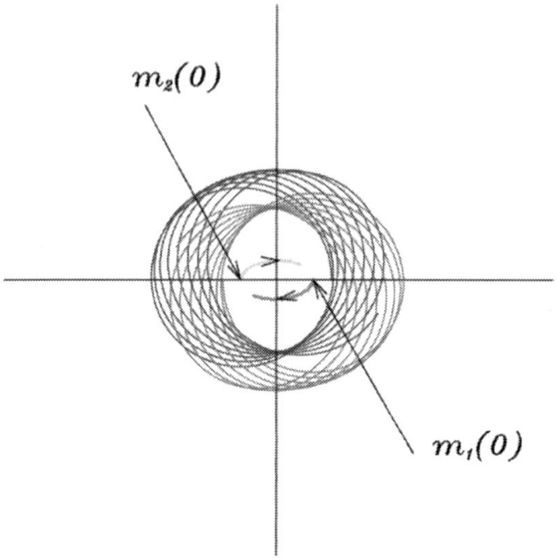

Figure 3. Bound motion of the CC's (red and blue) and CM's (light red and blue) of two spinning electrons with parallel spinsut. This bosonic system is stable under external electric fields but it is destroyed with a transversal magnetic field.

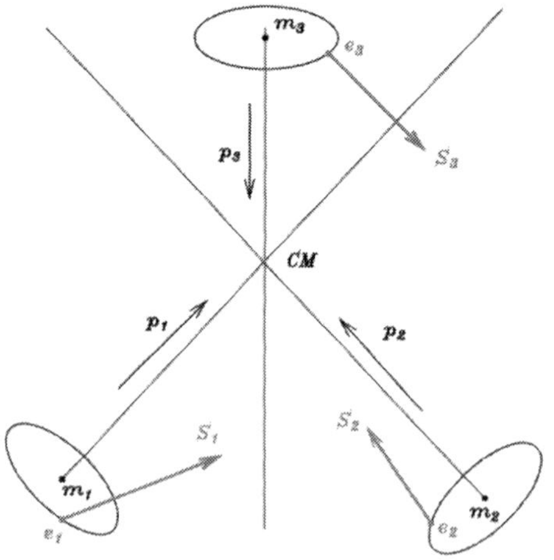

Figure 4. Proton at rest showing its CM and the coplanar motion of the CM's of the three quarks where the $\boldsymbol{S}_i$ (in red) represent the angular momenta of each quark with respecto to the corresponding CC, $\boldsymbol{r}_i$. Since the $\boldsymbol{S}_i$ are represented by the Dirac spin operators in the quantum case, it is impossible that the addition of the three Dirac spin operators $\boldsymbol{S}_i$ gives rise to the total angular momentum of the quarks with respect to the CM of the proton. The addition of the three terms $\boldsymbol{r}_i \times \boldsymbol{p}_i$ is left

[7] Rivas M 2003 The dynamical equation of the spinning electron, *J. Phys. A: Math. and General*, **36**, 4703. (arXiv:physics/0112005).
[8] Rivas M 1998 Is there a classical spin contribution to the tunnel effect?, *Phys. Lett. A*, **248**, 279
[9] Rivas M 2011 A lacking term in the proton spin, *J. Phys. Conf. Series*, **295**, 012040

IARD 2014 IOP Publishing
Journal of Physics: Conference Series **615** (2015) 012018 doi:10.1088/1742-6596/615/1/012018

Relativity and quantum mechanics

B.G. Sidharth

G.P. Birla Observatory & Astronomical Research Centre, B.M. Birla Science Centre, Adarsh Nagar, Hyderabad - 500 063, India

Abstract. We first deduce Special Relativity from Quantum Mechanics. Then we discuss the implications and ramifications.

It is well known that both Special and General Relativity, and quantum theory had independent origins and development. This brand of quantum mechanics was the non-relativistic theory. It was only around 1930 that Paul Dirac could "unite" Special Relativity and Quantum Theory through his famous Dirac equation.

Can we go the reverse way? Let us see how Special Relativity can be "derived" from Quantum Theory. We first define a complete set of base states by the subscript $\imath$ and $U(t_2, t_1)$ the time elapse operator that denotes the passage of time between instants t_1 and t_2, t_2 greater than t_1. We denote by, $C_\imath(t) \equiv \langle \imath|\psi(t)\rangle$, the amplitude for the state $|\psi(t)\rangle$ to be in the state $|\imath\rangle$ at time t, and [1, 2]

$$\langle \imath|U|j\rangle \equiv U_{\imath j};\; U_{\imath j}\,(t+\Delta t, t) \equiv \delta_{\imath j} - \frac{\imath}{\hbar}H_{\imath j}(t)\Delta t.$$

We can now deduce from the superposition of states principle that,

$$C_\imath(t+\Delta t) = \sum_j [\delta_{\imath j} - \frac{\imath}{\hbar}H_{\imath j}(t)\Delta t]C_j(t) \tag{1}$$

and finally, in the limit $\Delta t \to 0$,

$$\imath\hbar\frac{dC_\imath(t)}{dt} = \sum_j H_{\imath j}(t)C_j(t) \tag{2}$$

where the matrix $H_{\imath j}(t)$ is identified with the Hamiltonian operator.

Starting from equation (2) Feynman had deduced a version of the Schrodinger equation using the probability amplitude for a particle to diffuse from one point to another. We had argued earlier at length that (2) leads to the Schrodinger equation [1, 2]. In the above we had taken the usual unidirectional time to deduce the non relativistic Schrodinger equation. If however we consider in (1) that time is oscillating between $t - \Delta t$ and $t + \Delta t$, then we will have to consider instead of (2)

$$C_\imath(t-\Delta t) + C_\imath(t+\Delta t) = \sum_j \left[2\delta_{\imath j} - \frac{\imath}{\hbar}H_{\imath j}(t)\right] C_j(t). \tag{3}$$

Journal of Physics: Conference Series **615** (2015) 012018 doi:10.1088/1742-6596/615/1/012018

Equation (3) in the limit was seen to lead to the relativistic Klein-Gordon equation rather than the Schrodinger equation [3].

This follows if we consider a simple Taylor expansion:

$$f'(x) = \lim_{h\to 0}\frac{f(x+h)-f(x)}{h} \tag{4}$$

$$\begin{aligned} f''(x) &= \lim_{h\to 0}\frac{f'(x)-f'(x-h)}{h} = \lim_{h\to 0}\frac{f(x+h)-f(x)-f(x)+f(x-h)}{h^2} \\ &= \lim_{h\to 0}\frac{f(x+h)+f(x-h)-2f(x)}{h^2} \end{aligned} \tag{5}$$

so that by analogy

$$U(t+\Delta t,t) = 1-\frac{\imath}{\hbar}H(t)\Delta t+\frac{1}{2}\left(\frac{\imath}{\hbar}H(t)\Delta t\right)^2+0(\Delta t^3) \tag{6}$$

$$C_\imath(t+\Delta t)+C_\imath(t-\Delta t) = \sum_j 2\delta_{\imath j}C_j(t)+\left[\left(\frac{\imath}{\hbar}H(t)\Delta t\right)^2\right]_{\imath j}C_j(t)+0(\Delta t^3) \tag{7}$$

$$C_\imath(t+\Delta t)+C_\imath(t-\Delta t)-2C_\imath(t) = \sum_j\left[\left(\frac{\imath}{\hbar}H(t)\right)^2\right]_{\imath j}C_j(t)\Delta t^2+0(\Delta t^3) \tag{8}$$

$$\frac{d^2}{dt^2}C_\imath(t) = \left[\left(\frac{\imath}{\hbar}H(t)\right)^2\right]_{\imath j}C_j(t) \tag{9}$$

From this expression, the Klein-Gordon equation is motivated, and it is already evident that the generator of time translations $H(t)$ is the square root of the operator in the Klein-Gordon equation, in the limit rather than the Schrodinger equation. In other words considering a process in (1) that is time oscillating between $t-\delta t$ and $t+\delta t$ we reach the relativistic equation [3]. This has been discussed in great detail over the years, (cf [5] and references therein). In this case we have $(\Delta t)^2 = \beta\Delta x$ exactly as in a typical Wiener process. We should bear in mind, that here we are dealing with a non-differential spacetime manifold as in the Fokher-Planck theory (cf. [6]). This is the beginning of some modern Quantum Gravity approaches. More specifically, if we can neglect the fundamental scale **a** which stands for a minimum scale as in (14) below, we have classical theory. If $\bar{\mathbf{a}}^2$ can be neglected, we have the usual Quantum Theory and Quantum Field Theory. But if $\bar{\mathbf{a}}^2$ has to be retained, we have the Quantum of area and the above approach which encompasses fractal spacetime.

As it is well known the Klein-Gordon equation is obtained from the energy momentum dispersion relation of Special Relativity

$$E^2 = p^2+m^2 \qquad (c=1=\hbar). \tag{10}$$

Using the usual Quantum Mechanical substitutions for energy and momentum, (10) becomes

$$\left(\frac{d^2}{dt^2}-\nabla^2-m^2\right)\psi = 0. \tag{11}$$

Conversely, from (11) we can deduce (10) with substitutions of the type

$$p^\mu = \frac{\hbar}{\imath}\nabla^\mu.$$

It is well known that the problem with (11) was that there is no fixed eigenvalue, that is loosely the negative energy states appear because of the second time derivative. On the other hand Dirac linearized the second time derivative in (11) to obtain his well-known equation

$$\{\gamma^\mu p_\mu - m\}\,\psi = 0 \qquad (12)$$

where the symbols have their usual well known meanings. But there was a price to pay for this linearized relativistic Quantum Mechanical equation. Firstly we encounter spin half which had (and has) no place in the Classical Theory. Secondly, though (12) was linear in the time derivative, it still leads to the very same negative solutions. This then led Dirac to propose his Hole Theory.

We can now see the reason for the appearance of negative energy solutions in the derivation from (3): time as it were could run forward, this in the usual sense. But now we allow it to run backward (negative energy).

Let us now take another approach, starting from the coordinate for a Dirac electron as derived by Dirac himself [7] (cf. Appendix):

$$x = (c^2 p_1 H^{-1} t) + \frac{\imath}{2} c\hbar(\alpha_1 - c p_1 H^{-1}) H^{-1} \qquad (13)$$

The first term in (13) is known by the usual (Hermitian) position. The imaginary part of (13) is of the order of the Compton wavelength.

It is at this stage that a proper physical interpretation begins to emerge. Dirac himself observed, that to interpret (13) meaningfully, it must be remembered that Quantum Mechanical measurements are really averaged over the Compton scale [7]: Within the scale there are the unphysical zitterbewegung effects represented by the second term of (13): for a point electron the velocity equals that of light.

Once such a minimum spacetime scale **a** is invoked, then we have a noncommutative geometry as shown by Snyder [8, 9]:

$$[x,y] = (\imath \mathbf{a}^2/\hbar) L_z \qquad [t,x] = (\imath \mathbf{a}^2/\hbar c) M_x \qquad [x,p_x] = \imath\hbar[1 + (\mathbf{a}/\hbar)^2 p_x^2] \qquad \text{etc.} \qquad (14)$$

In this analysis, time too is a matrix (operator) — this is because we are allowing it to run backward and forward, or alternatively, we are allowing for the possibility of positive and negative energies.

The relations (14) are compatible with Special Relativity as shown by Snyder himself. Indeed such minimum spacetime models were studied for several decades, precisely to overcome the divergences encountered in Quantum Field Theory which originate from the point electron [2], [9]-[14], [15, 16].

Before proceeding further, it may be remarked that when $\mathbf{a}^2$, which we will take to be the squared Compton wavelength (including the Planck scale, which is a special case of the Compton scale for a Planck mass viz., 10^{-5} gm), in view of the above comments can be neglected, then we return to ordinary Quantum Theory.

It is interesting that starting from the Dirac coordinate in (13), we can deduce the noncommutative geometry (14), independently:

We observe that the first term on the right hand side is the usual Hermitian position. For the second term which contains α, we can easily verify from the commutation relations of the σ's that

$$[x_\imath, x_j] = \beta_{\imath j} \cdot l^2 \tag{15}$$

where l is the Compton scale which comes from the factor $(c\hbar/H)$ in (13) and $\beta_{\imath j}$ are suitable 2×2 matrices.

There is another way of looking at this. As can be seen the one dimensional coordinate in (13) is complex. We now try to generalize this complex coordinate to three dimensions. Then as is known we encounter a surprise - we end up with not three, but four dimensions,

$$(1, \imath) \to (I, \sigma),$$

where I is the unit 2×2 matrix and σs are the Pauli matrices. We get the special relativistic Lorentz invariant metric at the same time. (In this sense, as noted by Sachs [17], Hamilton who made this generalization would have hit upon Special Relativity, if he had identified the new fourth coordinate with time).

That is,

$$x + \imath y \to Ix_1 + \imath x_2 + jx_3 + kx_4,$$

where $(\imath, j, k)$ now represent the Pauli matrices; and, further,

$$x_1^2 + x_2^2 + x_3^2 - x_4^2$$

is invariant. Before proceeding further, we remark that special relativistic time emerges above from the generalization of the complex one dimensional space coordinate to three dimensions.

While the usual Minkowski four vector transforms as the basis of the four dimensional representation of the Poincare group, the two dimensional representation of the same group, given by the right hand side in terms of Pauli matrices, obeys the quaternionic algebra of the second rank spinors (cf [18, 19, 17] for details).

To put it briefly, the quarternion number field obeys the group property and this leads to a number system of quadruplets as a minimum extension. In fact one representation of the two dimensional form of the quarternion basis elements is the set of Pauli matrices. Thus a quarternion may be expressed in the form of the representation $SL(2, C)$,

$$Q = -\imath\sigma_\mu x^\mu = \sigma_0 x^4 - \imath\sigma_1 x^1 - \imath\sigma_2 x^2 - \imath\sigma_3 x^3 = (\sigma_0 x^4 + \imath\vec{\sigma} \cdot \vec{r})$$

This can also be written as

$$Q = -\imath \begin{pmatrix} \imath x^4 + x^3 & x^1 - \imath x^2 \\ x^1 + \imath x^2 & \imath x^4 - x^3 \end{pmatrix}.$$

It is well known that, there is a one to one correspondence between a Minkowski four-vector and Q [17]. The invariant is now given by $Q\bar{Q}$, where $\bar{Q}$ is the complex conjugate of Q.

However, as is well known, there is a lack of spacetime reflection symmetry in this latter formulation. If we require reflection symmetry also, we have to consider the four dimensional representation,

$$(I, \vec{\sigma}) \to \left[\begin{pmatrix} I & 0 \\ 0 & -I \end{pmatrix}, \begin{pmatrix} 0 & \vec{\sigma} \\ \vec{\sigma} & 0 \end{pmatrix} \right] \equiv (\Gamma^\mu)$$

(cf also [20] for a detailed discussion). The motivation for such a reflection symmetry is that usual laws of physics, like electromagnetism do indeed show the symmetry.

We at once deduce spin and Special Relativity and the geometry (14) in these considerations. This is a transition that has been long overlooked [21]. It must also be mentioned that spin half itself is relational and refers to three dimensions, to a spin network in fact [22, 23].

While a relation like (15) above has been in use recently, in non-commutative models, we would like to stress that it has been overlooked that the origin of this non commutativity lies in the original Dirac coordinates.

This can be seen directly from the Dirac theory itself where we have [24] that the velocity operator is given by

$$\frac{c^2\vec{p}}{H} - \frac{2\imath}{\hbar}\hat{x}H \tag{16}$$

In (16), the first term is the usual velocity. The second term is the extra contribution due to zitterbewegung.

In fact we can easily verify from (16) that

$$\vec{p} = \frac{H^2}{\hbar c^2}\hat{x} \tag{17}$$

where $\hat{x}$ has been defined in (16).

We finally investigate the angular momentum $\sim \vec{x} \times \vec{p}$ — that is, the angular momentum at the Compton scale. We can easily show that [25]

$$(\vec{x} \times \vec{p})_z = \frac{c}{E}(\vec{\alpha} \times \vec{p})_z = \frac{c}{E}(p_2\alpha_1 - p_1\alpha_2) \tag{18}$$

where E as usual in the continuum case [26] is the eigenvalue of the Hamiltonian operator H. Equation (18) shows that the usual angular momentum but in the context of the minimum Compton scale cut off, leads to the "mysterious" Quantum Mechanical spin. In the above approach, spin emerges from the Compton scale [27].

In the above considerations, we started with the Dirac equation and deduced the underlying non commutative geometry of spacetime.

Interestingly, starting with Snyder's non commutative geometry, based solely on Lorentz invariance and a minimum spacetime length, which we have taken to be the Compton scale, (14), it is possible to deduce the relations (18), (17) and the Dirac equation itself.

We have thus established the correspondence between considerations starting from the Dirac theory of the electron and Snyder's (and subsequent) approaches based on a minimum or fuzzy spacetime interval and Lorentz covariance.

Special Relativity however gets "modified" due to Quantum Effects. In fact the relation [28]

$$(\Delta x)^2 - (\Delta t)^2 = 0 \tag{19}$$

goes over to, as is well known [29],

$$(\Delta x)^2 - (\Delta t)^2 \leq \frac{1}{m^2} \quad (\hbar = 1 = c) \tag{20}$$

where term on the right side arises from the uncertainty in the Compton wavelength region as discussed in detail by Weinberg [10]. Effectively, the velocity of light $c(=1)$ becomes c', given by from (20),

$$c'^2 \leq 1 + \frac{1}{m^2(\Delta t)^2} \leq 2 \tag{21}$$

as $\Delta t \geq \frac{1}{m}$, the Compton time. Alternatively, the energy momentum relation gets modified.

So, the usual energy $E = mc^2$ becomes E' where

$$E < E' \leq 2E. \tag{22}$$

There is an increase in energy in the extreme relativistic case. The extra energy is the contribution of the background dark energy as argued by the author elsewhere [29].

The actual situation is that within the Compton time there are superluminal "velocities" or more realistically the Zitterbewegung effects. These effects are interpreted as the appearance of anti particles [30].

There is however another perspective: Wigner and Salecker [31] have argued that no physical measurement is possible within the Compton scale. In other words all our physics originates outside the Compton scale. There would thus be a minimum spacetime interval within which there would be no physics (cf [5] for a detailed description). This is exactly what we were discussing earlier. Such minimum spacetime theories are now in vogue in Quantum Gravity approaches, including the author's own.

Interestingly this interplay between Special Relativity and Quantum Mechanics manifests itself in graphene [32, 33]. As is known graphene is a two dimensional form of carbon with a honeycomb structure. The particles in graphene obey the equation

$$v_F \vec{\sigma} \cdot \vec{\nabla} \psi(r) = E\psi(r) \tag{23}$$

This is the two dimensional Dirac equation that describes neutrinos, except that $v_F \sim 10^6$ m/s replaces the velocity of light c. A finite graphene sheet cannot be expected to be Lorentz invariant, by the very nature of its finiteness. However we have argued that an infinite sheet reproduces Special Relativistic effects [34]. In fact in this case we have

$$v_F^2 = \left(\frac{\hbar \pi}{m}\right)^2 \cdot \frac{1}{A} \tag{24}$$

where A the area is $\sim l^2$, $l$the inter lattice distance.

Whence we get

$$\frac{m^2 v_F^2}{\hbar^2} \cdot l^2 \sim 0(1) \tag{25}$$

We can see from (25) that if v_F tends to the velocity of light c, h/mv_F tends to the Compton wavelength consistently. That is spacetime can be modelled consistently with an infinite lattice sheet.

Indeed following this train of thought it has been argued [35] that the Hall Effect parallels the Lorentz Force of electromagnetic theory.

Finally, it may be mentioned that graphene resembles a wind tunnel in the sense that there are Reynold number type scalings that lead to 3D Minkowski space [34]. Specifically, $L \sim 10^3 l$, $c = 300 v_F$ and $m_{graphene} \sim 10^{-2} m$ where m is the electron mass.

This suggests that graphene can be used as a test bed for High Energy experiments in view of the fact that it can be used to harvest protons, as confirmed by Andre Geim [36]. We must also remember that its mysterious minimum conductivity gives a source of protons themselves.

Appendix (Based on *The Principles of Quantum Mechanics* by P.A.M. Dirac [7])

The x_1-component of the velocity is

$$\dot{x}_1 = [x_1, H] = c\alpha_1. \tag{26}$$

The $\dot{x}_1$ given by (26) has as eigenvalues $\pm c$, corresponding to the eigenvalues ± 1 of α_1. That is a measurement of a component of the velocity of a free electron is $\pm c$. This conclusion is easily seen to hold also when there is a field present. We must remember that the velocity in above is the velocity at one instant of time while observed velocities are always average velocities through appreciable time intervals. Actually the velocity is not at all constant, but oscillates rapidly about a mean value which agrees with the observed value. This can also be seen from an application of the Uncertainty Principle.

To measure the velocity we must measure the position at two slightly different times and then divide the change of position by the time interval. "...In order that our measured velocity may approximate to the instantaneous velocity, the time interval between the two measurements of position must be very short and hence these measurements must be very accurate. The great accuracy with which the position of the electron is known during the time-interval must give rise, according to the principle of uncertainty, to an almost complete indeterminacy in its momentum. This means that almost all values of the momentum are equally probable, so that the momentum is almost certain to be infinite. An infinite value for a component of momentum corresponds to the value $\pm c$ for the corresponding component of velocity.

Let us now examine how the velocity of the electron varies with time. We have

$$\imath\hbar\dot{\alpha}_1 = \alpha_1 H - H\alpha_1.$$

Now since α_1 anticommutes with all the terms in H except $c\alpha_1 p_1$,

$$\alpha_1 H + H\alpha_1 = \alpha_1 c\alpha_1 p_1 + c\alpha_1 p_1 \alpha_1 = 2cp_1,$$

and hence

$$\imath\hbar\dot{\alpha}_1 = 2\alpha_1 H - 2cp_1 = -2H\alpha_1 + 2cp_1. \tag{27}$$

Since H and p_1 are constants, it follows from the first of equations (27) that

$$\imath\hbar\ddot{\alpha}_1 = 2\dot{\alpha}_1 H. \tag{28}$$

This differential equation in $\dot{\alpha}_1$ can be integrated immediately, the result being

$$\ddot{\alpha}_1 = \dot{\alpha}_1^0 e^{-2\imath Ht/\hbar}, \tag{29}$$

where $\dot{\alpha}_1^0$ is a constant, equal to the value of $\dot{\alpha}_1$ when $t = 0$. The factor $e^{-2\imath Ht/\hbar}$ must be put to the right of the factor $\dot{\alpha}_1^0$ in (29) on account of the H occuring to the right of the $\dot{\alpha}_1$ in (28). The second of equations (27) leads in the same way to the result

$$\dot{\alpha}_1 = e^{2\imath Ht/\hbar}\dot{\alpha}_1^0.$$

We can now easily complete the integration of the equation of motion for x_1. From (29) and the first of equations (27)

$$\alpha_1 = \frac{1}{2}\imath\hbar\dot{\alpha}_1^0 e^{-2\imath Ht/\hbar}H^{-1} + cp_1 H^{-1}, \tag{30}$$

and hence the time-integral of equation (26) is

$$x_1 = -\frac{1}{4}c\hbar^2\dot{\alpha}_1^0 e^{-2\imath Ht/\hbar}H^{-2} + c^2 p_1 H^{-1} t + a_1, \tag{31}$$

a_1 being a constant."

This is the desired result.

Acknowledgements

I am grateful to the referee for pertinent comments and suggestions.

References

[1] Sidharth, B.G. (1997). *Ind.J. of Pure and Applied Physics* 35, pp.456ff.
[2] Sidharth, B.G. (2001). *Chaotic Universe: From the Planck to the Hubble Scale* (Nova Science, New York).
[3] Sidharth, B.G. (2002). *Chaos, Solitons and Fractals* 13, pp.189–193.
[4] Feynman, R.P. *Feynman Lectures on Physics* Vol.3, pp.16-2-16-4.
[5] Sidharth, B.G. (2008). *The Thermodynamic Universe* (World Scientific, Singapore).
[6] Nottale, L. (1993). *Fractal Space-Time and Microphysics: Towards a Theory of Scale Relativity* (World Scientific, Singapore), pp.312.
[7] Dirac, P.A.M. (1958). *The Principles of Quantum Mechanics* (Clarendon Press, Oxford), pp.4ff, pp.253ff.
[8] Snyder, H.S. (1947). *Physical Review* Vol.72, No.1, July 1 1947, pp.68–71.
[9] Snyder, H.S. (1947). *Physical Review* Vol.71, No.1, January 1 1947, pp.38–41.
[10] Kadyshevskii, V.G. (1962). *Translated from Doklady Akademii Nauk SSSR* Vol.147, No.6 December 1962, pp.1336–1339.
[11] Schild, A. (1948). *Phys.Rev.* 73, pp.414–415.
[12] Dirac, P.A.M. (1978). *Directions in Physics* (John Wiley, New York).
[13] Bombelli, L., Lee, J., Meyer, D. and Sorkin, R.D. (1987). *Physical Review Letters* Vol.59, No.5, August 1987, pp.521–524.
[14] Finkelstein, D.R. (1996). *Quantum Relativity A Synthesis of the Ideas of Einstein and Heisenberg* (Springer, Berlin).
[15] Wolf, C. (1990). *Hadronic Journal* Vol.13, pp.22–29.
[16] Lee, T.D. (1983). *Physics Letters* Vol.122B, No.3,4, 10 March 1983, pp.217–220.
[17] Sachs, M. (1982). *General Relativity and Matter* (D. Reidel Publishing Company, Holland), pp.45ff.
[18] Sidharth, B.G. (2003) *Found.Phys.Lett.* 16, (1), pp.91–97.
[19] Shirokov, Yu. M. (1958). *Soviet Physics JETP* 6, (33), No.5, pp.929–935.
[20] Heine, V. (1960). *Group Theory in Quantum Mechanics* (Pergamon Press, Oxford), pp.364.
[21] Sidharth, B.G. (2002). *Found.Phys.Lett.* 15, (6), pp.577–583.
[22] Penrose, R. (1971). *Angular Momentum: An approach to combinational space-time* in *Quantum Theory and Beyond*, Bastin, T. (ed.) (Cambridge University press, Cambridge), pp.151ff.
[23] Misner, C.W., Thorne, K.S. and Wheeler, J.A. (1973). *Gravitation* (W.H. Freeman, San Francisco), pp.819ff.
[24] Gueret, P. and Vigier, J.P. (1982). *Found.of Phys.* Vol.12, (11), pp.1057ff.
[25] Sidharth, B.G. (2002). *Foundation of Physics Letters* 15(5),2002,501ff.

[26] Schiff, L.I. (1968). *Quantum Mechanics* (McGraw Hill, London).
[27] Sachidanandam, S. (1983). *Physics Letters* Vol.97A, No.8, 19 September 1983, pp.323–324.
[28] Weinberg, S. (1972). *Gravitation and Cosmology* (John Wiley & Sons, New York), p.61ff.
[29] Sidharth, B.G. (2010). *Int.J.Mod.Phys.E* Vol.19, No.1, 2010, pp.79-90.
[30] Weinberg, S. (1972). *Gravitation and Cosmology* (John Wiley & Sons, New York), p.61ff.
[31] Salecker, H. and Wigner, E.P. (1958). *Physical Review* Vol.109, No.2, January 15 1958, pp.571-577.
[32] Sidharth, B.G. (2014). *Int.J.Mod.Phys.E* Vol.23, No.4, 2014, pp.1450025-1-1450025-8.
[33] Sidharth, B.G. (2014). *New Advances in Physics* Vol.8, (2), 2014.
[34] Sidharth, B.G. (2014). *Int.J.Mod.Phys.E* Vol. 23, (11), 2014.
[35] Sidharth, B.G. *PSTJ* 3 (11), 2012.
[36] Sidharth, B.G. *An addendum to the paper, Graphene and High Energy Physics* IJMPE Volume 23, Issue 05, May 2014

AUTHOR INDEX